MW00838267

SAFETY
DULUTH
MISSABE
&
IRON RANGE
FIRST

DULUTH MISSABE & IRON RANGE RY.
DMIR 20143
DMIR 20231
DMIR 20389
31268
31429
31556
31039
25951
27156
WT 43800 1956

DULUTH, MISSABE & IRON RANGE EQUIPMENT, 1883–2004

Daniel P. Holbrook

Signature Press
Wilton and Berkeley, California

Duluth, Missabe & Iron Range Equipment, 1883–2004

Published by Signature Press
11508 Green Road
Wilton, CA 95693
www.signaturepress.com

Publisher's Cataloging-in-Publication
(Provided by Cassidy Cataloguing Services, Inc.)

Names: Holbrook, Daniel P., author.
Title: Duluth, Missabe & Iron Range equipment, 1883-2004 / by Daniel P. Holbrook.
Description:
Wilton and Berkeley, California : Signature Press, [2019] | Includes bibliographical references and index.
Identifiers: ISBN: 9781930013414 | LCCN: 2019931181

Subjects:
LCSH: Duluth, Missabe, and Iron Range Railway--History. | Duluth, Missabe and Northern Railway Company--History. | Frieght cars--United States--History. | Freight cars--Minnesota --History. | Railroads--Minnesota--History | Railroad equipment industry--United States-- History. | Railroads--Freight--United States--History.
Classification:
LCC: TF470 .H64 2019 | DDC: 625.2/4--dc23

Book design and typography by John Signor and Jonathan Signor

Printed in the United States of America

Frontispiece: Proctor Yard September, 1974. A virtual sea of ore cars. Unusual is caboose C-135 normally assigned to T-Bird service at Virginia, Minnesota, and a string of leased former B&LE ore cars re-stencilled DM&IR tied on to MofW service passenger cars. (D.P. Holbrook collection)

Opposite: DM&IR Proctor Yard on September 27, 1946. (R.V. Nixon photo, Museum of Rockies collection)

Pages 6-7: It is the twilight of steam on June 16, 1959. DM&IR 502, a Class E 2-10-2 acquired by DM&N in 1919, is at the north end of Proctor Yard working one of many day ore-sorter jobs. Tied on to the drawbar is caboose C-97, one of nine acquired from EJ&E in 1948 during a severe caboose shortage. The first car is a 50-ton ore car, one of 2,985 left on the roster which would all be retired by 1964. All remaining cars in this cut are 70-ton capacity, many of which would soldier on past the year 2000, obtaining FRA exemptions for their continued use. (Bruce Meyer)

Pages 8-9: Extra 234 North, behind a 2-8-8-4 Yellowstone, is trailing an seemingly endless string of 50- and 70-ton ore cars at Kelsey, Minnesota, on June 16, 1959. (Bruce Meyer)

CONTENTS

INTRODUCTION

Iron Ore! Lifeblood of the American industrial revolution was found from coast to coast and border to border. Major deposits were concentrated in northern Wisconsin and the upper peninsulas of Michigan and Minnesota with the largest of these deposits found in the Arrowhead region of Northern Minnesota. First on what would be called the Vermilion iron range in early 1863, and in 1866 on the future Mesabi iron range. Mesabi, a Chippewa Indian word for "sleeping giant" was truly an indication of what had been discovered.

Once the industrial revolution began the demand for high-grade iron ore rapidly increased and so did the need for finding additional deposits of iron ore to feed the steel mills. Exploration and development of the Minnesota iron ranges was at hand. Scotsman

Andrew Carnegie, who would later head U.S. Steel, constructed the first modern Bessemer process steel mill at Braddock, Pennsylvania, between 1873 and 1875. The steel industry that developed around this process and would consume the iron ore was located predominantly in the Eastern United States involving over a 1,000 mile movement of iron ore from Northern Minnesota. The only transportation solution was to construct railroads from the mines to ports on Lake Superior for boats to forward ore to the mills. Two railroads were built, initially, the Duluth & Iron Range to serve the Vermillion iron range and the Duluth, Missabe & Northern to service the Mesabi iron range. Both, eventually, would extend their railroads into the competitor's territory. Both initially acquired flat cars for early construction and a large fleet of purpose-built, wooden ore cars to service the mines. As Northern Minnesota developed, this car fleet expanded to include all types of equipment. ■

ACKNOWLEDGEMENTS

Multiple railroads, heavy industry and marine traffic combined to create a lifelong interest in the Twin Ports of Duluth, Minnesota, and Superior, Wisconsin. The synergy of these created a location with few equals. One of the commodities involving all three was iron ore. This dense, heavy commodity created a need for heavily built railroads, boats and docks, but also the need for specialty type equipment to carry the ore. Starting in 1985, I began a serious quest to research the equipment of the Duluth & Iron Range, Duluth, Missabe & Northern and Duluth, Missabe & Iron Range. A 1975 diagram book and my own equipment photos were the beginnings of what is now in your hands.

This book would not have been possible without the assistance of the following organizations. The Missabe Railroad Historical Society allowed access to their vast collection of AFE (Approved For Expenditure) files, diagrams, maps and photos. The society leadership of Brian Hoag, Tim Schandel, Hans Kremer, Dave Schauer, Damian Kostron, Doug Buell and Tim Vitelli gave their support and time to make this book possible. The author strongly encourages readers to join the society by going to www.missabe.com. Additional equipment diagrams, photos, history and operation information is available with their quarterly publication *Ore Extra* and from their on-line store.

Pat Maus, curator of the Northeast Minnesota Historical Collection at the University of Minnesota Duluth campus, provided advice and direction as I reviewed their collections. Additional sources were: Minnesota Historical Society, Museum of the Rockies, Jim Singer with the Burlington Route Historical Society, Kay Peterson of the Smithsonian Institution Archives Center, Nicholas Fry, curator, John W. Barriger III National Railroad Library, St. Louis Mercantile Library, St. Louis, Missouri (UMSL), and the Transportation Library at Northwestern University in Evanston, Illinois.

I would also like to acknowledge the assistance of freight car experts: the late Richard Hendrickson, Ed Hawkins, Eric Neubauer, Jim Kinkaid, Jim Eager, Keith Jordan, Jerry Stewart, Bill Welch, Ray Kucaba and Dave Lehlbach.

Many other individuals provided photos and documentation. These include: R.C. Anderson, Glen Blomeke, Dick Bradley, the late Hank Brower, Doug Buell, Ron Christensen, Joe Collias, Gene Collora, Chuck Corwin, the late Ed DeRouin, Joe Economy, Jeff Eggert, Martin Fair, Paul Faulk, Larry Goolsby, Art Griffen, John Gruber (Mainline Photos), Wes Harkins, the late Lee Hastman, Ed Hawkins, Tim Johnson, Jim Kinkaid, Llloyd Keyser, the late Frank A. King, Bob Liljestrand (Bob's Photo), Ray Kucaba, Bruce Kuitunen, George LaPray, John C. LaRue, the late Owen Leander, Todd Lindahl, Steve Lorenz, Jim Maki Sr., Brian Memmot, Arnold Menke, the late Bruce

Meyer, the late Wayne C. Olsen, Bill Polard, the late Bill Raia, D. Repetsky, D.K. Retterer, Dave Schauer, Ted Schnepf, Andy Sharp, Shorpy Photo Archives, the late Mark Simonson, Jim Singer, Perry Sugerman, Johnathan Sugerman, Mike Urie, the late Harold K. Volrath, Duke Wahl, Bill Welch, Wilbur C. Whittaker, and Craig Wilson. Photos from Minnesota Historical Society are identified as MNHS in photo captions. Photos from the Missabe Railroad Historical Society Collection are identified as MRHS in photo captions.

I especially want to thank Guy Wilber for his guidance with ICC, ARA and AAR rules and laws relating to railroad equipment.

A number of people provided inspiration, guidance and research assistance. Special thanks to Perry Sugerman for data processing assistance, Jon Sugerman for guidance with the fine points with Photoshop and Krystal Flowers for photo restoration efforts. Steve Lorenz and the late Lee Hastman accompanied me on multiple trips over the years to the research institutions mentioned above and always found a tidbit I had missed.

One of the shortcomings of smaller, regional, out of the way railroads, like the D&IR and DM&N, is photographic coverage. Up until DM&IR became popular in the late 1950s for operating steam, very few people ventured to the Arrowhead region of northern Minnesota and most did not photograph equipment. Early coverage of equipment is mostly through builder photographs and supplemented with equipment diagrams.

Most books on railroad equipment leave the unanswered question of: What were the cars used for? I've attempted, in Chapter 5 of the book, to give a broad overview of the commodities handled and industries served by the DM&IR and its predecessors in the hopes of filling that void.

Data were compiled from railroad diagrams, railroad AFE (Approved For Expenditure) files, railroad annual reports, auditor's reports, ledgers, Minnesota Warehouse Commission reports, *Poor's*, *Moody's*, *Official Railway Equipment Registers* (*ORER*), and by comparing photos. Sometimes conflicts arose between various sources. Whenever conflicts appeared, railroad documents were used in preference to other sources.

Lastly, none of this would be possible without my wife Linda. We met working for Burlington Northern in 1978, got married shortly after, and worked five years together at the railroad. I'm not aware of too many wives that both understand railroading, can talk about it and can assist with research. She accompanied me on many trips to Duluth, Two Harbors, St. Paul, Minnesota and Evanston, Illinois and assisted in going through company records. My efforts are only reflected in your patience as I pursued this project. Thanks for being my better half.

To those that assisted, thank you. Hopefully I've thoroughly covered the equipment of the DM&IR and any mistakes are entirely mine. ■

DM&IR Proctor Yard in 1943. On the left is the ore sorting yard and engine terminal. The bottom of the photo is the empty yard for assembling departing trains for the iron range and sandwiched between them is the small commercial yard for merchandise traffic. (D.P. Holbrook collection)

CHAPTER ONE
HISTORY

As with most railroads, the Duluth Missabe & Iron Range Railroad was the product of merger, consolidation and reorganization of a number of component companies over the years.

THE DULUTH & IRON RANGE

The Duluth and Iron Range (D&IR) was incorporated on December 21, 1874, however no track was built until 1883. On March 1, 1882, the original D&IR incorporation was transferred to a group of investors with construction starting on June 20, 1883. Charlemagne Tower incorporated the Minnesota Iron Co. on December 1, 1882, for $10 million dollars to develop mining operations on the Vermillion iron range. He quickly formed the D&IR to transport iron ore to Agate Bay, which was latter platted as Two Harbors in 1885. The first iron ore was shipped from Tower, Minnesota on July 31, 1884, and arrived at Agate Bay at 11:00 p.m. where it was dumped into the pockets of the new 1,200-foot long ore dock. The first train consisted of ten cars including the first car loaded, which was 25-ton capacity wood ore car 406.

The first locomotive arrived in Duluth during July 1883. Orders for equipment were immediately placed with Northwestern Car & Manufacturing of Stillwater, Minnesota, in 1883 for six 34-foot combination flat cars, followed quickly by an additional 24 identical cars for use in the initial construction of the railroad from Two Harbors to Tower, Minnesota. It appears that these cars were possibly numbered 31–80, odd numbers only. Additional orders were placed with Northwestern Car & Manufacturing in January 1884 for 300 24-foot wood ore cars, 251–550, and coal cars

D&IR Engine 1 with train at Tower Jct. Minnesota in 1885. (University of Minnesota Duluth, Kathryn A. Martin Library Archives and Special Collection, Collection S3742 Box 71 Folder 20)

236–250. All of these cars were delivered by June 1, 1884. At the same time, car shops were established at Two Harbors to construct additional equipment

Until December 20, 1886, when the Two Harbors-to-Duluth trackage was completed, all new equipment was delivered by barge from Duluth. After this date, connections at Duluth allowed interchange of equipment between D&IR and other railroads.

Troubled times were ahead. H.H. Porter, head of Illinois Steel Co., took notice. A group of prominent capitalists, including John D. Rockefeller, had acquired a number of mining deposits in the Vermillion Range. They soon made it clear to Charlemagne Tower that they desired to acquire the D&IR or they would build their own railroad to compete with it. Reluctantly Tower sold his railroad and mining interests to Illinois Steel Co. in April 1887 for $6.4 million. Illinois Steel provided the finances to expand the railroad.

The main line was extended from Tower to Winton between 1888 and 1894. Two additional ore docks were built at Two Harbors in the 1890s. The discovery of iron ore on the Mesabi Range in 1890 caused the D&IR to build an 18-mile branch from Wyman to McKinley, and then to Virginia, Minnesota, in 1894 to service the new iron range.

Two Harbors shop was converting combination flat cars to box cars in 1884 and by 1886 they were capable of building their own equipment with flat cars and cabooses being the first cars built in the shops. The shops would go on to build log cars, box cars, ore cars, refrigerator cars, cabooses, and also to rebuild flat cars to box cars, and convert box cars to stock cars, vegetable and oil cars. During the 'teens many cars received steel center sills. The last cars built were refrigerator cars 8032–8041 in 1918.

Vermillion iron range tonnage shipped through Two Harbors rose to 10,735,853 tons in 1916 but five years later in 1921 fell to 3,286,338 tons. 1916 saw the last 50-ton steel ore cars purchased. Outside of 10 refrigerator cars constructed at Two Harbors in 1918, no additional equipment would be purchased by D&IR. The underground mines of the Vermillion range were expensive to operate, while the neighboring Mesabi iron range had a combination of underground and cheaper to operate open pit mines. While DM&N tonnages continued to increase, the Vermillion range was falling on hard times. Making matters worse, old-growth timber that had contributed to the bottom line was playing out and the last major lumber mill would close in 1925. U.S. Steel, seeing the writing on the wall, began to make plans to merge the DM&N and D&IR to obtain the economy of operating one railroad.

THE DULUTH, MISSABE & NORTHERN

The Duluth, Missabe and Northern (DM&N) was incorporated on February 11, 1891, using the 1882 charter for the Lake Superior and Northwestern Railway. Prior to this the Merritt brothers of Duluth—Lewis, Leonidas, Alfred, Cassius and Napoleon—had taken notice of the developments on the Vermillion iron range. They discovered even higher grade iron ore south and west of the Vermillion. Quickly they acquired 140 land leases on what was to become the Mesabi iron range. However they lacked the capital to invest in sinking mine pits, constructing a railroad and erecting an ore dock. The brothers found investors in St. Paul, Minnesota, to capitalize their dreams.

Construction of the railroad from the Mountain Iron Mine on the Mesabi iron range to Stony Brook Junction was begun January 28, 1892. Initially the DM&N hauled iron ore to a connection with the Duluth and Winnipeg (D&W) at Stony Brook Junction and the D&W forwarded the ore to docks on Lake Superior at Allouez, Wisconsin. DM&N and D&W had earlier reached an agreement with each supplying ore cars on a 50/50 basis for this new business. D&W never did comply with this agreement.

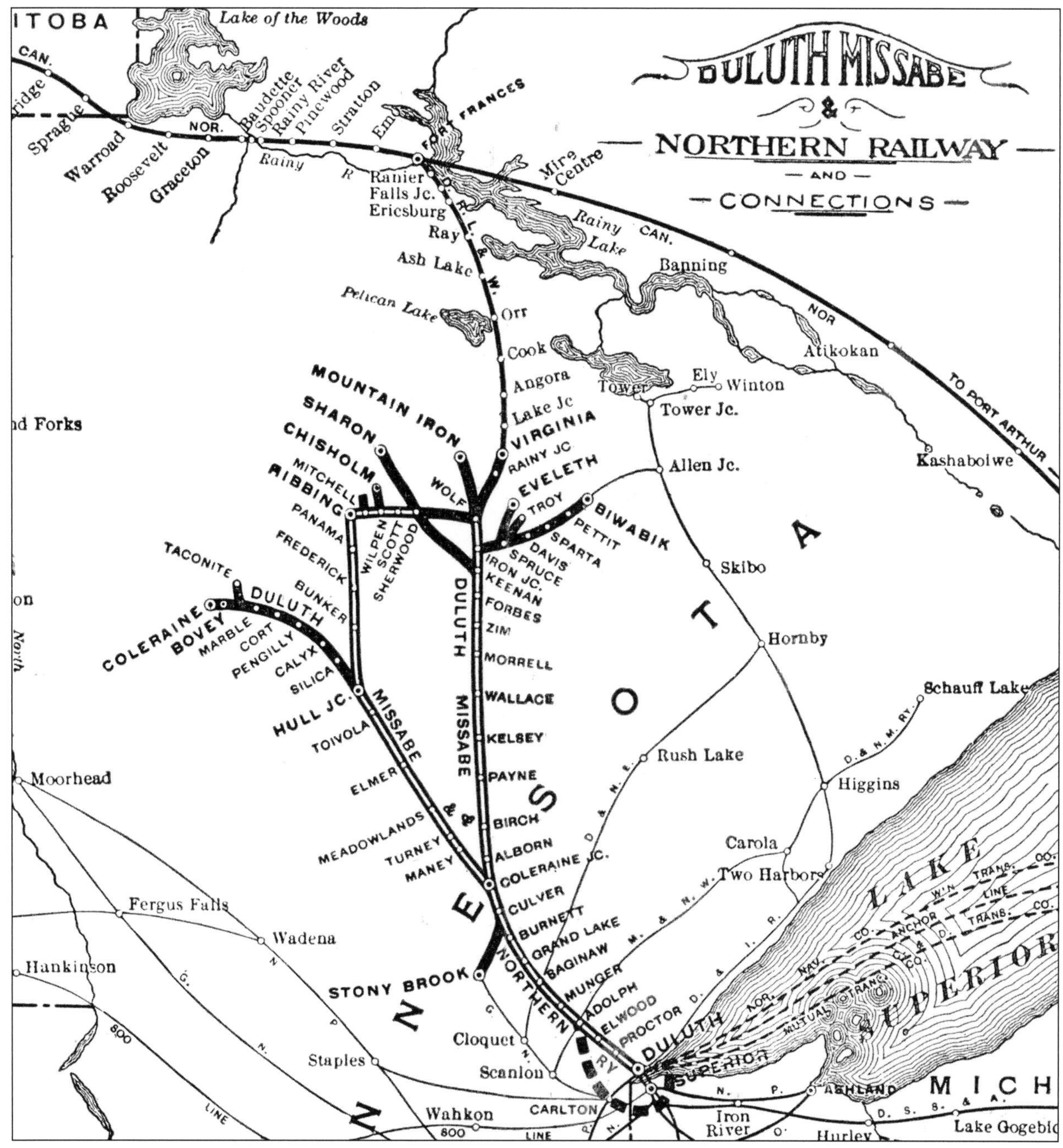

DM&N System Map 1911. Note D&IR trackage Duluth to Two Harbors, Hornby, Allen Jct, Biwabik and Winton. (D.P. Holbrook collection)

Needing additional consumers for the ore, the Merritts convinced Henry W. Oliver, a steel mill owner and associate of Andrew Carnegie, to visit the Mesabi iron range in the summer of 1892. Impressed, Oliver committed to delivery of 4,000 tons of ore within two years. The first iron ore was shipped from Mountain Iron Mine on October 17, 1892, in 25-ton capacity ore car 342, arriving Duluth at 11:15 a.m. on October 18th. This car was loaded and run as one car train and proudly displayed in Duluth. The ore was eventually moved to the D&W interchange and Allouez, Wisconsin, for loading on a lake boat. Regular train service was begun on November 1, 1892, with a train of nine cars each carrying 22 tons of ore. Conflicts with D&W not providing their portion of ore cars, and the Merritts wanting to control the entire movement of ore from mine to dock caused DM&N to extend their main line to

Duluth, with the first train arriving on Saturday, July 22, 1893. Construction of the first ore dock had begun in January, 1893 and the first boat was loaded in October.

The first equipment orders were placed in May 1892 with Duluth Manufacturing Co. for 200 24-foot wood ore cars 200–399, 50 34-foot flat cars 2000–2049 and 10 36-foot box cars 3000–3009 for delivery during 1892. Before the first cars were delivered an additional 16 36-foot box cars, 3010–3035, were added to the order for 1893 delivery. Unlike the D&IR, DM&N did not use odd and even number series to distinguish between house and flat cars.

All cars on DM&N were equipped with air brakes by 1893, except for the 50 Russell-type log flats. Sometime before 1899 these cars disappeared from the roster and all cars on the roster were equipped with air brakes.

The DM&N fell on hard times during the financial panic of 1893. John D. Rockefeller, through investments in American Steel Barge Co., came to agreements with the DM&N for pooling of mine, railroad and ship properties. Rockefeller quickly saw that he would have to invest large sums of money to keep the Merritts from failing. At the DM&N board meeting on February 6, 1894, which included principal mining partners, the Merritts lost voting control of their company and Rockefeller was in full control

During 1894 both railroads were now controlled by "big steel," the D&IR by Illinois Steel Co. and the DM&N by Lake Superior Consolidated Iron Mines, which would later become part of Andrew Carnegie's Carnegie Steel Co. in 1896.

CONSOLIDATION BEGINS

In 1901, Federal Steel, successor to Illinois Steel Co. and Carnegie Steel Co., were the basic components in the consolidation that created United States Steel. This brought both the D&IR and DM&N under U.S. Steel ownership. The 1901 shipping season found D&IR serving 12 Mesabi Range and five Vermillion Range mines and DM&N serving nine Mesabi Range mines. Competitor Eastern Railroad of Minnesota (ERM), later absorbed by Great Northern, was serving 10 Mesabi Range mines. Initially D&IR and DM&N were operated as individual railroads. Slowly, cost savings and efficiencies were realized and operations started to be consolidated. From 1901 to 1914 when World War I began, both railroads continued to grow their traffic base with no financial concerns.

World War I broke out in Europe in 1914 and initially the U.S. maintained neutrality. Finally, in April 1917, the U.S. declared war on Germany. Almost immediately there was a dramatic traffic increase. Ore tonnage surged from 13,358,264 tons in 1914 to 33,334,925 tons in 1916. Nationwide traffic increased by 30 percent in 1916 and 43 percent in 1917, creating a national shortage of freight cars. Railroad labor unions by this time had an established foothold on railroad employment with the assistance of political support in Congress. The traffic increase, congestion and rapid inflation in 1916 and 1917 all contributed to the threat of a nationwide rail strike. Congress reacted quickly and created the United State Railroad Administration (USRA) on December 17, 1917, to operate the nation's railroads. This lasted for 28 months until March 1, 1920.

Major changes to DM&N during this period of time were the construction of all-steel ore Dock No. 6 in Duluth and for D&IR, the construction of the Wales Branch, which tapped rich timber resources north and east of Two Harbors. Under USRA control, D&IR rostered 5,718 steel ore cars during January 1919, of these, 1,698 were being used by sister road DM&N.

The depression of 1921 caused traffic levels to fall by over 50 percent. This rebounded back to normal levels in 1923. The Great Depression began on October 29, 1929 and bottomed out during the spring of 1933. Ore tonnage fell to 1,458,711 tons in 1932 and slowly increased to

Duluth circa 1905. Visible in the right center foreground is a DM&N box car and a flat car converted to a gondola. (Shorpy Photo Archives)

27,764,490 tons in 1937. Traffic levels in 1932 were so bad that no ore trains were operated with all ore traffic moving on local trains. Recovery began in earnest during 1938 and 1939 with ore tonnage reaching pre-Depression levels of 28,005,441 tons in 1940.

Management and operation of the DM&N and D&IR were unified in 1930 as the precursor to the eventual merger, with equipment accounting consolidated on February 1, 1930. The D&IR was a frugal railroad and used its extensive shop capacity at Two Harbors to continually build and rebuild equipment. Two Harbors built 388 cars, rebuilt over 1,500 cars, and purchased 6,220 new cars between 1901 and 1937. The DM&N had the shop capacity to build equipment but only built 153 cars between 1901 and 1937, instead purchasing 12,269 new cars.

Steel 50-ton ore cars had become the standard starting in 1899 on DM&N and 1900 on D&IR. Advances in freight car construction and design saw the introduction of 95- and 81-ton capacity ore cars in 1925 and 1928. Numerous car builders supplied these cars, which were sampled in small orders by DM&N and by 1937 the design had evolved to a 70-ton ore car which would become the standard for all future orders. DM&N would purchase 70-ton cars but D&IR, the weak sister, would not.

The Interstate Commerce Commission D&IR Lease, 158 ICC 373 was decided on

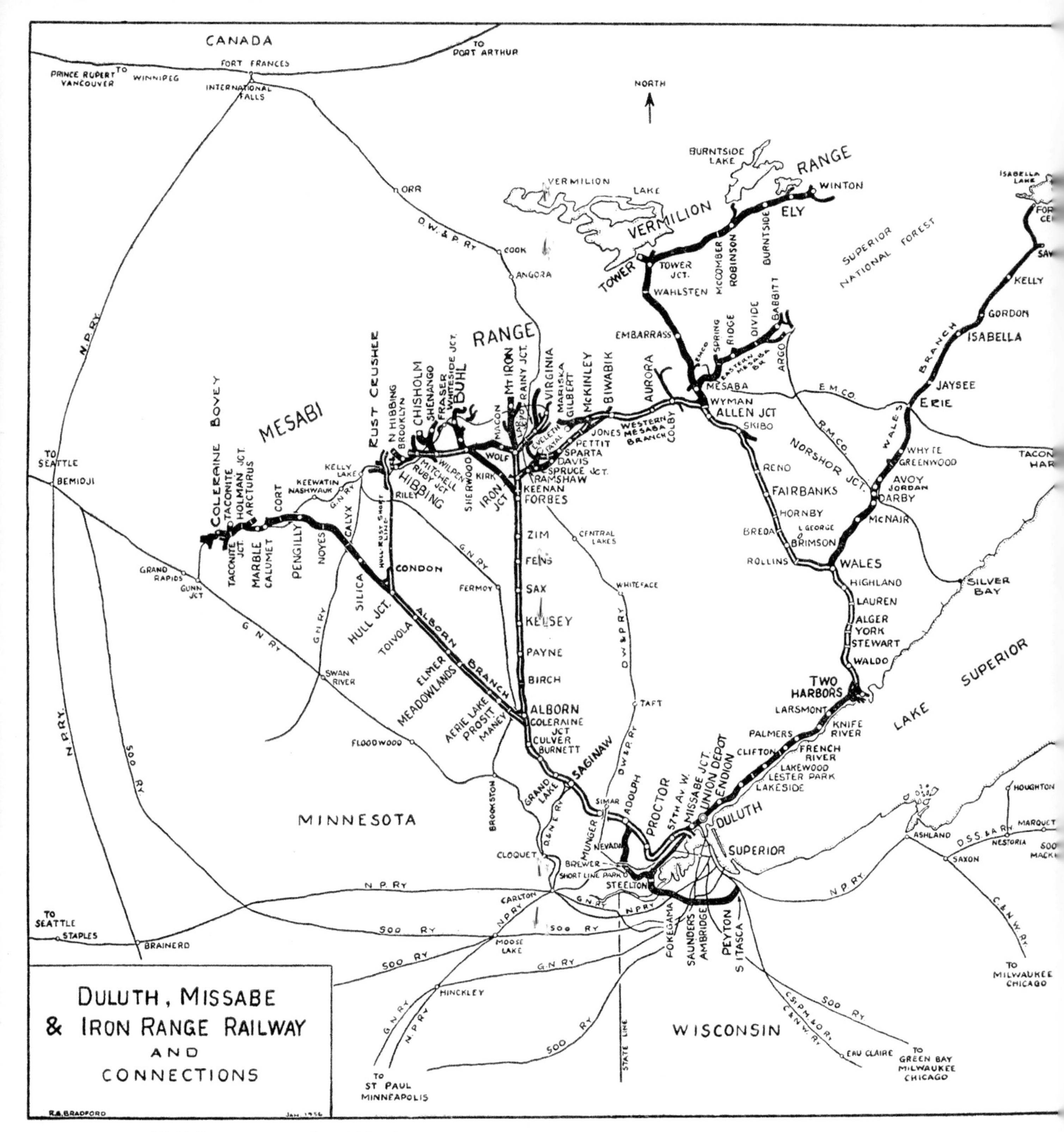

DM&IR system map July, 1956. (D.P. Holbrook collection)

December 11, 1929. On January 10, 1930, all of the railroad property of the D&IR was leased to the DM&N. The *Official Railway Equipment Register* (ORER) began listing both owners equipment under the DM&N listing after this date.

The DM&N and the Spirit Lake Transfer railroad were merged in July 1, 1937, under DM&N Railway Co. consolidation, 221 ICC 539, to become the Duluth, Missabe & Iron Range (DM&IR).

ICC DMIR Docket 11763 for merger

DM&IR 503, a Class E 2-10-2, is working an ore "sorter" job on June 16, 1959. Tied to its drawbar is 42758, a Class E11 50-ton ore car built in 1916 by Standard Steel Car as D&IR 12758. In the foreground is 33032 a Class U31 70-ton ore car built by American Car & Foundry in 1957. Note the "stencil group" letters located in the middle of side sills, "X" on 33032 and "D" on 42758. These groups were established in 1952 to allow easier identification of cars for air brake testing. In the background two EMD SD9s are preparing to depart with ore empties for the iron range. In a little over a year the last fires would be dropped on steam. (Bruce Meyer)

of DM&IR, the D&IR, and the Interstate Transfer Railway was submitted on November 30, 1937, and approved on December 28, 1937. On March 21, 1938, the D&IR and DM&IR were consolidated into the DM&IR. At the time of the merger the traffic base was 81 percent iron ore that was produced by Oliver Mining Co., nine percent by independent ore operators, and ten percent general freight. A small amount of ore was going to American Steel & Wire at Steelton, Minnesota. One of the main reasons for the merger was to allow a single line-haul movement of traffic for ore producers which were increasingly mixing ore from different mines on both railroads. Significant economies in operation included the joint use of equipment, ore docks and other facilities.

By 1937, the two carriers were serving four mines on the Vermillion and 36 on the Mesabi iron ranges. Total shipments for 1937 were 16,721,205 tons over the Duluth docks, 9,743,476 tons over the Two Harbors docks, and 40,484 tons all-rail to on-line and off-line customers. Many felt that the Depression was over, but this would prove to be a false recovery. The following year would find ore tonnages fall from 27,764,490 tons in 1937 to 8,265,732 tons in 1938. War fever would start to grip the nation in 1939 and ore tonnages soared through the 1940s.

POST-MERGER YEARS, 1938 UNTIL 1965

The economy continued to improve as the storm clouds in Europe began to affect America's industry. Production of planes, boats, tanks, and other munitions of war required continually increasing steel production. Even without the involvement of the United States, the war in Europe was creating a demand for the munitions of war. The bombing of Pearl Harbor caused the United States to enter World War II on December 8, 1941.

On January 1, 1942, World War II found the newly formed DM&IR with 11856 freight cars. Between 1939 and 1942 ore tonnage shipped more than doubled from 18,489,768 to 44,788,189 tons. DM&IR was not well situated for handling this increase of tonnage and needed additional 70-ton steel ore cars. The transportation of iron ore was declared by the War Production Board (WPB) as a necessity for winning the war. This allowed the DM&IR to receive 1,500 70-ton steel ore cars in 1942 and an additional 500 cars dur-

Fifteen EMD SW9s were delivered in 1953. Good for switching but too light for road ore trains, they would all be off the roster by 1963. Shown is the Endion Transfer crew at the NP Rice's Point Yard, Duluth, Minnesota, on August 3, 1957 with SW9s 21 and 18, former D&IR caboose C-180 and ore cars loaded with coal. Note the raised brackets supporting early VHF radio antennas on the SW9s. (D.P. Holbrook collection)

ing 1943 to handle the swelling tonnage of ore. On January 1, 1946 a total of 12,088 ore cars were on the roster out of 13,602 total cars. Only 254 ore cars were retired during the war, a testament to the shops at Proctor and Two Harbors.

The war years took their toll on the older 50-ton ore cars and replacements were needed. From 1948 to 1957, 6,000 70-ton ore cars were acquired from four builders. All were identical in construction, with Enterprise bottom doors, and rode on ASF A-3 Ride Control trucks. They would become the standard ore car for the DM&IR. A total of 10,087 50-ton ore cars were on the roster in 1948 declining to 4,037 cars in 1957, a net reduction of almost 60 percent. Reduced ore shipping caused no 50-ton ore cars to be used in ore service during 1958. From 1959 to 1965 50-ton ore cars were stored and only placed in service during traffic surges. Eight years later in 1965 one 50-ton car remained. Except for work service, the 50-ton ore car era was over.

Economic changes on the iron ranges of Northern Minnesota began to have their effect on the DM&IR. The first diesels were delivered in 1956 and complete replacement of steam accomplished in 1960. Ore tonnages declined from 48,799,815 tons in 1951 to 18,418,952 tons in 1961. The DM&IR during 1953 was serving four mines on the Vermillion and 110 mines on the Mesabi iron ranges. This would decline to three mines on the Vermillion and 58 mines on the Mesabi iron ranges in 1963. Notable in 1953 were two plants, Erie Preliminary Plant operated by Pickands Mathers at Aurora, Minnesota and Pilotac operated by U.S. Steel at Virginia, Minnesota shipping taconite which would eventually replace natural ore.

High-grade iron ore was playing out and steel companies began to obtain ore from Canada, Venezuela and other foreign countries. This caused consolidation of all ore shipping at Duluth and the closing of Two Harbors ore docks in the spring of 1963. DM&IR began analyzing every expense and purchase, trying to remain profitable. Cutting costs became a primary management focus. Two Harbors ore docks would reopen in 1966 as taconite plants opened and lake shipping distances were shorter from Two Harbors versus Duluth.

Pulpwood shipments remained high during the 1950s, but the age of equipment was

DM&IR Proctor Yard in 1959. From right to left we see two tracks containing empty ore cars assembled into departing trains. The third track begins with 21114, a Class U13 ore car built by Pullman Standard in 1925, with cast steel underframes supplied by General Steel Castings. The sixth car from the left is one of two cars, 23176 or 23177. The eighth car from the left is one of five cars, 23171–23175. All seven, with two different side sheet arrangements, were built as demonstrator 70-ton ore cars by Standard Steel Car in 1923 as STDX 171–177. DM&N would purchase and renumber them in 1931. The fifth track from bottom shows the first box car from left being 3151, a Class P1 box car built by Standard Steel car in 1951 as a single-door box car and rebuilt with a 12-foot door opening in 1910 for automobile service. The unique end platform coal hoppers in the background are from 50 Class Q2 cars in 4700–4749 series cars built by The Ryan Car Co. in 1937. (Basgen Photography, D.P. Holbrook collection)

Overleaf: Proctor, Minnesota, Yard on September 3, 1953 finds steam everywhere. 1953 was a watershed year for DM&IR with the first diesels—EMD SW9s—arriving, and the highest ore tonnage ever shipped, 49,317,625 tons, over DM&IR during the shipping season. Noteworthy in this photo is the string of gondolas; PRR, B&O, MILW, EJ&E and others, loaded with iron ore being shipped all-rail to an eastern destination. Note the loaded ore car in the bottom left corner with pooled water from "washed iron ore." (John Szwajkart)

Caboose C-58 is at Rainy Junction Yard, Virginia, Minnesota, during July 1976 tied on to a string of loaded coal hoppers received from BN for movement to Minnesota Power & Light's Laskin power plant at Aurora, Minnesota. (George LaPray)

causing a rapid retirement of rack flats. No new flat cars or gondolas had been purchased since 1941. DM&IR had, for many years, requested empty cars from connecting carriers at Duluth to assist in car supply for pulpwood and steel loading. This caused the railroad to pay car hire and per-diem to foreign railroads for the use of their equipment, and the division of revenue was not covering the cost of these borrowing habits. A decision was made to purchase second-hand equipment to cover on-line loading that would allow DM&IR to earn per-diem when cars were off-line.

The first proposal in 1965 was to acquire second-hand: 150 rack-end gondolas for pulpwood, 50 billet gondolas for steel billet loading, 10 65-foot gondolas for reinforcing rod loading, two flat cars for moving mining equipment, and 100 open hoppers for coal shipments. All had been acted on by the late 1960s. This was followed in 1966 with purchase of 20 second-hand 50-foot box cars for wood chip and other loading. DM&IR had greatly reduced requests to connecting carriers to supply empties for everyday loading, reducing the daily per diem cost for using foreign-owned equipment.

By the late 1960s, the railroad had stemmed the flow of red ink.

THE TACONITE YEARS, 1964 UNTIL 2004

By the early 1960s, natural ore was beginning to play out. Additional sources of ore were needed to feed America's steel industry. During the 1930s and 1940s the University of Minnesota led efforts to find an economical method to extract iron from lower grade taconite rock. The first large scale efforts to process taconite using this new process were begun in 1949 by newly formed Erie Mining Co. This was a purpose-built plant and railroad between Hoyt Lakes, Minnesota, and docks at Taconite Harbor on Lake Superior. Followed quickly in October 1955, Reserve Mining Co. built a railroad between their mine at Babbitt, Minnesota, and a taconite plant and docks at Silver Bay on Lake Superior.

After two world wars, iron ore reserves in the Vermillion and Mesabi iron ranges, were declining and the Korean conflict would further deplete those reserves. The Vermillion Range ended shipments of ore with the clos-

Crude taconite was shipped via rail from T-Bird North and T-Bird South loading pockets to the Fairlane pellet plant located at Forbes, Minnesota. Here a T-Bird crude ore train with rebuilt SD-M 310 is unloading at the Eveleth Taconite Plant at Forbes on September 27, 2001. (D.P. Holbrook)

ing of the Soudan Mine in 1962 and the remaining mines at Ely, Minnesota, in 1963. U.S. Steel began testing taconite processing methods during 1953 with the experimental Pilotac plant near Mountain Iron, Minnestoa and companion Extaca plant at Virginia, Minnesota. Pilotac was constructed for the concentration of low-grade taconite. This material was shipped to the Extaca plant at Virigina for additional processing into nodules and sinter.

Construction of a large-scale taconite plant was a huge investment and U.S. Steel was reluctant to invest until voters in Minnesota passed the Taconite Amendment to the state constitution in 1964. This gave critical tax relief to steel companies for taconite mining.

Two taconite plants quickly located on the DM&IR, Eveleth Taconite Co. (EVTAC) at Forbes, Minnesota, which shipped their first taconite pellets in 1965. This was quickly followed by U.S. Steel's Minntac plant at Mountain Iron, which shipped their first pellets in 1967. Ten year later, in 1977, Inland Steel built their Minorca plant near, served by DM&IR via trackage rights on Duluth, Winnipeg and Pacific.

Between 1967 and 1968 taconite shipments surpassed iron ore shipments. December 31, 1978, showed a total of 6,982 ore cars on the roster: 1,560 in mini-quad sets for taconite pellets, 280 assigned to T-Bird crude taconite service, 2,506 in natural ore service, 2,061

DM&IR would acquire 20 rebuilt SD40-3s between 1996 and 1997 to replace SD9/18s and SDMs. Iron Junction, Minnesota, September 27, 2002, finds the Extra 419 North with Minorca empties meeting the Extra 406 South with Minntac pellets. Note the stylized University of Minnesota style "M" on the 419, introduced in late 1997. (D.P. Holbrook)

stored serviceable, 472 leased to other carriers and the remaining cars out of service for repairs. The natural iron ore era was coming to a close. The last natural ore was shipped in 2002.

Many of the early 70-ton ore cars were sold off to Lake Superior & Ishpeming, C&NW, Birmingham Southern, Bessemer & Lake Erie, U.S. Steel and other industrial concerns. Ownership of 70-ton ore cars went from 8822 in 1953 to 6685 in 1980. The newest ore cars were rebuilt as taconite cars to service the new taconite plants allowing DM&IR to service the taconite plants without spending additional money on new equipment. All-rail ore movements found DM&IR ore cars in Weirton, West Virginia, Birmingham, Alabama, Granite City, Illinois, Gary, Indiana, Pittsburgh, Pennsylvania and a host of other locations. Numerous railroads; PRR, GN, NP, CB&Q, ITC, MON, L&N, C&NW, EJ&E, SOO, MILW, PC, IHB and others hosted these movements. Leasing would find ore cars on RDG, N&W, ACL, NP, PRR, SCL, RI, SOO, MILW, EJ&E, LS&I, C&NW and other roads taking the cars from coast to coast.

U.S. Steel, owner of the DM&IR, changed their name to USX Corporation in 1986. The transportation division of USX was moved to a subsidiary company, Transtar, in early 1988 and in June 1988 the majority share of ownership would be sold to Blackstone Capital Partners who took over on December 28, 1988. Blackstone on March 23, 2001, placed DM&IR in a new company named Great Lakes Transportation that also owned, the B&LE Railroad, boats of the U.S.S. Great Lakes Fleet and the Pittsburgh and Conneaut Dock Co.

Through three major wars, the Great Depression, and the change from natural ore to taconite, the flood of ore would continue until present day, driving the economy of Northern Minnesota and the nation. The economics of size caught up with the DM&IR on May 10, 2004, when the railroad was acquired by the Canadian National. This is the story of the equipment used by the Duluth, Missabe and Iron Range Railroad up to 2004.

MANUFACTURERS AND CAR SHOPS

NORTHWESTERN CAR & MANUFACTURING

The firm Seymour, Sabin and Co. was merged in February 1882 with the Northwestern Construction Co. and the name was changed in March 1882 to Northwestern Car & Manufacturing Co. A promotional brochure showed the company was incorporated May 1882 with $5,000,000 of capital for the manufacturer of threshers, portable and traction farm engines, freight and passenger cars. Prior to the merger George Seymour had secured contracts to construct buildings for the state prison at Stillwater, Minnesota, in 1861 and 1870 and contracted with prison officials to use prison labor to produce barrels.

In 1882 they constructed 1,500 threshing machines, 350 agricultural engines, 1,493 freight cars, and 51 baggage cars and cabooses. The plant was also manufacturing doors, blinds and office furniture. Twelve hundred men were employeed and sales during that year were $2,272,000. The plant was capable of building per day: 25 freight cars, 8 threshing machines, 4 farm engines and 20 farm wagons. By late 1883 the company ensured a continued supply of lumber for building railroad equipment by purchasing 20,000 acres of hardwood timber in Northern Wisconsin.

A major fire occurred on January 7, 1884, destroying the shops of the Northwestern Car & Manufacturing Co. The main shop—300 feet long, 75 feet wide and four stories high—was destroyed. On January 25, 1884, another fire broke out destroying one of the prison buildings, 350 feet long and 60 feet wide. By January 31st a decision was made to rebuild the shops.

Two million dollars in debt and near bankruptcy, on May 10, 1884, it was announced that E.S. Brown of the firm of Hersey, Bean and Brown had been appointed receiver of Northwestern Car & Manufacturing. Financial troubles continued and the company struggled along until early 1887.

A North-Western district court decided to order the sale of the Northwestern Car & Manufacturing at Stillwater, Minnesota, in August 1887 and gave six weeks notice when entering the order. The plant was to be sold by a referee to be appointed by the court. Price was set at $1 million.

It should be noted that sometime between June 1884 and October 1887 the name of the company was changed from Northwestern Car & Manufacturing to Northwestern Manufacturing & Car.

On November 2, 1887. the assets of the Northwestern Manufacturing & Car Co. were sold to Henry D. Hyde for $1,105,000. The final transfer was made on January 5, 1888, when the sale was confirmed and the company name was changed to Minnesota Thresher Manufacturing Co. The new company discontinued building of railroad equipment and concentrated on farming equipment.

Many D&IR equipment diagrams show builders as "Stillwater" or "Stillwater Manufacturing," the actual builder was Northwestern Car & Manufacturing.

DULUTH MANUFACTURING

The Minnesota Car Co. was organized in Duluth, Minnesota, on August 29, 1888. A freight car shop was constructed between 1888 and 1889 and placed into operation in October 1889. The facility was located in West Duluth at Redruth and 55th Avenue West and consisted of four heating furnaces, five Smith gas producers, one 2,500-pound and one 60,000-pound pad hammer, one 10- and one 18-inch train of rolls. Products produced included bar iron, railroad fasteners, railroad axles and other forgings. Annual capacity of 12,000 tons of rolled iron and 5,000 tons of forgings. The de-

Duluth Manufacturing Co manufactured railroad equipment at West Duluth, Minnesota, from August 1891 until 1897 when plant was closed. D&IR and DM&N both purchased equipment from this builder. Photo circa mid. 1890s. (University of Minnesota Duluth, Kathryn A. Martin Library Archives and Special Collections, Collection S2421 Box 5 Folder 7)

sign of the plant allowed for the construction of 15 standard cars or 18 flats and gondolas per day. The general arrangement of the plant included a 58 x 362-foot paint shop with a capacity of 25 cars, an erecting shop, a planing mill, and a machine shop of 97 x 527 feet, one half of which had a capacity of 20 cars. The plant foundry had a capacity of 100 wheels per day. Power for the plant was furnished by two Corliss engines of 250 and 500 horsepower. When the plant was in full operation it employed 1,000 people. Principal officers of the company were President John F.T. Anderson, Superintendent, George H. White and General Manager, William E. Tanner.

In October 1889 Minnesota Iron Car Co. was incorporated as the successor to Minnesota Car Co. The property of Iron Car Co. failed in May 1890 for non-payment of a note for $50,000 and liabilities of $165,000. The Minnesota Iron Car Co. was also involved in the failure. An amicable agreement was worked out amongst creditors to reorganize the company over three months to a year. *Railway Age* reported on September 26, 1890, that the reorganized shops of Minnesota Iron Car Co. were running by October 1, 1890. At that time it was reported that they would be working on an order for 1,500 cars for the Iron Car Equipment Co. During August 1891 it was reported that the plant was sold to Duluth Manufacturing Co.

Duluth Manufacturing became a very busy car builder by March of 1892. At that time they were employing 450 men and building 25 iron cars for the Knoxville, Cumberland Gap and Louisville, 300 25-ton 34-foot box cars for the Lake Erie & Western and six cabooses for the Duluth & Iron Range.

By July 1892, 500 men were employed with the following orders being reported: 100 iron box cars for the Chicago, Hamilton & Dayton, 200 ore cars, 50 flat cars, 10 box cars, 50 logging cars and 6 cabooses for the Duluth, Missabe & Northern, 200 ore cars, 50 flat cars, 10 box cars and 6 cabooses for the Duluth and Winnipeg. Orders were also being filled for 50 log cars for Cranberry Lumber Co., 60 log cars for Mitchell & McClure and 22 log cars for Merritt & Ring. They also had contracts for cast and wrought iron work for several bridges and were selling large numbers of wheels. In August 1895 the Minneapolis & St.Louis ordered 200 box cars. Production of freight cars continued until 1897 when the company was closed.

It was reported in *Railway Age* on November 5, 1897, that Duluth, Missabe & Northern was considering purchasing the plant. This never took place.

By 1897, the New Jersey charter for Duluth Manufacturing was void for unpaid back taxes. The property sat idle from 1896 until 1899 when it was purchased by John E. Searles of New York, who was president of American Cotton Co., but he was unsuccessful in reviving steel production. In later years this site was occupied by American Carbolite Co.

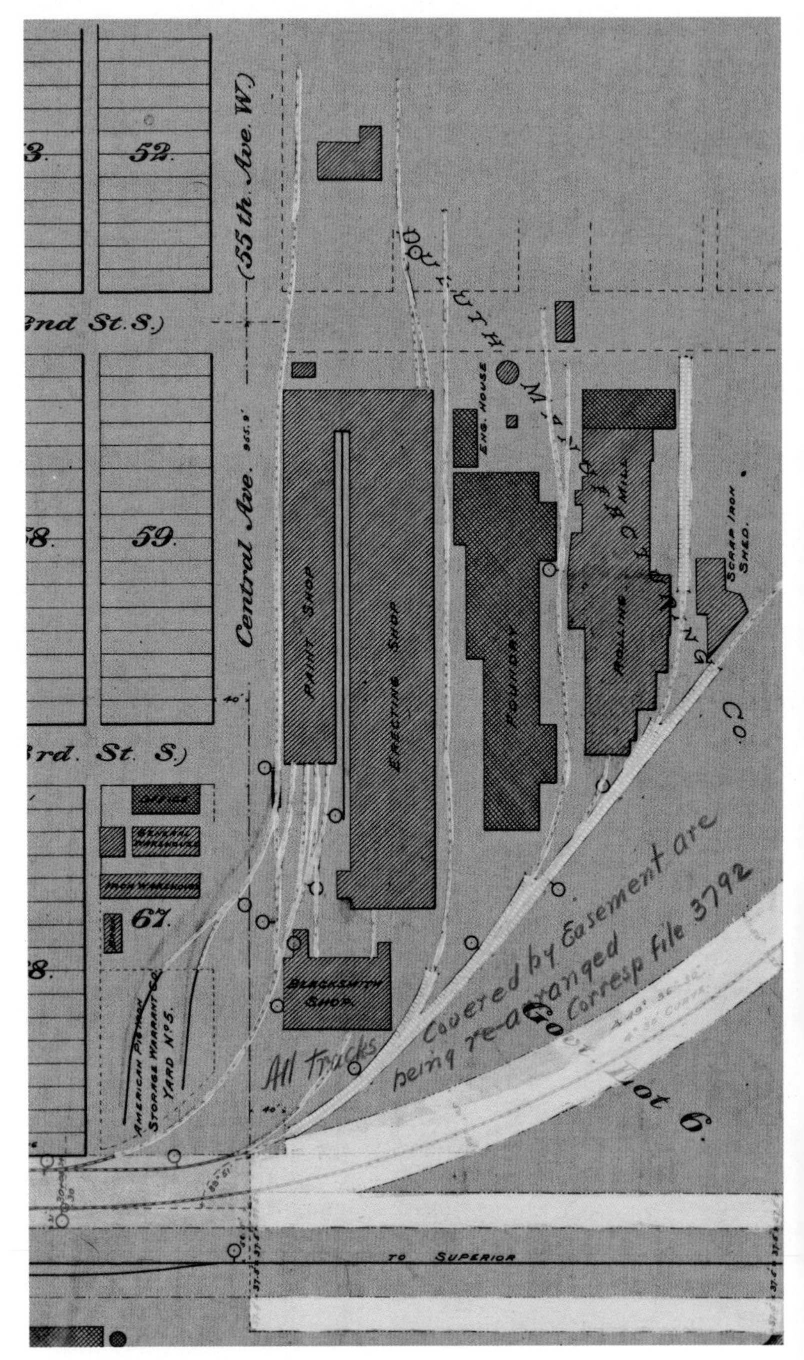

Duluth Transfer Co. map of Duluth Manufacturing Co. at West Duluth, Minnesota, (MRHS Collection)

TWO HARBORS SHOPS

Initially, the D&IR, when constructed, had no outside rail connections and all equipment arrived at Two Harbors by barge. This landlocked situation caused D&IR to quickly construct roundhouse and shop facilities at Two Harbors that would allow maintenance, rebuilding and construction of equipment. Two Harbors shop was unique in that cars spotted at the west end of the building could be moved by gravity through the car shop and when completed gravity could be used to allow cars to roll from the shop facility into the empty yard across from Two Harbors depot.

The first rebuilding of equipment took place here in 1884 when six of the combination flat cars delivered in 1883 were rebuilt to boxcars. Additional flat car-to-box car conversions took place with four more rebuilds in 1885, two in 1886 and two more in 1887.

During 1887 Two Harbors constructed five 25-ton 36-foot flat cars, 45 24-ton 36-foot flat cars and 25 20-ton 34-foot flat cars. Traffic continued to increase and Two Harbors built

D&IR Two Harbors, Minnesota, roundhouse with car shops at left about 1910. Note the string of early 50-ton steel ore cars. (University of Minnesota Duluth, Kathryn A. Martin Library Archives and Special Collections)

Two Harbors car shops and power plant about 1910. Note the three wood ore cars loaded with coal at the power plant. (University of Minnesota Duluth, Kathryn A. Martin Library Archives and Special Collections)

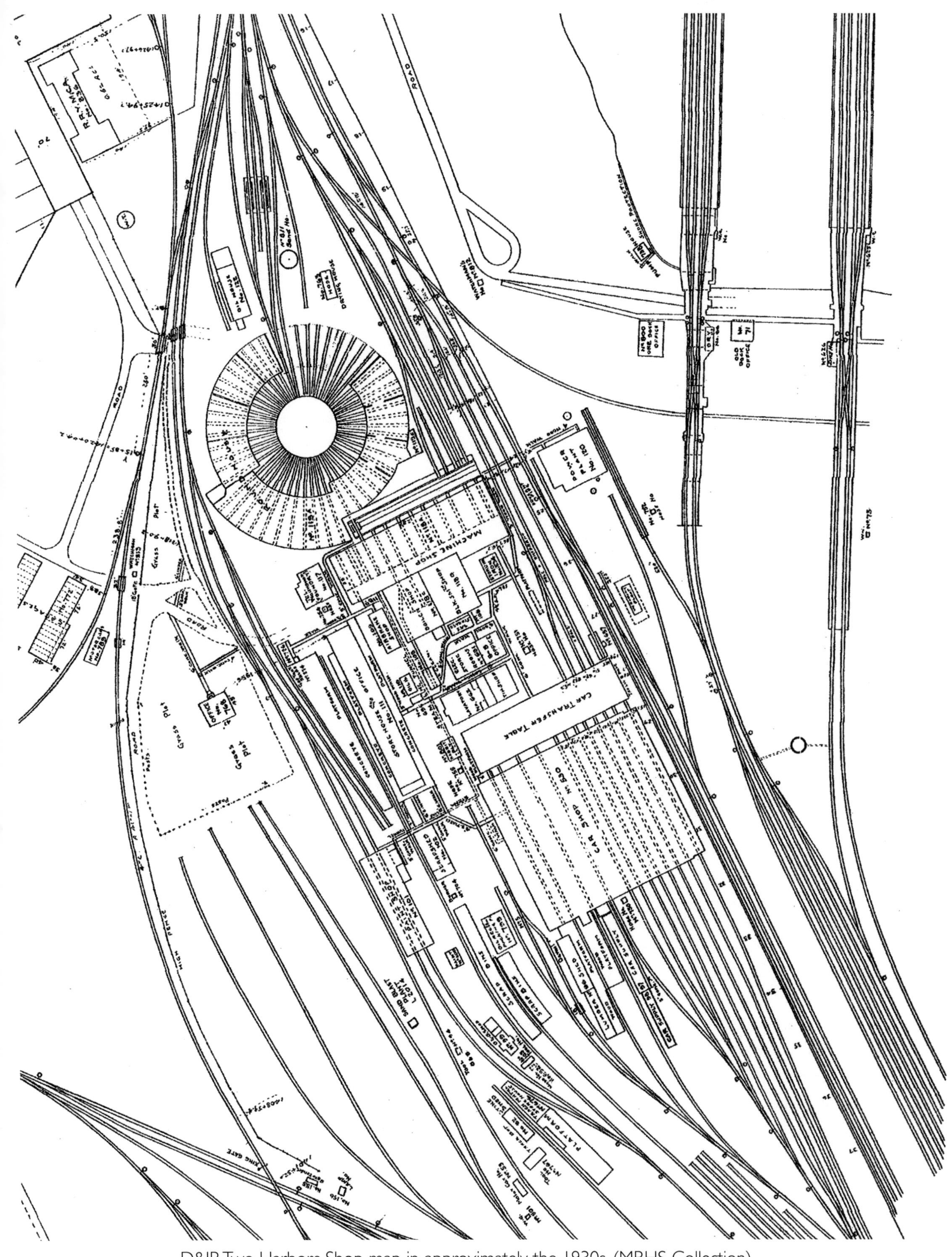

D&IR Two Harbors Shop map in approximately the 1930s. (MRHS Collection)

12 4-wheel cabooses C22–C29 and C35–C38 in 1888.

Railway Age announced on September 13, 1895, that D&IR had contracted with Barnett & Record of Minneapolis to add 23 stalls to the 33-stall Two Harbors roundhouse and further enlarge the shop complex at Two Harbors.

The need for additional box cars caused Two Harbors to rebuild, during 1897 and 1898, 12 additional flat cars to box cars. These were followed in 1899 by an additional 40 30-ton 36-foot flat cars

As logging in Minnesota increased, the D&IR found the need for log cars for hauling cut old-growth timber. Initially they built 50 30-ton 36-foot flat cars in 1900 without trucks. These were classified as "bunker flats" on equipment lists and equipped with chains for log loading. During the winter months trucks off stored wood ore cars were used to place these cars in service. Many of the logging railroads in Northern Minnesota had started to use 22-foot Russell-type log cars to haul logs. Taking notice, D&IR between 1901 and 1910 constructed a fleet of eventually 320 of these cars which were based on the Russell design. The first of these cars were rebuilt from retired wooden ore cars.

Between 1901 and 1905 Two Harbors rebuilt five box cars into stock cars. An additional eight 24-ton 34-foot box cars were constructed in 1904 and numbered 5138–5152(2nd). The need for additional flat cars to haul finished lumber created the need to build 50 additional 36-foot flat cars during the winter of 1905-06. Stock cars X5134–X5136 were rebuilt from 34-foot to 36-foot cars with the capacity changed from 1,865 cu. ft. to 2,358 cu. ft.

Additional cabooses 61–63 were built at Two Harbors in 1905, followed quickly by 64–68 and 83–84 built in 1906, and 80–81 rebuilt at Two Harbors in 1907. During 1905 Two Harbors also built for DM&N 4-wheel cabooses 27–41. These were followed in 1907 by cabooses 42–58 for DM&N. Cabooses 68–70 and 86–87 were built in 1908. Also during the 1908 shop program, D&IR rebuilt 39 36-foot flat cars to 36-foot box cars.

The 50 remaining 4-wheel cabooses were rebuilt during 1909 and 1910 to 8-wheel cabooses and the length changed from 20-feet 8-inches to 31-foot. The updating of the caboose fleet was now complete and Two Harbors returned to building and rebuilding in 1910. Ten 32-foot box cars from the 5150–5172 series were rebuilt to oil cars for the transportation of barrels of oil. This conversion created a car very similar to D&IR stock cars with an open area for ventilation at the top of the car. An additional 50 30-ton 36-foot "bunker flats" were constructed in 1910.

No additional cars were built between 1910 and 1918 but numerous rebuilding projects took place. Replacing wooden underframes with new steel center sills became a priority. Steel center sills were applied to automobile cars 5200 in 1910 and 5202 in 1912. Steel center sills were applied to box cars 5000–5120 during 1913, 5124–5220 and 5264–5340 during 1914.

Gondolas 6100–6124 originally built in 1907, were rebuilt with drop-bottom doors in 1914. Deteriorated ends and roofs were replaced by Murphy steel roofs and ends during 1914 on refrigerator cars 8000–8021. Delivered in 1916, 77 flat cars in series 6180–6279 were rebuilt with wooden gondola sides and ends. Two years later 14 of these cars were rebuilt again with steel gondola sides and ends. Two additional gondolas from the 6125–6179 were rebuilt with steel sides and ends during 1922. The remaining 53 cars in the 6125–6179 series were rebuilt in 1929 with wooden gondola sides and ends.

Appropriation 531 for $236,192.70 was approved in 1916 for construction of new shop facilities at Two Harbors. This included a "new" car repair shop and transfer table, new paint shop, converting existing steel car and tank shop into a two-story mill, carpenter, tin and pipe shop.

During 1916 the shops focused on rebuilding 100 of the 20-foot Russell-type log cars by reinforcing the underframes with pieces of used 60-pound rail and the application of pulpwood racks on all 100 cars. This was caused by the declining logging of old-growth timber and heavier shipping of pulpwood. Beginning in 1914 and completed by 1917, was the rebuilding of 47 flat cars with steel center sills.

Just before USRA took control of D&IR and DM&N, the shops at Two Harbors built ten 40-ton 37-foot refrigerator cars 8032–8041 in 1918. These would be the last "new" built cars at Two Harbors and after USRA control was relinquished, the shops would stick to rebuilding equipment.

Beginning on June 4, 1919, under AFE 54, 73 50-ton ore cars from series 3900–4399, 13 gondolas cars from 201–235 series and 14 from 251–310 series had the common wheel-type hand brake lever replaced with Blackall Ratchet brake levers. The Blackall Ratchet brake levers were unsuccessful and were all removed during June 1926. Also during 1919, Murphy metal roofs began to replace wood roofs on 5000–5340 series box cars. Murphy roofs were applied to 14 box cars in 1923, 15 box cars in 1924 and 10 box cars in 1925.

Experimenting with hand brakes, nine steel ore cars from 3450–9699 series were equipped with Klasing hand brakes during November 1920. Six box cars, one gondola and three flat cars were equipped with Universal hand brakes late in 1922. Between February 20, 1924, and May 31, 1924, 25 gondolas from 201–260 series had the wheel-type handbrakes replaced with Universal hand brakes.

Caboose 89 was rebuilt at Two Harbors in 1920. During 1922 box car 5216 was rebuilt to an automobile car. Wooden brake steps on 4400 and 9000 series 50-ton ore cars had deteriorated and 108 cars had these steps replaced with Irving Subway brake steps during December 1925 and January 1926.

During the winter shopping program of 1927-28, 28 Class E ore cars were rebuilt with longitudinal doors and 10-inch side extensions to increase cubic capacity and unloading speed. This project was not successful enough to be repeated. Previously rebuilt to drop side gondolas, 6100–6124, were converted to solid side gondolas in 1931.

From 1931 until 1964 Two Harbors shop continued to rebuild equipment. After World War II shop functions with Proctor were reorganized. Proctor became the primary ore car facility with remaining merchandise, passenger, and caboose equipment being shopped at Two Harbors. Notably in 1952 the shop rebuilt Class U21 ore car 25600 with new doors with a tire attached on the outside edge to operate the doors. This was the test car for Erie Mining's purposed automated dumping system on their new Taconite Harbor ore dock.

The economies of joint operation and the declining shipments of ore caused the roundhouse, engine terminal and office buildings at Two Harbors to be sold on August 11, 1964, to Universal Fiberglass Corp. for establishment of a plant for the production of fiberglass articles, such as replacement automobile fenders. Final removal of machines from the Two Harbors car shop was started on May 1, 1964 and completed in July. Effective July 10, 1964, the car shop at Two Harbors was closed.

Even with the closing of the shop, DM&IR continued to utilize Two Harbors for rebuilding equipment. The late 1960s found Proctor with a full heavy repair schedule when 200 secondhand gondolas were purchased from EJ&E. Cars required rebuilding and the application of end extensions for pulpwood service. A temporary outdoor production line rebuilding area was established at Two Harbors at the site of the old coal docks adjacent to the Two Harbors depot. The cars had reinforced draft sills and body bolsters applied, drop ends welded in upright position and new end extensions applied. Rebuilding of cars was started in October 1969 and lasted until December 1970. This same location was used starting in 1971 for rebuilding Class Q open hoppers. Offset sides, slope sheets and hopper bottoms

DM&N Proctor, Minnesota car shop interior in 1918. Note the Whiting overhead crane lifting up an entire 50-ton ore car. (University of Minnesota Duluth, Kathryn A. Martin Library Archives and Special Collections)

from these cars were removed and loaded as scrap into gondolas. The remaining ends and center sill had new ribbed bodies, slope sheets and hopper bottoms installed.

DM&N PROCTOR SHOPS

Not as prolific as Two Harbors, DM&N, being the stronger of the two carriers, normally purchased new equipment when needed. Initially the shops involved themselves with rebuilding equipment. Beginning in 1903 Proctor started building complete cars.

Three box cars were rebuilt to refrigerator cars in 1894 and one additional in 1895. One was destroyed in 1896 and the other three were rebuilt at Proctor back to box cars in 1905.

Needing an official car for the owners, Proctor rebuilt Class I caboose 13 to an official car 99 in August 1898. Car 99 would have number removed and be named *Olivette* in March 1907.

Strapped for cash, DM&N purchased 20 25-ton 34-foot box cars from Armour Refrigerator Line (ARL) in 1899 and numbered them 3043–3062. Two cars 3046 and 3062 were rebuilt to stock cars at Proctor Shops and given an "S" prefix to identify them as stock cars. Two additional stock cars were purchased from ARL in 1900 and numbered 4000 and 4001.

From 1899 to 1903, nine wooden ore cars of various classes were rebuilt from 25-ton to 35-ton capacity. The expansion of iron ore production on the Mesabi iron range was occurring so quickly DM&N could not order enough 50-ton steel ore cars to keep up with

Aerial photo of DM&IR car shops at Proctor, MN taken during the 1960s. (D.P. Holbrook collection)

the demand. Unable to purchase new steel ore cars quick enough to fulfill the increasing ore shipments, DM&N rebuilt the newest 1,700 wooden ore cars, 6000–7699 from 1905 to 1907, allowing them to be used for another 10 years, while new 50-ton steel ore cars were acquired.

During 1903 Proctor Shops built 50 30-ton 36-foot flat cars numbered 2229–2279 which were quickly followed by 100 identical flatcars in 1905 numbered 2280–2379.

Two automobile cars numbered 100 and 101 were constructed in 1905 and then quickly renumbered to 3073 and 3074 during the summer of 1905.

Returning to rebuilding, 12 box cars were rebuilt to oil cars between 1907 and 1908

During 1910 and 1911 all 50 4-wheel cabooses were rebuilt at Proctor to 8-wheel cabooses. Three new 36-foot center cupola side door cabooses were built at Proctor, one each in 1911, 1912 and 1913 numbered 59–61. New cabooses were built at Proctor over many years. Two in August 1918, six in 1923, seven 70-76 in 1924 and one additional identical car 77 built in 1926. This would end new construction of cars at Proctor.

After the merger, Proctor became the primary ore car shop and Two Harbors was used as the merchandise car shop. Proctor continued to rebuild equipment from 1924 until 2004 and became the primary DM&IR shop in 1964 when Two Harbors car shop closed.

The three through tracks in the car shop provided space for repairing 36 cars inside. During fall of 1974 the north end of the car shop had black top installed allowing outdoor space for holding 65 ore cars. ■

This view of the Duluth & Iron RangeTwo Harbors, Minnesota, yard and shops in 1915 shows a mix of wood and steel ore cars in the foreground. In the background is a passenger train with, from left to right, Baggage-RPO either 2nd 8 or 2nd 9, and two coaches, either 12, 13 or 14. (University of Minnesota Duluth, Kathryn A. Martin Library Archives Special Collection)

CHAPTER TWO
DULUTH & IRON RANGE
ROLLING STOCK

Before delving into the extensive rosters connected with the Duluth, Missabe & Iron Range, it is necessary to lay a little ground work on car classification, general renumberings, jointly-ordered equipment and the development of the ore car itself, which was at the heart of the fleet.

DULUTH & IRON RANGE CAR CLASSES

The D&IR did not begin to apply car class letters until 1910, when drawings began to show car classes. Starting in January 1911, the Recapitulation of Equipment lists the D&IR compiled each year showed car classes in the summary for ore cars. Class C and D was for "slow-dumping" ore cars and Class E for fast dumping ore cars equipped with newer longitudinal bottom doors. By the time the 1929 D&IR diagram book was issued, car classes started to be applied to most D&IR equipment. By this time, the DM&N, who had started using car classes in the 1895 diagram book—possibly before—and the D&IR, had begun to combine

D&IR Car Classes

Class	*Type of car*
A	
B	
C	Steel slow dump ore cars: 3050–3399
D	Steel slow dump ore cars: 3400–4899 and 9000–9699
E	Steel fast dump ore cars: 9700–12499 and 3049 (former Summers 101)
F	Box Cars
G	Steel underframe Gondolas
H	Gondolas
I	36ft Flat Cars
J	
K	Pulpwood Flat Cars
L	

Class	*Type of car*
M	Stock Cars
N	Oil Cars
O	
P	
Q	
R	Refrigerator Cars
S	Automobile Cars
T	
U	
V	
W	
X	
Y	
X	

operations and mechanical department functions under the U.S. Steel umbrella.

Starting in 1931, numbers began to be added to the letters to denote specific groups of cars. The 1951 diagram book still shows no car class series assigned to former D&IR cabooses and passenger cars.

DM&N Car Classes

A	Official (Business) Cars
B	Passenger Cars
C	Passenger Cars
D	Combination Passenger Cars
E	Baggage-Mail Cars
F	Baggage Cars
G	Four-wheel wood Caboose
H	Eight-wheel wood Caboose
I	Eight-wheel wood Caboose. After 1899 Class I was assigned to stock cars
J	Refrigerator cars
K	Refrigerator cars. After 1907 Class K was assigned to stock cars. After DM&IR merger Class K assigned to former D&IR cabooses
L	Box Cars
M	Box Cars
N	Box Cars
O	Wood Ore Cars. After 1905 Class O was assigned to vehicle cars 3073, 3074
P	Wood Ore Cars. After 1906 Class P was assigned to 40-foot box cars and also used for stock cars
Q	Hart Convertible and open and covered Hoppers
R	Wood Ore Cars
S	Wood Ore Cars
T	Wood Ore Cars
U	Steel Ore Cars
V	36-foot Flat Cars. Starting in 1941 with Class V2, V was also assigned to gondolas
W	34-foot Flat Cars
X	Phelan Flat cars
Y	Flat cars
Z	

Beginning with 1903 car orders, Class N1 boxcars 3063–3072 and Class U1 steel ore cars 8005–8354, a number started to be added to the Class "letter" so succeeding car orders had distinctive class letter and number.

All DM&N and D&IR equipment retained their car classes when merged into DM&IR, with the following exceptions:

D&IR Car Class to DM&IR Car Class

D&IR Class I flat cars	5543–584 to DM&IR Class K 5543–5841
D&IR Class I flat cars	6000–6099 to DM&IR Class K1 6000–6099
D&IR Class G gondolas	6100–6179 to DM&IR Class G1 6100–6179
D&IR Class I flat cars	6180–6304 to DM&IR Class K2 6180–6304
D&IR Class I flat cars	6305–6306 to DM&IR Class K3 6305–6306

This created a conflict with Class H1 and G1 already being used for cabooses on DM&N and Class K through K3 also being used for former D&IR cabooses, but apparently this was overlooked. Freight cars were stenciled with the Class letter and number but cabooses and passenger cars were not.

DULUTH & IRON RANGE 1887 AND 1896 RENUMBERINGS

Two major re-numberings of early D&IR equipment took place between 1884 and 1896.

The first D&IR equipment delivered was 30 combination flat cars. All were ordered from Northwestern Car & Manufacturing and delivered in 1883. No documents have been found to support the numbering of this equipment when delivered. A photo taken at Ely, Minnesota, in 1884 shows a flat car, three box cars, a combination car and one early ore car, with the box cars having two-digit numbers. The number on the end of one of the box cars is 68. Records indicate that in 1884 six of the 30 combination flat cars had been rebuilt into box cars. This process would continue with four more rebuilds in 1885, two in 1886 and two more in 1887. The first cabooses were delivered in 1884 and numbered from 21 through 38. It appears that revenue freight cars, other than ore cars, delivered before 1888 were numbered between 39 and 250, below the 1884-built wood ore cars 251–550.

The first renumbering of this equipment was to place all box cars in a 2200 series with even numbers only and flat cars in the same 2200 series, odd numbers only, and coal cars in a 2000 series. It appears this coincided with the equipping of these cars with air brakes and took place between 1887 and 1888. Ore cars were not renumbered. According to ICC records, the D&IR was the first railroad in the United States to fully equip all of their equipment with air brakes in 1888.

During 1896 and early 1897, another renumbering occurred, possibly as a result of cars having their link and pin couplers replaced with automatic couplers. The 5000-series numbers were chosen to free up the number series from 1999 up for additional ore cars. This placed following orders of ore cars in one continuous number series. Renumbering was as follows:

Coal, side dump	2000–2014 to 236–250
Box cars (even)	2200–2226 to 5122–5132 and 5138–5152 (even)
Stock cars (even)	X2248, X2250 to X5134 and X5136 (even)
Box cars (even)	2228–2302 to 5050–5120 (even)
Flat cars (odd)	2201–2211 to 5001–5011 (odd)
Flat cars (odd)	2213–2259 rebuilt to box cars in 2200–2226 series before 1895
Flat cars (odd)	2261–2269 to 5053–5061 (odd)
Flat cars (odd)	2271–2359 to 5063–5145 (odd)
Flat cars (odd)	2361–2409 to 5013–5051 (odd)
Flat cars (odd)	2411–2809 to 5147–5541 (odd)

This freed up the number series between 2000 and 4999 to number the ore cars.

The renumbering was accomplished during 1896 and early 1897. Both renumberings would have been to make it easier for operating department employees to determine, by car number, if a car had hand brakes only, automatic air brakes, link and pin couplers or automatic couplers. Box cars, flat cars, and coal cars remained in service during winter months making the renumbering necessary for easy identification. Ore cars would not have been renumbered, because the shops at Two Harbors would have concentrated on making these conversions during the winter months when mines were not shipping.

By 1906, the series for ore cars was filled up to 4899 and the next group of ore cars delivered in 1907, 9000–9699 were numbered above the highest-numbered refrigerator cars which occupied the 8000 series.

D&IR AND DM&N RENUMBERING TO DM&IR

The initially application for the consolidation of the D&IR and DM&N was made to the ICC during July 1929. Various conflicts existed between D&IR and DM&N freight car numbers. The first cars renumbered were DM&N refrigerator cars 5034–5058 which were renumbered to 7034–7058 to avoid D&IR boxcars in the 5000 series. This renumbering took place starting in May, 1931 and was completed November, 1931, keeping their DM&N reporting marks.

The lease of the D&IR by the DM&N was approved on January 10, 1930, but the first cars began to be renumbered and receive DMIR reporting marks on August 21, 1937. Recap of equipment reports for both railroads were consolidated by 1934. The actual merger was approved on March 21, 1938.

The merged carriers equipment retained their numbers when re-stenciled DM&IR, but a handful of D&IR cars requiring renumbering. The 8000-series D&IR refrigerator cars were placed in the 7100 series. This placed all refrigerator cars of the combined roads between 7000–7140 numbers. D&IR GS-type gondolas 201–235 and 251–310 were renumbered by placing a "4" in front of the current D&IR numbers. For example, D&IR 201 became DM&IR 4201. All D&IR ore cars conflicted with numbers assigned to DM&N equipment. All D&IR

ore cars with 4-digit numbers: 3900 to 4399, 4400 to 4899 and 9000 to 9699 were renumbered by placing a "3" in front of their D&IR numbers. For example, D&IR 9000 became DM&IR 39000. All D&IR ore cars with 5-digit numbers: 10500 to 11249, 11250 to 11399, 11500 to 12299, 12300 to 12749 and 12750 to 13249 were renumbered by replacing the "1" with a "4". For example, D&IR 11500 became DM&IR 41500.

As late as 1957, there were still D&IR freight cars that had not been changed to DM&IR. By early 1959 all cars remaining from the merger of the two roads had been restenciled and renumbered.

JOINTLY ORDERED EQUIPMENT

After the 1901 takeover of both properties by U.S. Steel, equipment buying remained the same, with both railroads operating independently based on need except for the purchasing of 50-ton steel ore cars. The economics of scale and the need to replace the large wood ore car fleets caused them to place joint orders for the same type of 50-ton cars from various builders, beginning with 1903 order from Pressed Steel Car for 800 50-ton cars being split 350 to DM&N and 450 to D&IR. These buying practices would continue until 1911 with the purchase of 10 Rakowsky 50-ton ore cars, five for DM&N and five for D&IR. After 1911, with U.S. Steel under much scrutiny by the government as a monopoly, the buying habits returned to more independent and non-joint purchases.

It is interesting to note that with the 1910 orders it appears that U.S. Steel began to set aside blocks of numbers to renumber D&IR equipment with the same type and builder with DM&N equipment. Blocks of numbers were set aside in 1910 for D&IR equipment and any additional equipment orders from the same builders to be grouped in one number series. AC&F equipment was assigned 12000–13999, Western Steel Car and Foundry 14000–15999 and Standard Steel Car 16000–17999. This thought process was short lived and by 1911 individual allocation of numbers without regard to D&IR or DM&N ownership continued. This resulted in the DM&N 13000–13999 and 17000–17999 series never being utilized for equipment numbering. The one exception was the 1913 order for 1500 50-ton ore cars that was placed in the 14500–15499 series grouped with identical cars in the 14000–14149 series, but apparently as an afterthought. This kept the same builder within this series.

Joint ordering of equipment was not restricted to freight cars. Both roads in 1908 jointly ordered baggage cars from ACF.

THE NEED FOR SPEED

Shipments of iron ore increased from 7.5 million tons in 1899 to over 21.5 million tons in 1907. Mines had the ability to ship as fast as headframes and shovels could load. Steel mills could store in stockpiles and use ore at whatever rate was needed. Loading and unloading the lake carriers was the sticking point in the movement of iron ore. Beginning in early 1907 the process became a focal point for improved productivity.

Original wood and steel ore cars had 4 small transverse drop doors, causing iron ore railroads to employ "punchers" to poke the sticky ore by standing on end platforms and poking at the ore with long poles to get the ore to flow out of the car into ore dock pockets. During the summer it took an average of 15 to 20 minutes to empty each car utilizing six men on each carload. Besides being a slow and labor-intensive process it also placed employees in harm's way with the possibility of falling off an ore car or worse yet, into a dock pocket. In the summer the computed cost of labor in unloading was 0.75 cents per ton and during freezing periods it increased to 2 cents or upwards of 5 cents for cars having sticky frozen ore.

The earliest efforts to increase speed were

the development of large longitudinal door openings. Two of the earliest developers of these doors were the Summers Car Co., which had a drop-bottom which separated longitudinally, and the Clark Car Co., which had a door which would slide on rollers. Charles H. Clark, president of Clark Car Co., developed the Summers type ore car while working as an engineer for the Summers Steel Car Co. He left Summers in 1907 and was involved in organizing Clark Car Co. with his newly developed Clark doors. Both were dropped or raised by hand as in a standard car and took from two to three minutes to open and close.

The first Summers self-clearing ore car, Number 101, was tested beginning in the spring of 1908, when it was placed in experimental service on the Duluth and Iron Range. This car would later be purchased by the D&IR and renumbered 3049 in 1914. Before this car arrived the wooden and steel ore cars had a 45-degree hopper slope and two bottom door openings 42-inch wide by 38-inch long, a total of 11.2 square feet. These small openings prevented the clear, unassisted dumping of ore on the docks. Car No. 101 had a single bottom door opening of 5-foot 6-inch by 8-foot 6-inch square for a total of 40 square feet. When the doors were open they had a 50-degree inclination that matched the 50-degree slope sheet inclination. This large door opening, both transversely and longitudinally had the effect of breaking the bridging effect of the ore causing it to flow easily out of the car. Worm gears on the door opening mechanism allowed the operator to open the doors slowly to any angle desired to prevent the ore from dropping out all at once which might cause damage to the ore dock. The tests proved successful and the D&IR ordered 800 cars numbered 9700–10499 that were placed in service in the spring of 1909. A number of loading records were established with the new cars including the loading of 9,311 tons of ore in 39 minutes into the steamer *William E. Corey* at Two Harbors on October 10, 1909.

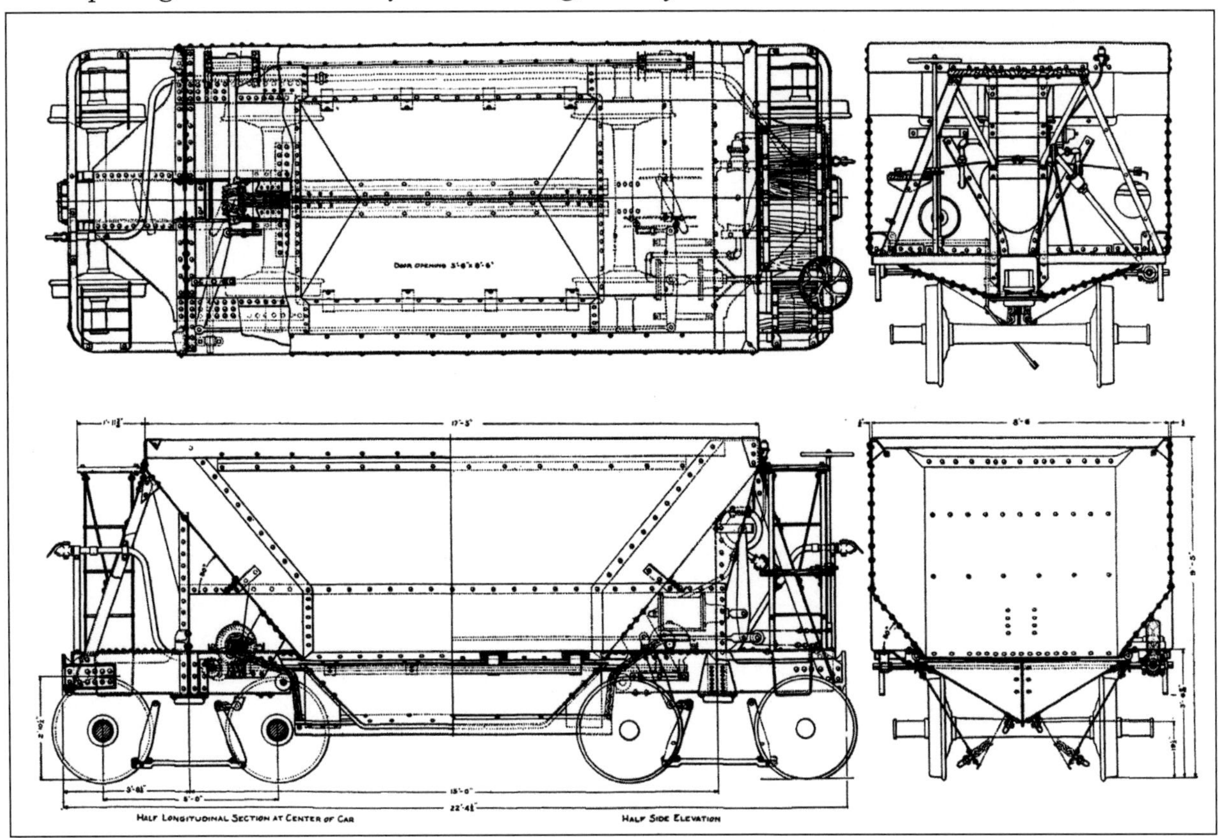

Plan and Elevations Summers Ore Car – DIR 9700-10499. (D.P. Holbrook collection)

The Clark Car Company, which was organized in 1907, developed the first balanced-door ore car and delivered the first of these cars to the DM&N in 1909. As mentioned, the car was developed by Charles H. Clark, while he was an engineer for Summers Steel Car Company.

During the spring of 1909 DM&N President W.A. McGonagle invited various car companies to a competitive test of their respective cars on the DM&N. The following cars were supplied for testing:

Ore Cars Dumping Tests on DM&N: 1909

			Per car avg. time		Per man per car		
Car	No. of tests	Avg. men per car	Min	Sec	Min	Sec	Failure to operate
Summers # 101	12	5.00	4	21	21	43	None
Summers # 9983	14	2.80	2	41	7	54	None
Summers # 10384	12	2.25	1	39	3	43	None
Rakowsky # 100	32	1.47	2	00	2	56	1
Clark # 170	25	4.50	3	18	14	51	2
Clark # 210	5	1.00	1	37	1	37	None

The Clark Car No. 170, the first of its kind built with a 45-degree end slope sheet, compared favorably with the cars entered. The Summers Car No. 101 with a 50-degree end slope was also a sample car. The Clark car 170 was damaged at the mines and withdrawn from further testing on August 1, 1909, and on September 9, 1909, Clark No. 210 was entered in the competition. The Summers Car No. 10384 and the Rakowsky car No. 100 each had doors that were operated by air. The Clark Car No. 210 was hand-operated on two tests and air on the remaining three, with manual and air taking the same amount of time to unload. This proved that the balanced doors on the Clark car required very little effort to close. The doors on the Clark car also locked together in the center and did not gap in the center like other ore cars. All of the newer quick-dumping cars had a hopper bottom opening from 40 to 50 square feet of area.

It was noted that the most time consuming effort in unloading these cars was the unwinding and winding of the doors. A sample car was built by Standard Steel Car, Hammond, Indiana in 1909. This car was extensively tested at the builder's shop while under load and given a series of bumping test from 4 to 18 mph, when two cars were damaged. This car, and one additional Summers car with air-operated doors, one each for service on the D&IR and DM&N, were placed in in service in July 1909.

Dumping of ore using the geared mechanism took from 1 minute 24 seconds to 2 minutes 14 seconds. The air-operated doors allowed dumping of the entire load within 12 to 14 seconds. Cars equipped with the air-operated doors could be unloaded either manually or by air. D&IR equipped an unknown number of additional Summers cars with air-operated doors.

A later advance was the Rakowsky 50-ton automatic unloading ore car manufactured by the Automatic Unloading Car Co., Division of Northwestern Steam Boiler Co. of Duluth. Their plant was located on Garfield Ave. and covered a street frontage of 400 feet. They manufactured boilers, engines, stand pipes, refuse burners, smokestacks and mining equipment. Construction of the first Rakowsky Patent Ore Car was announced in the *Duluth Evening Monogram Magazine* on April 29, 1909. It opened instantly and freely by having two halves (buckets) that when opened struck against the car axles causing the ore to fall out of the halves. This car was developed by Gustav A. Rakowsky, a mechanical engineer, from Duluth, who applied for patents for his unique design in 1910 and was awarded patents 973208, 973209, and 973210.

To open the halves there was a 12-inch vertical air cylinder at each end which, when open, created a 52½-degree slope having an opening of 5 feet to discharge the ore. The buckets rotated on trunions located on the main frame top members. They were set so the weight of the load assisted in the dumping. Total time to unload was about 10 seconds and in most cases did not require any manual labor to finish unloading the car. A hand device was available for manual unloading.

The DM&N began testing the prototype Rakowsky car during the summer of 1908. This testing showed various weak points in the car design. A redesigned version of the car was delivered to DM&N in 1909. The 50-ton, 22-foot car had a capacity of 900 cu. ft. with a tare weight of 33, 900 pounds.

Comparison of ore cars				
	Present Type	Summers (SSC)	Clark (ACF)	Rakowsky
Cubic Capacity	689	650	650	700
Weight of Car	32,500	32,500	32,090	33,900
Slope of Surface (dumping) degrees	45	51	50	52½
Center of gravity from rail	6 ft. 7½ in.	6 ft. 9½ in.	6 ft. 9½ in.	6 ft.
Pay-load to total weight	76%	76%		80%

The D&IR purchased five of the Rakowsky type cars in 1911 numbering them 3045 to 3049. DM&N purchased five of the Rakowsky type cars in 1911 numbering them from 18000 to 18004. Loading of iron ore was unforgiving on the Rakowsky cars and they had a short service life.

Until 1940, opening and closing of doors was done mechanically. During 1940, the first of the gasoline-powered trapping machines to open and close doors, was placed in service at Duluth and Two Harbors. This allowed quicker opening and closing of the ore car doors. These greatly assisted with the war effort during World War II.

Starting by 1942, many of the Class U3 8405–10554, U4 10555–11704, and Class D 3900–4399, 4400–4899, 9000–9699 began to have carbody interiors modified by adding a corner slope sheet between the end and side slope sheets. The change is visible in photos, with the end slope sheet ending at the side sheet between the first and second rib on the left and the fifth and sixth rib on the right.

The new slope sheet arrangement eliminated the rivets on the side sheets with all plates being welded together. This made the cars better at self-clearing when being unloaded. No paper documentation has been found for this program but it is supported by photographs from the period and it is unclear how many cars were modified. The War Production Board (WPB) during the war had strict restrictions on use of steel and speculation is that DM&IR hid the rebuilding process from the WPB. Over 150 of the modified cars were retained for company ballast service in later years.

The next speed modification took place in February 1969 when two Class U-30 cars,

Class U3 and U4 ore cars as built had ends of side sheets slope to the first C-channel rib as depicted on the right two cars, W8513 and W39218 (former 9218). Starting in early 1940, an unknown number of Class U3 and U4 ore cars were rebuilt with corner slope sheets and side sheets were modified as depicted on the left hand car, W8993. Proctor, August, 1978. (D.P. Holbrook collection)

Two Class U30 ore cars were sent to Pullman Standard in February 1969 for rebuilding with air operated bottom dump doors. DMIR 32880 at Michigan City with new doors applied. (Pullman-Standard photo, Jim Kinkaid collection)

Three regular ore cars were rebuilt by Trinity, Miner Industries and Johnstown America in 1999–2000 to test air operation. DMIR 30412 at Proctor on September 30, 1999 was rebuilt by Johnstown America and painted in this gray paint scheme. (D.P. Holbrook)

32880 and 32675, were returned to Pullman Standard under job order JO 40053 and equipped with Parker-Hannifin air cylinders and hand valves as part of an air-operated bottom dump door arrangement. Both of these cars were rebuilt to taconite cars with 19½ inch-high sides on July 28, 1969, repainted and renumbered 52880 and 52675 and retained the option of air-operated doors.

During 1997 the DM&IR was looking to replace the aging ore car fleet. Various car builders were contacted to design and manufacturer new cars with quicker opening and closing doors. Three manufacturers, Miner, Johnstown America and Trinity Industries assisted DMIR in converting two Class U29 (31056 and 31456) and one Class U-27(30412) cars with air-activated door dumping systems. DMIR 31456 was converted by Miner Industries in 12/99 and painted blue. DMIR 31056 was converted by Trinity Industries in 9/2000 and painted maroon. DMIR 30412 was converted by Johnstown America in 1999 and painted gray with maroon lettering. This car was transported to Johnstown by truck and returned to Proctor on August 16, 1999. All three cars underwent extensive testing in both pellet and natural ore service during 1997 thru 2003. The CN merger would intervene and the new taconite cars would come painted for CN with DMIR reporting marks. ■

Miner Industries installed air-operated doors on 31456 in December 1999. It was repainted into this blue paint scheme to identify it as a test car. Proctor, December 26, 1999. (D. Schauer)

Trinity Industries converted DMIR 31056 with air-operated doors in September 2000 as a test car for future ore car orders. Specially painted, it is at Proctor, September 30, 2000. (D.P. Holbrook)

DULUTH AND IRON RANGE

D&IR Roster: 1883-1886

Number Series	Class	Description	Total	O.L.	Cu.Ft	Lbs.	Builder	Built	Trucks	Notes
31–89o?	–	Combination Flat car	30	34'0"			Northwestern Car & Mfg.	1883	DIR Std	1
40–42e, 48–70e?	–	Combination Box	14	34'0"			Northwestern Car & Mfg.	1883	DIR Std	2
44–50e?	–	Combination Stock Car	4	34'0"			Northwestern Car & Mfg.	1883	DIR Std	3
236–250	–	Coal Car	15	34'6"	1022	50,000	Northwestern Car & Mfg.	1884	DIR Std	4
251–550	–	Wood Ore Car	300	28'0"	259	44,000	Northwestern Car & Mfg.	1884	DIR Std	

1. 30 flat cars delivered in 1883 possibly numbered 31-89 odd. Re-numbered to 2201-2259(odd) in 1887. Rebuilt to box cars; 6 in 1884, 4 in 1885, 2 in 1886, 2 in 1887. Rebuilt to stock cars; 4 in 1885. 2 were numbered 48 and 50 other two had unknown numbers.

2. Re-numbered to 2200–2226 (even) in 1887

3. Re-numbered to 2248–2250 (even)in 1887

4. Re-numbered to 2000–2014 in 1887

D&IR Roster: 1887-1895

Number Series	Class	Description	Total	O.L.	Cu.Ft	Lbs	Builder	Built	Trucks	Notes
251–550	–	Wood Ore Car	300	28'0"	259	44,000	Northwestern Car & Mfg.	1884	DIR Std	1
551–650	–	Wood Ore Car	100	28'0"	259	44,000	Wells & French	1887	DIR Std	1
651–750	–	Wood Ore Car	100	28'0"	250	44,000	Haskell & Barker	1888	DIR Std	1
751–1020	–	Wood Ore Car	270	22'0"	269	50,000	Haskell & Barker	1889	DIR Std	1
1021–1145	–	Wood Ore Car	125	22'0"	270	48,000	Wells & French	1892	DIR Std	1
1201–1500	–	Wood Ore Car	300	22'0"	270	54,000	Pullman Car Co. Lot 2095	1895	DIR Std	1
1501–1541	–	Wood Ore Car	41	22'0"	269	54,000	Lafayette Car Works	1890	DIR Std	1
1542–1666	–	Wood Ore Car	125	22'0"	270	48,000	Wells & French	1892-1893	DIR Std	1
1667–1991	–	Wood Ore Car	325	22'0"	270	48,000	Wells & French	1893	DIR Std	1
1992–2191	–	Wood Ore Car	200	22'0"	305	60,000	Duluth Mfg Co.	1895-1896	DIR Std	1
2000–2014	–	Coal, Side dump	15	34'6"	910	50,000	Northwestern Car & Mfg.	1884	DIR Std	2
2200–2226e	–	Box Car	14	34'0"	1865	48,000	Northwestern Car & Mfg.	1883	DIR Std	3
2228–2302e	–	Box Car	38	34'0"	1865	50,000	Wells & French	1887	DIR Std	4
X2248, X2250	–	Combination Stock car	2	34'0"	–	50,000	Northwestern Car & Mfg.	R1887	DIR Std	
2201–2211o	–	Combination Flat Car	6	34'0"	–	48,000	Northwestern Car & Mfg.	1883	DIR Std	5
2213–2259o	–	Combination Flat Car	24	34'0"	–	50,000	Northwestern Car & Mfg.	1883	DIR Std	6
2261–2269o	–	Flat Car	5	34'0"	–	50,000	D&IR Two Harbors	1887	DIR Std	7
2271–2359o	–	Flat Car	45	34'0"	–	48,000	D&IR Two Harbors	1887	DIR Std	8
2361–2409o	–	Flat Car	25	34'0"	–	40,000	D&IR Two Harbors	1887	DIR Std	9
2411–2809o	–	Flat Car	200	36'0"	–	60,000	Lafayette	1888	Thielson	10

1. Not renumbered in 1896

2. Renumbered to 236–250 in 1896

3. Renumbered to D&IR 5122–5132e and 5138–5152e series in 1896

4. Renumbered to D&IR 5050–5120 series even numbers in 1896. Two cars destroyed by 1896 renumbering

5. Renumbered to 5001–5011o in 1896

6. Entire series rebuilt to box cars 2200–2226 and MofW support cars before 1896 renumbering

7. Renumbered to 5053–5061o in 1896

8. Renumbered to 5063–5145o in 1896

9. Renumbered to 5013–5051o in 1896

10. Renumbered to 5147–5541o in 1896

D&IR Roster: 1895-1937

Number	Class	Description	Total	O.L	Cu.Ft.	Lbs	Builder	Built	Trucks	Notes
201–220	H	Steel Gondola	20	41'0"	1532	100,000	Pressed Steel Car Co	1908	Arch Bar	
221–235	H	Steel Gondola	15	41'0"	1532	100,000	Pressed Steel Car Co	1907	Arch Bar	
236–250		Coal Car	15	36'0"	906	60,000	Northwestern Car & Mfg.	1884	DIR Std	1
251–260	H	Steel Gondola	10	41'8"	1532	100,000	Pressed Steel Car Co	1908	Arch Bar	
261–310	H	Steel Gondola	50	41'8"	1532	100,000	Western Steel Car & Foundry	1917	A. U-section	2
251–550		Wood Ore Car	300	28'0"	259	44,000	Northwestern Car & Mfg	1884	DIR Std	
551–650		Wood Ore Car	100	28'0"	259	44,000	Wells & French	1887	DIR Std	
651–750		Wood Ore Car	100	28'0"	250	44,000	Haskell & Barker	1888	DIR Std	
751–1020		Wood Ore Car	270	22'0"	269	50,000	Haskell & Barker	1889	DIR Std	
1021–1145		Wood Ore Car	125	22'0"	270	48,000	Wells & French	1892	DIR Std	
1146–1200		Wood Ore Car	55	22'0"	454	60,000	THC&M Co	1895-1896	DIR Std	3
1201–1500		Wood Ore Car	300	22'0"	270	54,000	Pullman Car Co. Lot #2095	1895	DIR Std	
1501–1541		Wood Ore Car	41	22'0"	269	54,000	Lafayette Car Works	1890	DIR Std	
1542–1666		Wood Ore Car	125	22'0"	270	48,000	Wells & French	1892-1893	DIR Std	
1667–1991		Wood Ore Car	325	22'0"	270	48,000	Wells & French	1893	DIR Std	
1992–2191		Wood Ore Car	200	22'0"	305	60,000	Duluth Mfg, Co	1895-1896	DIR Std	
2192–2548		Wood Ore Car	450	22'0"	454	60,000	THC&M Co	1896	DIR Std	3
2549–2848		Wood Ore Car	300	22'0"	454	60,000	Illinois Car & Equipment Co.	1898	DIR Std	
2849–3048		Wood Ore Car	200	22'0"	454	60,000	Illinois Car & Equipment Co.	1899	DIR Std	
3045–3049		Steel Ore Car	5	22'1"	700	100,000	Northwestern Steam Boiler Co	1911	A. L-Section	4
3049		Steel Ore Car	1	22'	580	100,000	Summers Car Co (Std Steel Car)	1908	Arch Bar	5
3050–3399	C	Steel Ore Car	350	22'	670	100,000	Pressed Steel Car Co	1900	PSC Arch Bar	
3400–3449	D	Steel Ore Car	50	22'0"	680	100,000	ACF Lot #2517	1903	Arch Bar	6
3450–3899	D	Steel Ore Car	450	22'0"	680	100,000	Pressed Steel Car Co	1903	Arch Bar	
3900–4399	D	Steel Ore Car	500	22'0"	710	100,000	Standard Steel Car Co	1905	Arch Bar	
4400–4899	D	Steel Ore Car	500	22'0"	710	100,000	Pressed Steel Car Co	1906	Arch Bar	
4400 / 4656	E	Steel Ore Car	22	22'0"	710	100,000	Pressed Steel Car Co	R1926/27	Arch Bar	7
5000–5048e	F	Box Car	25	34'0"	1865	50,000	Duluth Manufacturing Co	1895	DIR Std	8
5050–5120e	F	Box Car	36	34'0"	1865	60,000	Wells & French	1886	DIR Std	9
5050–5120e		Vegetable cars	11	34'0"	1865	60,000	Rebuilt D&IR Two Harbors	1889	DIR Std	10
5122–5132e	F	Combination Box	6	34'0"	1865	48,000	Northwestern Car & Mfg	1884	DIR Std	11
5122–5130e	F	Combination Oil Car	5	34'0"	1865	48,000	Rebuilt D&IR Two Harbors	1896 to 1908	DIR Std	12
X5132		Combination Stock	1	34'0"	1865	48,000	Rebuilt D&IR Two Harbors	R1901	DIR Std	13
X5134–X5136e		Combination Stock	2	34'0"		40,000	Northwestern Car & Mfg	1887	DIR Std	14
X5134–X5136e	M	Stock Car	2	36'2"	2358	60,000	D&IR Two Harbors	1905	DIR Std	15
5138–5152e (1st)		Combination Box	8	34'0"	1865	48,000	Northwestern Car & Mfg	1884	DIR Std	16
5138–5152e(2nd)	F	Box Car	8	36'2"	2358	60,000	D&IR Two Harbors	1904	DIR Std	17
5150	N	Oil Car	1	36'2"	2358	40,000	Rebuilt D&IR Two Harbors	R1910	DIR Std	18
5154–5172e	F	Box Car	10	34'0"	1579	40,000	Rebuilt D&IR Two Harbors	R1896	DIR Std	19
5154	N	Oil Car	1	32'5"	1579	40,000	Rebuilt D&IR Two Harbors	R1910	DIR Std	20
5156	N	Combination Oil Car	1	32'5"	1579	40,000	Rebuilt D&IR Two Harbors	R1910	DIR Std	21
5158	N	Combination Oil Car	1	32'5"	1579	40,000	Rebuilt D&IR Two Harbors	R1910	DIR Std	22
5162–5172e	N	Oil Car	6	32'0"	1579	40,000	Rebuilt D&IR Two Harbors	R1910	DIR Std	23
5174–5222e	F	Box Car	25	36'10"	2448	60,000	Western Steel Car & Foundry	1903	DIR Std	24
5174–5180e	M	Stock Car	4	36'2"	2358	60,000	D&IR Two Harbors	R1921	DIR Std	25
5182		Stock Car	1	36'2"	2358	60,000	D&IR Two Harbors	R1920	Barber	26
5184		Stock Car	1	36'2"	2358	60,000	D&IR Two Harbors	R1920	Barber	27

Number	Class	Description	Total	O.L	Cu.Ft.	Lbs	Builder	Built	Trucks	Notes
5200	S	Automobile Car	1	36'10"	2448	60,000	Western Steel Car & Foundry	R1910	Barber	28
5202	S	Automobile Car	1	36'10"	2448	60,000	Western Steel Car & Foundry	R1912	Barber	29
5216	S	Automobile Car	1	36'10"	2448	60,000	Western Steel Car & Foundry	R1922	Barber	30
5224–5262e	F	Box Car	20	37'3"	2448	60,000	Western Steel Car & Foundry	1908	Arch Bar	31
5244		Stock Car	1	37'3"	2448	60,000	D&IR Two Harbors	R1920	Arch Bar	32
5264–5340e	F	Box Car	37	36'2"	2358	60,000	D&IR Two Harbors	R1908	DIR Std	33
5342–5390e	F	Box Car	25	36'11"	2448	70,000	Standard Steel Car Co	1918	A. U-section	
5001–5011o	I	Combination Flat	6	36'0"		48,000	Northwestern Car & Mfg.	1883	DIR Std	34
5013–5051o	I	Flat Car	20	36'0"		60,000	D&IR Two Harbors	1887	DIR Std	35, 61
5053–5061o	I	Flat Car	5	36'0"		48,000	D&IR Two Harbors	1887	DIR Std	36, 61
5063–5145o	I	Flat Car	45	36'0"		60,000	D&IR Two Harbors	1887	DIR Std	37, 61
5147–5541o	I	Flat Car	200	36'0"		50,000	Lafayette Car Works	1888	DIR Std	38, 61
5543–5641o	I	Flat Car	50	36'0"		60,000	Illinois Car & Equipment Co	1899	Arch Bar	61
5643–5741o	I	Flat Car	50	36'0"		60,000	Western Steel Car & Foundry	1903	Arch Bar	61
5743–5825o	I	Flat Car	42	36'0"		60,000	D&IR Two Harbors	1905	DIR Std	61
5827–5841o	I	Flat Car	8	36'0"		60,000	D&IR Two Harbors	1906	DIR Std	61
6000–6049	I	Log Car	50	36'0"		60,000	D&IR Two Harbors	1900	DIR Std	39
6050–6099	I	Log Car	50	36'0"		60,000	D&IR Two Harbors	1910	DIR Std	40
6100–6124	I	Flat Car	25	37'3"		60,000	Western Steel Car & Foundry	1907	Arch Bar	41
6125–6179	I	Flat Car	55	37'3"		60,000	Western Steel Car & Foundry	1907	Arch Bar	42
6100–6124	G	Gondola	25	37'3"	940	60,000	Western Steel Car & Foundry	R1914	Arch Bar	
6125–6179	G	Gondola	55	37'3"	934	60,000	Western Steel Car & Foundry	R1929	Arch Bar	
6135	G	Gondola	1	37'3"	1050	60,000	Western Steel Car & Foundry	R1930	Arch Bar	43
6144	G	Gondola	1	37'3"	1050	60,000	Western Steel Car & Foundry	R1922	Arch Bar	44
6180–6279	I	Flat Car	100	36'9"		60,000	ACF Lot #7934	1916	Arch Bar	45
6280–6304	I	Flat Car	25	36'9"		60,000	ACF Lot #8406	1917	A. U-section	46
7000–7074	K	Log Car	75	20'		40,000	D&IR Two Harbors	1901	DIR Std	47
7075–7124	K	Log Car	50	20'		40,000	D&IR Two Harbors	1902	DIR Std	
7125–7131	K	Log Car	7	20'		40,000	D&IR Two Harbors	1905	DIR Std	
7132–7159	K	Log Car	28	20'		40,000	D&IR Two Harbors	1906	DIR Std	
7160–7319	K	Log Car	160	20'		40,000	D&IR Two Harbors	1910	DIR Std	
8000–8007	R	Refrigerator Car	8	34'2"	1500	40,000	Pullman Car Co	1894	Thielson	48
8008–8011	R	Refrigerator Car	4	34'2"	1500	40,000	Pullman Car Co	1894	Thielson	49
8012–8021	R	Refrigerator Car	10	36'2"	1885	50,000	ACF Lot #4653	1907	Arch Bar	50
8022–8031	R	Refrigerator Car	10	37'2"	1995	80,000	Peteler Car Co	1912	Arch Bar	
8032–8041	R	Refrigerator Car	10	37'2"	1995	80,000	D&IR Two Harbors	1918	A. T-Section	51
9000-9699	D	Steel Ore Car	700	22'0"	710	100,000	Pressed Steel Car Co	1907	Arch Bar	
9156/9636	E	Steel Ore Car	6	22'0"	710	100,000	Pressed Steel Car Co	1907	Arch Bar	52
9700–10499	E	Steel Ore Car	800	22'0"	650	100,000	Standard Steel Car Co.	1909	A. L-Section	53
10500–11249	E	Steel Ore Car	750	22'0"	650	100,000	Standard Steel Car Co.	1910	Arch Bar	54
11250–11399	E	Steel Ore Car	150	22'0"	715	100,000	Pressed Steel Car Co	1910	Arch Bar	55
11400–11499	E	Steel Ore Car	100	22'0"	650	100,000	ACF Lot #5809A	1910	Arch Bar	56
11500–12299	E	Steel Ore Car	800	22'0"	695	100,000	Standard Steel Car Co	1913	Arch Bar	57
12300–12499	E	Steel Ore Car	200	21'9"	685	100,000	ACF Lot #7068	1913	Arch Bar	58
12500–12749	E	Steel Ore Car	250	21'9"	685	100,000	ACF Lot #7907	1916	A. T-Section	59
12750–13249	E	Steel Ore Car	500	21'11"	695	100,000	Standard Steel Car Co.	1916	A.T-Section	60

1. Renumbered from 2000–2014 in 1896. Rebuilt D&IR Two Harbors 1901 to 1904 with new sides and ends
2. Not delivered until 2-3, 1918
3. American Engineer and Railroad Journal announced orders had been received in April 1896 for 450 wood ore cars from Terre Haute Car Co. 55 cars were numbered 1146–1200, 357 cars were numbered 2137–2548 and remaining 38 cars were used to replace wreck and retired cars in 751–2136 series
4. Rakowsky ore cars(Patents 973208,973209,973210 issued to Gustav A. Rakowsky in Oct 1910
5. Received as Summers demo car #101 in 1909, renumbered 3049 in 1914
6. 100 cars ordered by D&IR on 10-15-1902. Order split on delivery in 1903 50 cars to D&IR and 50 cars to DM&N 8355-8404 ACF Detroit plant
7. 4400-4419, 4596, 4656 rebuilt with longitudinal doors and 10" side extensions D&IR Two Harbors during winter 1926-1927
8. Steel Underframes applied Two Harbors 1913
9. Steel Underframes applied Two Harbors 1913, Renumbered from 2228–2302 in 1895-96
10. 5062, 5078, 5106 rebuilt D&IR Two Harbors 1889 by adding interior stove next to door and vent stack to roof 5068 and 5092 in 1916, 5078 and 5084 in 1917, 5078,5026,5068,5106 in 1918 (5026 or 5062 in 1918)?
11. Renumbered from 2200–2210e in 1896
12. Rebuilt D&IR Two Harbors between 1896 and 1908 into oil cars
13. Rebuilt from box car 5132 D&IR Two Harbors in 1901; back to box car in 1912
14. Renumbered 1896 from X2248 and X2250. Retired1905
15. X dropped from reporting marks by Jan 5, 1923
16. Renumbered from 2212–2226e in 1896
17. New cars built 1904 at Two Harbors to replace same numbers
18. Rebuilt D&IR Two Harbors 1910 from box car 5150
19. Rebuilt D&IR Two Harbors from flat cars in 2361–2409 series
20. Rebuilt D&IR Two Harbors 1910 from box car 5154
21. Rebuilt D&IR Two Harbors 1910 from box car 5156
22. Rebuilt D&IR Two Harbors 1910 from box car 5158
23. Rebuilt D&IR Two Harbors 1910 from 5162–5172 series box cars
24. Steel Underframes applied Two Harbors 1914
25. Steel Underframes applied Two Harbors 1914; back to box cars in 1938, 1927, 1924 - 5176 retired in 1961
26. Steel Underframes applied Two Harbors 1914; back to box car 5182 in 1927
27. Steel Underframes applied Two Harbors 1914; back to box car 5184 in 1927
28. Rebuilt from 5200 to double-door auto car in 1910, Steel Underframe applied Two Harbors 1914
29. Rebuilt from 5202 to double-door auto car in 1912, Steel Underframe applied Two Harbors 1914
30. Rebuilt from 5216 to double-door auto car in 1922. Steel underframe applied Two Harbors 1914. Converted back to single-door box car winter 1924/25
31. Built with Steel Underframes
32. Rebuilt D&IR Two Harbors 1920; back to box car 5244 in 1927
33. Rebuilt D&IR Two Harbors 1-4-1908 from flat cars 5001, 5003, 5005, 5007, 5009, 5011, 5015, 5017,5023, 5025, 5029, 5043, 5047, 5049, 5053, 5055, 5057, 5061, 5067, 5069, 5071, 5081, 5093, 5095, 5103, 5107, 5109, 5111, 5115, 5117, 5123, 5125, 5127, 5129, 5133, 5135, 5139, 5141, 5143. Steel underframes applied, Two Harbors 1914
34. Renumbered from D&IR 2201–2211o in 1896
35. Renumbered from D&IR 2361–2409o in 1896
36. Renumbered from D&IR 2261–2269o in 1896
37. Renumbered from D&IR 2271–2359o in 1896
38. Renumbered from D&IR 2411–2809o in 1896
39. Bodies built in 1900 (no trucks) classified as "Bunker Flats" by D&IR.
40. Classified as "Bunker Flats" by D&IR
41. Rebuilt D&IR Two Harbors 1914 with dump doors and wood gondola sides; built with steel underframes
42. Rebuilt D&IR Two Harbors 1926 with wood gondola sides (except 6144, see below); built with steel underframes
43. Rebuilt D&IR Two Harbors 1930 with steel gondola sides
44. Rebuilt D&IR Two Harbors 1922 with steel gondola sides
45. Ordered 12-22-1915 - Blt ACF Chicago plant. 77 equipped with sideboards when delivered: 6180-6187, 6189-6198, 6200-6202, 6204-6206, 6208, 6209, 6211, 6212, 6214, 6216, 6217, 6219, 6220, 6222-6226, 6228-6234, 6236, 6237, 6239, 6241, 6242, 6245 to 6250, 6252, 6254, 6256, 6258, 6260, 6261, 6263, 6264, 6266 to 6278. - 14 equipped with steel side boards in 1918: 6231, 6251, 6221, 6228, 6240, 6241, 6243, 6244, 6251, 6255, 6257, 6259, 6265, 6279. 50 cars converted to rack flats: 10 cars 6269-6278 during 1934. 40 cars 6212, 6214, 6216, 6217, 6219, 6220, 6222, 6223, 6224, 6225, 6226, 6228, 6229, 6230, 6231, 6232, 6233, 6234, 6237, 6239, 6241, 6242, 6245, 6246, 6247, 6248, 6249, 6250, 6254, 6256, 6258, 6260, 6261, 6263, 6264, 6266, 6267, 6268 during 1935
46. Ordered 5-3-1917 - All delivered in 12/17 except 6301 delivered in 1/18 - Blt ACF Chicago plant
47. Starting in 1916, 20-ft log cars were rebuilt with six pieces of used 60-lb. rail to reinforce the underframes. A total of 96 cars were completed during 1916: 7005, 7012, 7014, 7015, 7016, 7020, 7026, 7030, 7031, 7035, 7036, 7041, 7055, 7056, 7065,7066, 7076, 7077, 7078, 7080, 7089, 7090, 7106, 7113, 7115, 7116, 7119, 7120, 7124, 7129, 7130, 7140, 7147, 7148, 7151, 7153, 7154, 7156, 7157, 7162, 7166, 7171, 7179, 7181, 7182, 7183, 7184, 7185, 7193, 7196, 7198, 7200, 7203, 7205, 7207, 7215, 7216, 7217, 7220, 7222, 7224, 7228, 7230, 7231, 7232, 7137, 7340, 7241, 7245, 7246, 7250, 7154, 7255, 7258, 7259, 7260, 7265, 7268, 7269, 7270, 7274, 7279, 7280, 7281, 7285, 7288, 7292, 7294, 7299, 7301, 7305, 7309, 7310, 7313, 7315, 7316. All of these cars were also equipped with pulpwood racks when the rail was added to the underframes. An additional 10 cars were completed in 1917: 7006, 7040, 7047, 7052, 7067, 7111, 7133, 7197, 7308, 7319 with all of these also equipped with pulpwood racks at the same time
48. Formerly ATSF 720–969 series refrigerator cars built by Pullman in 1894 - Purchased 2nd hand from ATSF 1901 - Murphy roof, steel underframe, Moore refrig/heating system applied D&IR Two Harbors 1914 - 8005 retired 11/05/29 used to build Steam Shovel car 800
49. Formerly ATSF 720–969 series refrigerator cars built by Pullman in 1894 - Purchased 2nd hand from ATSF 1902 - Murphy roof, steel underframe, Moore refrig/heating system applied D&IR Two Harbors 1914 - Show on equipment lists arriving in 1903 - 8010 and 8011 sold to D&NE 01/04/30
50. Equipped with meat racks - Steel Underframe applied D&IR Two Harbors 1914 - ACF Lot Number 4653 Ordered 1907 - Delivered Sept 8, 1907 -
51. Equipped with meat racks - Murphy roofs
52. 9156, 9195, 9348, 9426, 9566, 9636 rebuilt with longitudinal doors and 10" side extentions D&IR Two Harbors during winter 1926-1927
53. Summers type car
54. Summers type car
55. Built at Western Steel Car and Foundry plant at Chicago, IL
56. Clark type ore car
57. Summers type car
58. National type car - Built at ACF Peninsular shop Detroit, MI
59. National type car
60. Summers type car
61: Starting in 1914, steel center sills began to be applied to the fleet of flatcars Converted in 1914: 5161,5163,5507. Converted in 1915: 5089, 5137, 5163, 5165, 5169, 5183, 5199, 5201, 5317, 5447, 5459, 5463, 5537, 5615, 5649, 5667, 5671, 5717, 5725, 5737, 5789, 5825, 5833, 5835, 6001, 6002, 6007, 6009, 6011, 6016, 6021, 6022, 6023, 6026, 6027, 6028, 6037, 6038, 6039, 6047. Converted in 1916: 5633, 6042. Converted in 1917: 6041.

<u>Abbreviations</u>

/ = non-continuous series
e = even numbers only
o = odd numbers only
R = rebuilt
ACF = American Car & Foundry
THC&M = Terre Haute Car & Manufacturing
A = Andrews trucks

D&IR REFRIGERATOR CARS

The first refrigerator cars were eight cars leased in 1896. These were leased through 1898 and reduced to seven cars in 1899. In 1900 and 1901, a total of 12 refrigerator cars were on lease. Four cars were still on lease in 1902 when the leasing contract was terminated. Photos from this period indicate that cars were leased from Armour Refrigerator Lines.

During 1901 eight 20-ton 34-foot refrigerator cars with 1500 cu. ft. capacity were acquired from ATSF. These were former ATSF 720–969 series cars built by Pullman in 1894 and D&IR numbered them 8000–8007. They had Murphy roofs and truss-rod underframes. Four additional identical cars, 8008–8011 were acquired from ATSF in 1902. D&IR Two Harbors shop applied Murphy steel roofs, steel underframes and Moore refrigeration and heating systems in 1914. Two of these cars, 8010 and 8011, were sold to the Duluth & Northeastern Railroad on January 4, 1930. All remaining cars were off the roster by 1934.

The first new built refrigerator cars were 10 25-ton 36-foot 2-inch cars 8012–8021 with 1,885 cu.ft. capacity ordered from American Car & Foundry under Lot 4653 in 1907. Cars were ordered with meat rails for hauling hang-

Starting by 1895 D&IR leased reefers from Armour and ATSF, then purchased in 1901-1902 12 cars from ATSF 720–969 series and numbering them D&IR 8000–8011. A growing population and related traffic growth caused D&IR to eventually acquire a fleet of 42 refrigerator cars between 1894 and 1918. Builders photo of ATSF 960 represents the appearance of this group of cars before purchase by D&IR. (ACF builder photo, Art Griffen collection)

Satisfied with performance of second-hand ATSF reefers and expanding traffic base caused D&IR to purchase its first reefers in 1907 from American Car & Foundry in 1907. Ten cars D&IR 8012–8021 were built with meat racks and received on September 8, 1907. Renumbered to DM&IR 7112–7121, steel center sills would be applied in 1914 and Murphy roofs and Moore heating equipment during late 1916. (American Car & Foundry Collection, John W. Barriger III National Railroad Library at UMSL)

Left: DM&IR Class R1 7122 former D&IR 8022 is at Duluth on July 19, 1957. Originally built in series 8022–8031 by Peteler Car Co. in 1912. Delivered with four Moore Patent ice tanks, car was rebuilt with two Bohn Patent ice tanks in 1940. Note the large hatch platform and Moore heating stack on the roof. Car is preserved at Mid-Continent Railway Historical Society at North Freedom, Wisconsin. (Bob's Photo)

Below: Three years after being repainted at Two Harbors car shop, DM&IR 7138 is at Endion, Minnesota, during July, 1962 (Joe Collias)

ing meat and delivered on Sept 8, 1907. Steel center sills were applied at D&IR Two Harbors Shop in 1914. All ten cars were rebuilt with Murphy roofs and Moore heating systems at Two Harbors during late 1916 under Appropriation 544 for $4,136.40. All were off the roster by 1936.

The board of directors appropriated $12,500 on July 11, 1912, for purchase of 10 40-ton 37-foot 2-inch cars 8022–8031 with 1,995 cu. ft. capacity from Peteler Car Co. After the merger they were renumbered DMIR 7122–7131. All were retired by 1972.

The final group of refrigerator cars were 10 40-ton 37-foot 2-inch cars 8032–8041 with 1995 cu. ft. capacity built by D&IR Two Harbors shops in 1918. Purchase was approved on Appropriation 630 at the Board of Directors meeting on June 11, 1917. for a cost of $35,500. These 40-ton cars were almost identical to the Peteler cars built in 1912 and were equipped with Murphy roofs. After the merger they were renumbered DMIR 7132–7141. All were retired by 1972.

AFE 1781 was begun on January 10, 1926, for the application of racks and meat hooks in all refrigerator cars. This project was completed on March 11, 1926.

All refrigerator cars were placed in Class R starting in 1910. After the D&IR/DM&N merger cars 7122–7131 were placed in Class R and 7132–7141 in Class R1.

On September 29, 1939, AFE 4095 for $150.00 was approved for the purchase and installation of "Frigiwarm" curtains on five refrigerator cars. Cars 8022, 8024, 8025, 8028, 8029 in Two Harbors shop for heavy repairs at the time, had the curtains installed. These curtains allowed the icing of one end of the car only to allow merchandise to be loaded in one end and perishable traffic in the other. This project was completed on January 8, 1940.

End and roofs had deteriorated by 1950, and six of the remaining Class R and R-1 refrigerators required rebuilding. AFE 6869 for $3,600 was approved on September 22, 1950, to purchase and apply Murphy steel roofs, ends and Apex Type A running boards to six cars: 7131, 7132, 7133, 7134, 7135 and 7140. Rebuilding of the cars took place at Two Harbors between April 17, 1951 and April 20, 1951.

Declining LCL perishable movements and increased icing cost caused DM&IR in 1964 to look into dry ice for cooling needs on LCL refrigerator cars. AFE 9882 for $3,915 was approved on January 5, 1965, to modify 23 of the remaining 29 refrigerator cars by installing new insulated curtains and dry ice racks. Twenty-two cars would be modified on one end only and seven with both ends. Ten of the cars were former D&IR cars. Modified on one end only were: 7128, 7129, 7132, 7133, 7134, and 7140. Modified on both ends were 7122, 7124, 7135, and 7139. "High Kold Pak" insulated quilts (curtains) for project were supplied by CanPro Corporation of Fond Du Lac, Wisconsin. Rebuilding started on January 26, 1965 with all work completed by June 18, 1965.

D&IR BOX CARS

The first D&IR box cars appear to have been converted from combination flat cars in series 41–70 in 1884 that had originally been built in 1883 by Northwestern Car & Manufacturing. Quickly, after they were delivered, six of the cars were rebuilt to box cars in 1884, another four in 1885, two in 1886 and two more in 1887. No known photographs have been found of this flatcar series. Only one photo of this group of

D&IR scene at Tower Jct., Minnesota, circa 1885 showing three box cars, caboose and one ore car. The first boxcar, 68, is the only known photo of a combination flat car converted to box car showing the two-digit numbers. (University of Minnesota Duluth, Kathryn A. Martin Library Archives and Special Collections, Collection S3742 Box 7 Folder 20)

DM&IR 5132(2nd) at Proctor, July 28, 1969 was built at D&IR Two Harbors shops in 1916 and is preserved at Lake Superior Museum of Transportation at Duluth. (Owen Leander)

Three D&IR boxcars at Virginia, Minnesota, during 1912; left to right, 5220 built by Western Steel Car & Foundry in 1903 and 5032, 5008 built by Duluth Manufacturing in 1895. As built, cars were lettered with D&IR reporting marks to left of door and road number to the right. Around 1910, the road number was shifted to below the reporting marks left of the door. Note the large end door on 5220. Left-mounted grab irons have yet to be installed on 5032, as mandated by Congress under the Safety Appliance Act in 1911. By the end of 1919 all D&IR and DM&N cars would be in compliance with the law. (Minnesota Historical Society)

D&IR 5020, one of 25 25-ton 34-foot box cars 5000–5048, even numbers only, by Duluth Manufacturing Co. in 1895. Outside framed wood doors would be replaced by interior framed doors by 1920. 5020 was retired December 14, 1933. Duluth, circa 1905. Note the Fitger Refrigerator Line refrigerator car FRL 200 in the background. Fitger's brewing company of Duluth was in business from 1881 until 1972. (Shorpy Photo Archives)

box cars is available showing it carrying number 68. It was 24-ton 34-foot car with 1,865 cu. ft. capacity. Doors were inside framed wood with 5-foot wide door opening. It appears that the rebuilt boxcars carried numbers 40–42, even numbers, and 48–70, even numbers, but no confirming documents have been found. These cars were renumbered during the 1888 renumbering to 2200–2226, even numbers only. At the time of the 1895 renumbering all 14 cars on the roster were renumbered to 5122–5132 even numbers only and 5138–5152 even numbers only. During 1910 the D&IR placed these in Class F.

The next group was 38 25-ton 34-foot boxcars numbered with unknown 2 or 3 digit numbers purchased from Wells & French in 1884 as 25-ton cars with 1,865 cu. ft. capacity. They had a 5-foot by 6-foot 6-inch high side door opening and end doors 2-foot wide by 1-foot 4½-inches high on one end of the car. Cars were renumbered during the 1888 renumbering to 2228–2302, even numbers only. At the time of the 1895 renumbering a total of 36 of these cars remained on the roster and where renumbered 5050–5120, even numbers. During 1910 cars were placed in Class F. All were rebuilt with steel center sills at Two Harbors in 1913. Eleven of these cars were rebuilt to vegetable cars, between 1889 and 1918, with the addition of a stove just inside the door opening. During 1924 and 1925, car 5084 was in assigned service for hauling cement.

Another 25 25-ton 34-foot boxcars 5000–5048, even numbers only, with 1865 cu. ft. capacity were purchased from Duluth Manufacturing in 1895. Cars had 5-foot wide by 6-foot 6-inches high side door opening and equipped with end doors 2-foot by 1-foot 4½ inches high on one end of the car. Wood doors were an outside-framed, 3-panel design. These doors were replaced with interior framed doors by 1920 because rain water collected around the outside framing causing the doors to rot out. All were placed in Class F in 1910 and rebuilt with steel center sills at Two Harbors in 1913.

D&IR Two Harbors shop built the next 10 20-ton 34-foot box cars 5154–5172 even numbers only. Cars were rebuilt from combination flat cars in the 2361–2409 series with 1490 cu. ft. capacity and 5-foot wide by 6-foot high side door openings. All were placed in Class F in 1910. Nine of these cars were rebuilt into "oil cars" in 1910 at Two Harbors by placing small openings near the top edge of the sides and ends to allow ventilation.

In 1901 the American Railway Association (ARA) recommended a 36-foot boxcar design with 2,442 cu. ft. capacity which was adopted by the Master Car Builders' Association in 1904 as recommended practice. Many railroads and car builders built cars closely following this design. D&IR placed an order in September 1902, with Western Steel Car & Foundry for the purchase of 25 30-ton 36-foot 10-inch boxcars 5174–5222, even numbers only, with 2,448 cu. ft. capacity with truss-rod underframe. All were delivered in 1903 with a 6-foot wide by 7-foot 9-inches high side door opening and end doors

Right: Late in its service life DM&IR 5192 is at Duluth, on March 31, 1963. The notation "LOAD FOR DMIR POINTS ONLY" was stenciled on almost every car when it became overage or restricted to on-line service, account being equipped with arch bar trucks. (D. Repetsky, John C. La Rue, Jr. collection)

DM&IR 5208 at Proctor in July 1956 is one of 25 Class F 5274–5222 even numbers built by Western Steel Car & Foundry in 1903. (D.P. Holbrook collection)

2-feet wide by 3-foot 6-inches high on both ends. They were equipped with Barber trucks, Chicago couplers, Bryan tandem draft rigging, McCord journal boxes, Chicago roofs, Dunham door fixtures and Westinghouse brakes. DM&N received 10 identical cars 3063–3072 at the same time. Car 5214 was wrecked in transit in 1903 before being delivered. The replacement car was not delivered until 1905. All were placed in Class F in 1910. Steel center sills were applied when built and replaced at Two Harbors shops in 1914 with steel underframes. Three of these cars would be rebuilt to automobile cars; 5200 in 1910, 5202 in 1912 and 5216 in 1922.

DM&IR 5152 is one of eight 30-ton Class F boxcars built at Two Harbors shops in 1904 to replace eight cars with identical numbers built by Northwestern Car & Manufacturing in 1884. The NC&M cars had 1,865 cu. ft. capacity but the new cars were built with 2,358 cu .ft. capacity to a D&IR standard boxcar design that was developed by Two Harbors shops. (Bob's Photo collection)

Two Harbors shop built the next eight 20-ton 34-foot box cars 5138–5152, even numbers, only with 1,865 cu. ft. capacity in 1904. Door openings were 6-foot wide by 7-foot 6-inch high. These were the second cars to carry these numbers. Original D&IR 5138-5152(1st) were retired by 1904. Cars were placed in Class F in 1910.

Returning to 2448 cu. ft. capacity box car design, D&IR in 1908 placed an order for 20 30-ton 37-foot 3-inch long, double-sheathed box cars 5224–5262, even numbers only, from Western Steel Car & Foundry. Door openings were 6-foot wide by 7-foot 9-inch high with a small end door 2-foot wide by 2-foot 6-inches high. WSC&F had improved on the 1902 truss-rod underframe by constructing this order with girder underframes consisting of fishbelly side sills and fishbelly center sills. Cars were

DM&IR 5242, shown at Endion, Minnesota, freight house on July 19, 1957, is one of 20 5224–5262 even numbers only 30-ton Class F1 box cars built by Western Steel Car & Foundry in 1908. The cars were built with girder underframes consisting of two fishbelly side sills and two fishbelly center sills. For the time, the underframes were considered over-designed. (Bob's Photo collection)

DM&IR 5250 at Virginia, Minnesota, on July 27, 1969. Even at this late date, the outlawed arch bar trucks are still on this car and being used for on-line loading only. Note the sill step below the door applied under AFE 1329 beginning on March 4, 1926, to 250 box cars to allow easier access when unloading LCL. (Owen Leander, John C. La Rue, Jr. collection)

Two Harbors shops rebuilt 37 flat cars to box cars in 5264–5340 series in 1908. These would be the last group of cars lettered with reporting marks to left of door and road number to the right of the door. In the right corner of the photo is D&IR 5164, one of 15 oil cars rebuilt from box cars. Note the small framed openings on the ends and sides of the carbody. The rebuilding from flat cars is obvious with the stake pockets and single sheathing design (Minnesota Historical Society)

placed in Class F in 1910

An additional 37 30-ton 36-foot 2-inch long, double-sheathed truss-rod box cars 5264–5340, even numbers only with 2,358 cu. ft. capacity, were rebuilt at Two Harbors from flat cars in the 5001–5145 series, odd numbers only, during January through April of 1908. Door openings were 6-foot wide by 7-foot 6-inch high, with small end doors 2-feet wide by 1-foot 4½-inches wide on the A-end of car. All were placed in Class F in 1910

Appropriation 630 was approved at the June 11, 1917 Board of Director meeting for $70,625.00 to purchase 25 35-ton, 36-foot 11½-inches long, all-steel box cars 5342–5390, even numbers only, with 2448 cu. ft. capacity. This final D&IR boxcar order was placed with Standard Steel Car Co. and built in March 1918 at their Hammond, IN plant. Door openings were 6-foot wide by 7-foot 9-inches high with a small end door 1-foot 8-inches wide by 1-foot 2-inches high on the A-end of car. Cars were built with fishbelly side sill, riveted plate steel ends and doors. As-built plate steel doors were replaced with Youngstown doors by late 1940s. AFE 3235 was approved February 3, 1932 for reinforcing ends of six cars within the series.

D&IR, like most railroads, had retirements, cars destroyed and others rebuilt. Starting in 1916, retirements and cars placed in company service, caused many renumberings, rebuildings and new construction, which at first glance appear confusing. Not following ARA/MCB recommended practice, D&IR had in 1905 developed a standard plan for a 2358 cu. ft. capacity box car which was used to build replacement cars.

During 1916 the 5122–5136 series cars had

the following changes. Cars 5122 and 5126 were renumbered to shop cars 171 and 172. Cars 5124 and 5132 were retired. From series 5138–5152, cars 5146 was renumbered to the second 5124. A new second 5132 was constructed from D&IR 2358 cu. ft. Standard Boxcars plans at Two Harbors. Series 5138–5152 and 5154–5172 had following re-numberings: 5160 to the new 5150 and car 5150 to the new 5160.

Two cars from series 5174–5222, 5204 and 5214 were retired and new cars with the same number with 2358 cu. ft. capacity were constructed in 1916. At the same time one car from 5000–5124 series, 5024, was retired and a new 5024 was built with 2358 cu. ft. capacity. Two additional cars from this series, 5034 and 5102 were retired and new cars carrying the same numbers with 2,358 cu. ft. capacity were built as replacements. After 1916 the 5000 and 5100 series boxcars consisted of mismatched sizes of cars.

Wooden roofs on box cars began to be replaced with Murphy roofs with 20 cars completed in 1917 and an additional 30 in 1918.

DM&IR 5284, shown at Endion on July 19, 1957 is a Class F 30-ton 36-foot box car, one of 37 rebuilt from Class I flatcars at D&IR Two Harbors shops between January and April 1908. Six were originally built by Northwestern Car & Manufacturing in 1883 and the remainder at D&IR Two Harbors shops in 1887. Diagrams show these cars equipped with steel underframes in 1914. However, Two Harbors reinforced the underframes with 60-lb. salvage rail and declared them as having steel underframes. (Bob's Photo collection)

DM&IR 5306 at the Endion in 1961. The car was light weighed and painted at Two Harbors in June, 1958. This was originally flat car 5081 built at Two Harbors in 1887 and rebuilt to this box car in 1908. At 71 years old the car is in excellent shape. (Joe Collias)

Builders photo of D&IR 5342. Compare this to the photos below and note the as built door tracks, plate steel door, Andrews trucks and the single grab iron on the left corner. (D.P. Holbrook Collection)

Class F2 DM&IR 5344 is shown after 1959 having "Scotchlite" monogram, initials and numbers applied. Note that the as-built Andrews trucks have been replaced, also the bottom of the side sheathing has rusted out and been replaced by new steel along the floor line. (Mainline Photos, J.M. Gruber collection)

The only all-steel box cars purchased by D&IR were 25 Class F 5342–5390, even numbers only, from Standard Steel Car Co. in 1918. 5366 is at Virginia, Minnesota, in the Spring of 1961. (Jim Maki Sr.)

DM&IR 5344 at Biwabik, Minnesota, on July 28, 1975. One of 25 steel 36-foot Class F2 box cars, 5342-5390, it was built by Standard Steel Car in 1918 for D&IR. Note the placard for the Scotchlite monogram with the remaining lettering stenciled on the car. Maroon paint applied during early 1960s has faded to brown. (R.F. Kucaba)

The development of the automobile caused D&IR like many other railroads to convert box cars with wider doors to allow automobiles to be loaded inside the car. D&IR converted three Class F box cars, 5200 in 1910, 5202 in 1912 and 5216 in 1922 for automobile loading by equipping the cars with a single 12-ft. door and placing them in Class S. Most carriers opted for double doors, but D&IR equipped their auto cars as a single door. DM&IR 5202 is shown at Duluth during May, 1955. (W.C. Whitaker, Arnold Menke collection)

D&IR AUTOMOBILE CARS

The arrival of the Ford Model T in 1908 and initial highway development in Northern Minnesota created the need for box cars suitable to haul automobiles from docks at Duluth and Two Harbors. This need caused D&IR to rebuild box cars for this movement.

One Class F 30-ton 36-foot 10-inch box car in series 5174–5222, even numbers only, built by Western Steel Car & Foundry in 1903 was chosen to be rebuilt at D&IR Two Harbors in 1910 to a large single-side-door automobile car. Car numbered 5200 was placed in Class S after rebuilding. The door opening was enlarged from 6-foot wide by 7-foot 9-inches high side door feet to a 10-foot 11-inches wide by 7-foot 4-inches high single large side door. They also had small end door openings on both ends 2-foot wide by 3-foot 6-inches high.

Two additional cars were rebuilt from the same Class F box car series; 5202 in 1912 and 5216 in 1922. All three cars had steel center-sills applied at Two Harbors in 1914. Both were identical to 5200. Car 5216 was converted back to a single-door box car during the winter shopping season of 1924-25. The remaining two cars were retired by 1972.

D&IR car records show cars 5200 and 5202 being used mostly for on-line movement of automobiles, but 5216 showed being off-line on yearly annual reports and might have been used for the movement of automobiles from Detroit to northern Minnesota.

D&IR VEGETABLE CARS

A total of 11 box cars from Class F box cars in series 5050–5120, even, were rebuilt between 1889 and 1918 for vegetable service. The rebuilding involved placing a stove inside the door opening against the wall to allow cars to be heated in the fall, winter and spring to keep vegetables from spoiling. Initially, box cars 2240, 2256 and 2284 were rebuilt with the addition of the stove inside the right side door opening. These cars were renumbered 5062, 5078 and 5106 in 1896. Two additional cars, 5068 and 5092 were rebuilt in 1916, two more, 5078 and 5084 in 1917 and two additional, 5026, and 5106 in 1918. It is unknown when the stoves were removed, but the cars were no longer shown as individual diagrams by the 1929 D&IR diagram book.

D&IR STOCK CARS

Development of the area along the D&IR included the need to ship livestock, mules and horses, causing the need for stock cars. Mules and horses were used at the mines for movement of ore to the headframes for loading into ore cars. A total of four stock cars were shown on the D&IR roster beginning in 1885, and then reduced to two cars in 1887. Cars appear to have been rebuilt from combination flat cars in series 41–70 and numbered 44–50, even numbers. Two cars, 48 and 50, were renumbered to X2248 and X2250 in 1887. The other two cars, 42 and 46, were off the roster by the 1887 renumbering. They were 25-ton 34-foot cars with an unknown capacity. The X prefix was used account box cars 2228–2302 had been delivered from Wells and French in 1887. From 1887 to 1896, two box cars 2248, 2250 and two stock cars X2248, X2250 were on the roster. Both cars were renumbered during the 1895 renumbering to X5134 and X5136 placing them back in the series of rebuilt combination boxcars built by Northwestern Car & Manufacturing. Both cars were retired by 1905

Two replacement 30-ton 36-foot 2-inch stock cars, 2nd X5134 and X5136, were built at Two Harbors in 1905 based on the D&IR standard 2,358 cu. ft. box car design. The standard box car design was modified by the removal of a horizontal portion of the sides of a box car about 24 inches wide, about 18 inches below the top eave of the car. This left the internal bracing between the double sheathing to be visible. To reinforce the car, vertical steel rods were added about every 8 inches. Cars had 7-foot 6-inch wide by 6-foot high door openings and rode on 5-foot 4-inch archbar trucks. They were also equipped with a lumber type end door opening of 24-inches wide by 40-inches high located on the B-end of the car. D&IR applied Class M to these cars in 1910. Steel underframes were applied in 1914.

Because of confusion with car records, the ARA, in a letter to member railroads on October 21, 1920, advised the D&IR to drop the X prefix from cars 5134 and 5136. It appears that this was accomplished soon after this letter was received. Both cars were retired per AFE 8197 on January 8, 1957.

Class F box car 5132 built by Northwest-

Class M stock car DM&IR 5176 is shown at Duluth July 21, 1954. This was one of four Class F box cars rebuilt to stock cars at Two Harbors in 1921. It was retired in 1961. (Bill Raia collection)

ern Car & Manufacturing in 1884 was rebuilt as a stock car X5132 in 1901. It was rebuilt back to box car 5132 in 1912.

Class F box car 5182 built by Western Steel Car & Foundry in 1903 was rebuilt as a stock car in 1918 carrying the same number and then rebuilt back to box car in 1919. It was again converted in 1921 to a stock car with the same number and then rebuilt back to a box car in 1927. Per AFE 3380 car was retired on January 18, 1934.

Class F box cars 5174, 5176, 5178, 5180 built by Western Steel Car & Foundry in 1903 were all rebuilt to stock cars in 1921 at D&IR Two Harbors. These cars were 30-ton, 36-foot cars with 2,393 cu.ft. capacity. Door openings were 6-foot wide by 7-foot 6-inches with a small end door 24-inch wide by 16-inch high on the A-end of car. All rode on 5-foot 4-inch archbar trucks. In 1914 all were equipped with steel underframes. After 1931 they were equipped with Murphy steel roofs. Three were rebuilt back to boxcars with same numbers as follows: 5174 on February 16, 1938, 5178 in 1927, and 5180 in 1924. Stock car 5176 was retired in 1961.

Class F boxcar 5184 built by Western Steel Car & Foundry in 1903 was rebuilt as a stock car in 1920 carrying the same number and then rebuilt back to boxcar in 1927.

Class F boxcar 5244 built by Western Steel Car & Foundry in 1908 was rebuilt as a stock car in fall of 1920 carrying the same number and then rebuilt back to boxcar in 1922.

D&IR OIL CARS

A total of 15 double-sheathed box cars were converted to oil cars between 1896 and 1910. Their floors, interior sides and ends were lined with No. 16 gauge steel to allow cars to be used for hauling barrels and drums of oil.

The first cars rebuilt were five combination box cars in series 5122–5130, even numbers, that were originally built by Northwestern Car and Mfg. of Stillwater in 1884. Retaining their boxcar numbers, cars were rebuilt at D&IR Two Harbors shop between 1896 and 1908. When converted from combination flat cars, these had retained stake pockets on the carbody sides to mount vertical boards to hold the carbody. Rebuilt cars retained the distinctive stake pockets with vertical and diagonal bracing applied. Steel rods were used to support the interior sheathing. Cars in this series appeared to look like flat cars with a slatted stock car body applied.

The next ten cars looked more like traditional double-sheathed box cars. The outside sheathing was removed and replaced with stake pockets with vertical bracing and diagonals in between. Unlike the first five cars, these retained the full height interior sheathing and had small rectangular openings cut in sides and ends just below the roof line to provide ventilation.

The first four cars, 5150, 5154, 5156, 5158 were rebuilt in 1910 from box cars in series 2nd 5138–5152, even numbers only, originally built by D&IR Two Harbors in 1904 and box cars from series 5154–5172 that had been rebuilt from flat cars to box cars at Two Harbors in 1897. The last six cars, 5162–5172, even numbers only, were rebuilt in 1910 from box cars with the same numbers that had been rebuilt from flat cars to box cars at D&IR Two Harbors in 1897. All of the cars were placed in Class N in 1910.

The first five cars, 5122–5130, were renumbered to work train service in 1916. The remaining ten cars 5138–5172 were off the roster in late 1934.

D&IR WOOD ORE CARS

The first wood ore cars were ordered from Northwestern Car & Manufacturing Co. of Stillwater, Minnesota in January 1884. A total of 300 22-ton, 28-foot ore cars numbered 251–550. All were delivered by June 13, 1884. Car bodies consisted of a truss design with side sill (lower sill) and rail sill (upper sill) connected together by eight vertical corner and side post with angular side braces between the posts. A replica ore car to this design, 251, was rebuilt from a retired double spreader built in 1894 and rebuilt in 1899 from 28-foot ore car, for display at Two Harbors in 1934, however it is only an example of typical construction and not a copy of a specific wood ore car.

Increasing shipments of ore caused D&IR to order an additional 100 22-ton 28-foot wood ore cars 551–650 from Wells & French of Chicago in 1887. These were quickly followed by 100 more identical wood cars, 651–750, from Haskel & Barker of Michigan City, Indiana, in 1888.

It is unclear why, but the first two orders of ore cars were constructed as 28-foot cars but ore dock chutes and lake boat hatch spacing was 24 foot. To unload the 28-foot ore cars it was necessary to unload part of the car and then re-spot it to unload the remaining ore. Realizing the error in design beginning in 1889 length was standardized to 22 foot on all future ore car orders. Beginning in 1895, 200 of the 28-foot wood ore cars were rebuilt to 22 foot cars with the remainder completed by 1897.

Two hundred seventy 25-ton 22-foot wood ore cars, 751–1020 with 269 cu. ft. capacity, were ordered from Haskell & Barker in 1889. All were delivered that year. These were the first "1889 Standard" design and differed from the earlier 28-foot wood ore cars by having the upper and lower sills connected with six vertical supports with angular side braces. The center support post was replaced with two vertical rod supports.

The next group of 41 27-ton, 22-foot 1889 Standard wood ore cars 1501–1541 with 269 cu. ft. capacity were ordered from LaFayette Car Works in 1890. These were the first 27-ton cars and D&IR started a new number series to allow 24-ton and 27-ton cars to have their own blocks of numbers. Cars were equipped with Westinghouse air brakes.

Two years would elapse before additional 1889 Standard ore cars would be ordered. Built in 1892 by Wells and French Co. of Chicago were, 125 wood 24-ton, 22-foot 1889 Standard ore cars 1021–1145 with 270 cu. ft. capacity. All were equipped with Westinghouse air brakes, National hollow brake beams, and Safford drawbars. These were the first new built cars equipped with air brakes. These would be the last 24-ton cars built for D&IR.

The first ore cars built for D&IR were 350 by Northwestern Car & Manufacturing in 1883. This very early photo shows the 28-foot ore car with link and pin couplers and no air brakes. Development of standardized hatch spacing on lake boats created the need for a 22-foot standard car beginning in 1889. All D&IR 28-foot ore cars would eventually be rebuilt as 22-foot cars. (D.P. Holbrook collection)

Pre-1900 loading pocket on Vermillion iron range with a 28-foot wood ore car from 201–550 series built Northwestern Car & Manufacturing in 1884 spotted under loading spout. D&IR 992 one of 270 wood ore cars built by Haskell & Barker in 1889 is in the foreground. (D.P. Holbrook collection)

Left: D&IR 951 one of 270 25-ton 1889 Standard 22-foot wood ore cars 751–1020 built by Haskell & Barker in 1888 is at Canton Mine, Biwabik, Minnesota. Link and pin couplers date this prior to 1896 when D&IR had converted all cars to automatic couplers. Note the upper and lower sills extend beyond the end sills of the car. Later constructed cars had the upper and lower sills end at the end sill as shown D&IR 1548 coupled to the right. (MNHS Collection)

Below: LaFayette Car Works built 41 wood ore cars 1501–1541 in 1890. This builder photo shows the metal carbody details in gray. (University of Minnesota, Duluth, Kathryn A. Martin Library Archives and Special Collections Collection S3742)

D&IR 1548 one of 125 24-ton 1889 Standard 22-foot wood ore cars 1542–1666 built by Wells & French during 1892 and 1893. Canton Mine, Biwabik, about 1895.(MNHS Collection)

An additional 125 cars were ordered from Wells & French during March, 1892. These were 1889 Standard 27-ton, 22-foot wood ore cars 1542–1666 with 270 cu. ft. capacity. This order was delivered between November 1892 and January 1893. All cars were built with New York air brake equipment.

Railway Age and *Northwestern Railroader* issue of November 25, 1892, reported that Mr. H.S. Bryan, master mechanic of D&IR, had designed a standard ore car which became the D&IR "1892 Standard" wood ore car. Bids were requested for 450 to 500 of these cars with hopper bottom openings of 3-foot 7½ -inches by 7-foot 9¾-inches. The 24 gross ton capacity cars was to be constructed with white oak framing and side sills, side and end rail and deck beams from Norway pine. The four drop doors were each pair locked by a vertical

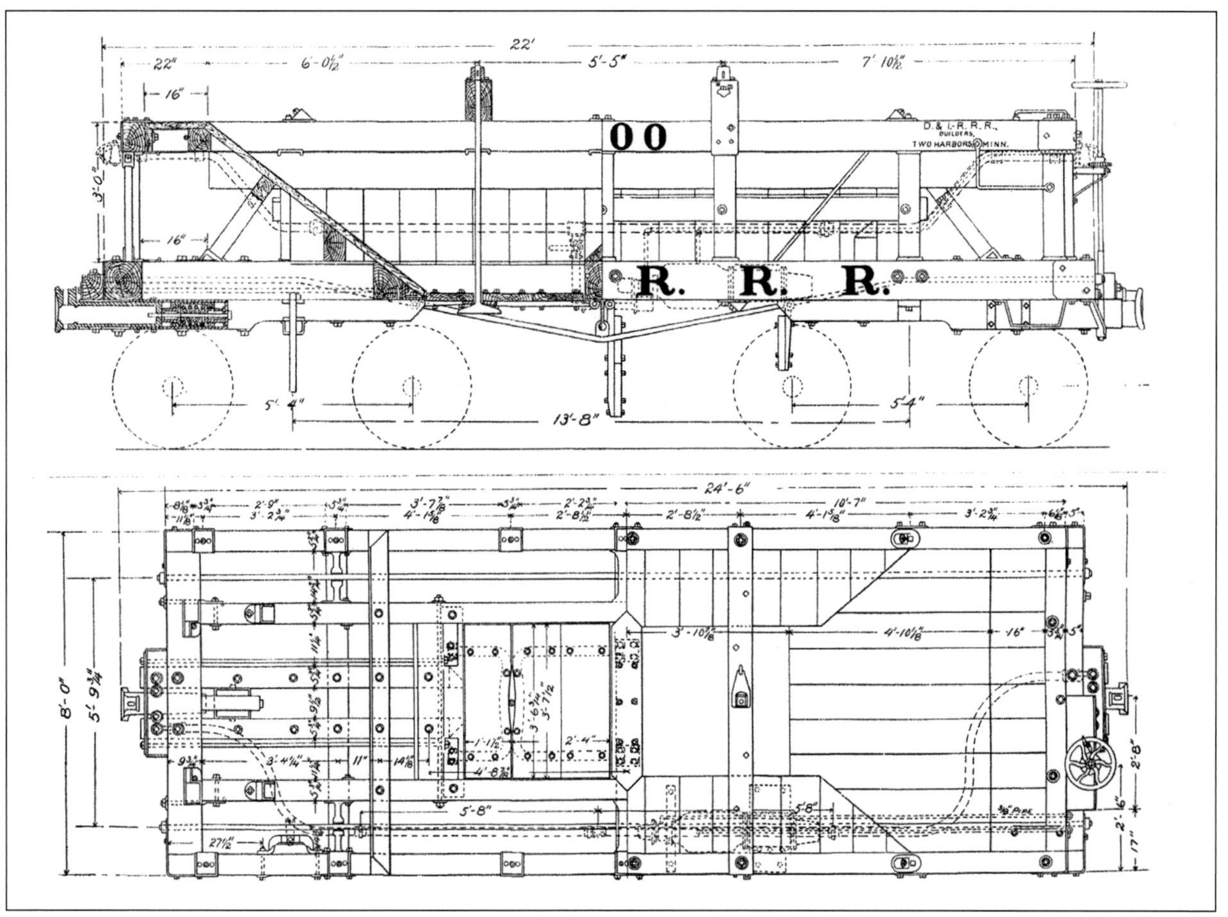

Duluth & Iron Range standard wood ore car – DIR 1667-3049. (D.P. Holbrook collection)

locking bar with a button on the bottom end and were equipped to be operated by a wrench from above. Cars were to be painted with two coats of D&IR standard freight car paint, with the end timbers being painted before car was assembled. This sample car was sent off-line as a demonstration car in an attempt to get orders for cars to be built at Two Harbors. It is unknown if any other railroads ordered cars from Two Harbors, but the design was used from 1892 to 1899 to build replacement cars for cars wrecked on the D&IR. Duluth Manufacturing Co. inspected the car and used the specifications for the 200 cars built in 1895 and 1896. The design was upgraded from a 24-ton to 27-ton car when the first orders were placed.

The first cars built to the new 1892 Standard design were 325 27-ton, 22-foot cars 1667–1991 with 270 cu. ft. capacity were ordered during December 1892 from Wells &

D&IR 1871, one of 325 24-ton 1892 Standard wood ore cars 1667–1991 built by Wells & French during 1893. Note in the upper background is D&IR 737 one of 100 22-ton 28-foot wood ore cars 651–750 built by Haskell & Barker in 1888. Canton Mine about 1895. (MNHS Collection)

Pullman built 300 22-foot ore cars 1201–1500 in 1895. Not mandated until 1911, there are no left corner grab irons or sill steps. (Pullman Palace Photographs, Archives Center, National Museum of American History, Smithsonian Institution)

French. Delivered in 1893, they were built with New York air brake equipment.

Three years later, 300 27-ton 22-foot 1892 Standard cars 1201–1500 with 270 cu. ft. capacity were ordered from Pullman Standard in February 1895 under Pullman Lot 2095. These cars were built during April and May 1895. Number series 1146–1200 had originally been left vacant to allow for additional 24-ton cars, if any where built, to be grouped with other 24-ton cars in the 201–1145 series. It was apparent in 1895 that this series would not be needed and D&IR started to fill in this gap in the wood ore cars by placing the Pullman 1892 Standard cars in series 1201–1500.

An additional 200 1892 Standard 27-ton 22-foot cars 1992–2191 were ordered from Duluth Manufacturing in May 1895 with 270 cu. ft. capacity. The cars were built in late 1895 and early 1896. While these were being delivered, coal cars 2000–2014 were being renumbered to 236–250 in late 1895.

During 1896, 450 1892 Standard 30-ton 22-foot wood ore cars were ordered from Terra Haute Car & Manufacturing with 454 cu. ft. capacity. A total of 357 cars were numbered 2192–2548 and 55 cars were numbered 1146–1200 filling in the remaining numbers left after the earlier 1895 Pullman order. The remaining 38 cars were used as replacements for cars destroyed in the 751–2136 series. These were the first wood ore cars built with an additional horizontal side extension applied above the rail sill (upper sill), causing the cars to have a larger "heaped capacity" then previous cars. All future wood ore car orders would have these additional extensions, called "side raising timbers by DM&N" applied. Many earlier wood ore cars also had this modification to allow more ore to be loaded in the cars.

An order for 300 1892 Standard 30-ton 22-foot wood ore cars 2549–2848 was placed in December 1897 with Illinois Car & Equipment Co. for $135,000 with 454 cu. ft. capacity. All were delivered in April and May 1898. The cars were equipped with Barber trucks, Westinghouse brakes, National hollow brakebeams, Tower couplers, Bryan tandem spring drawbar attachments, Scott springs, Barber roller side bearings and journal bearings split evenly between Acme and Ajax.

Washington state carrier Spokane Falls & Northern, later part of the Great Northern, thought enough of this design and builder that they ordered 25 identical cars with Barber arch-bar trucks.

D&IR returned again to Illinois Car & Equipment in February 1899 and ordered 200 1892 Standard 30-ton 22-foot wood ore cars 2849–3049 built to the same plans as the 1897 order. All were delivered between May 1899 and July 1899. These would be the last wood ore cars purchased by the D&IR.

By January 1908, 272 of the wood ore cars had been restricted to coal loading only, this increased to 486 cars in 1909. By 1910, with continued retirements of wood ore cars, only 1,029 wood ore cars remained on the roster, with 344 restricted to coal loading only. During 1911 there were only 329 wood ore cars remaining on the roster, almost all restricted to coal loading. Sixty seven remained in 1917 and all were off the roster before 1918.

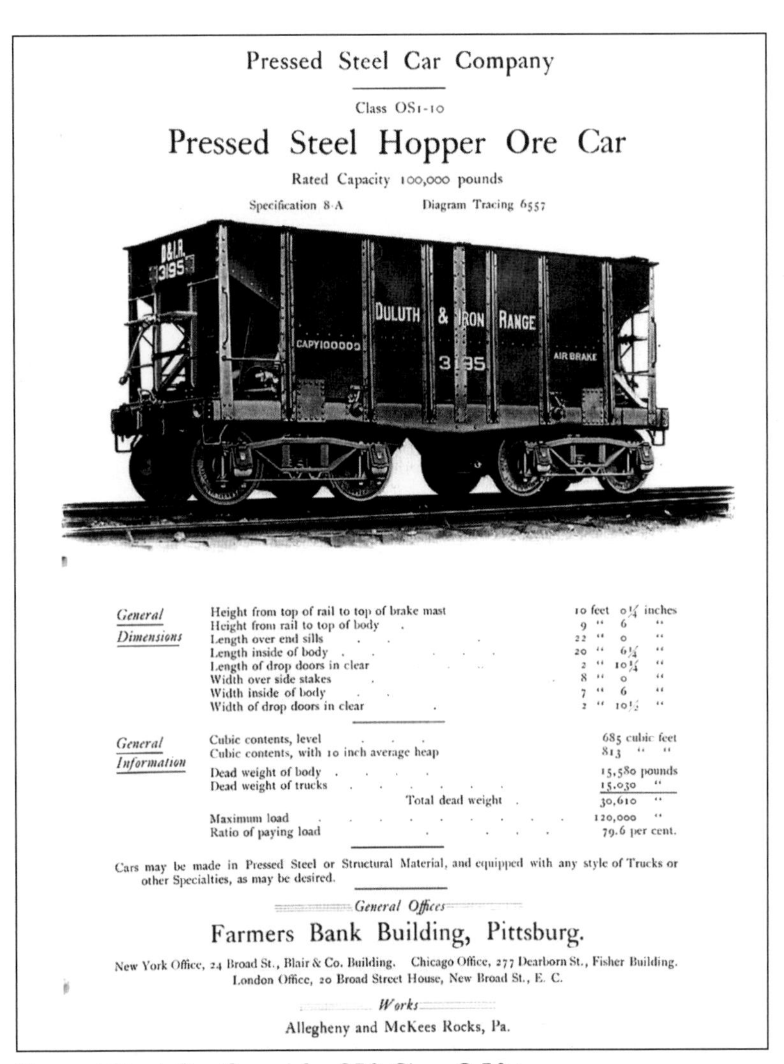

Pressed Steel Car Co. ad for 350 Class C 50 ton steel ore cars 3050–3399 built in 1900. The first ore cars were shortened versions of typical coal hoppers of the era and included, inside the rails, sloped coal hopper bays. This style of unloading bay made it difficult to unload the Vermillion iron range ore which consisted of larger hard chunks of ore. (D.P. Holbrook collection)

D&IR STEEL ORE CARS

The first steel ore cars acquired by D&IR were 350 cars ordered from Pressed Steel Car in September 1899. These 50-ton 22-foot ore cars 3050–3399 were delivered in 1900 with 670 cu. ft. capacity. The cars were a radical design change with the complete hopper and underframe constructed of steel. Design of these cars mimicked Pressed Steel Co. coal hoppers from the era, with eight feet of the middle section of car removed. Hopper doors were wood and equipped with a door trapping mechanism for operating the doors. They had 6 T-section pressed steel side ribs, fishbelly side sills and slope sheet bracing. Unique for ore cars, they had two hopper bottom doors much like coal cars from the era, but these were narrower, dumping only between the rails. Each door was only 2-foot 10½-inches square. The size of the opening would cause these cars to be difficult to unload the lump-type iron ore from the Vermillion iron range. Surprisingly, DM&N received five virtually identical cars numbered 8000–8004 in 1899, but did not order any additional cars of this type. During 1910 the D&IR placed these cars in Class C.

Once D&IR and DM&N came under the same corporate control of U.S. Steel, the economy of joint ordering of equipment was apparent. The next group of cars were ordered in October 1902 from American Car & Foundry under Lot 2517. A total of 100 50-ton cars were ordered with the order split with 50 cars going to D&IR and 50 to DM&N. The D&IR cars were delivered in 1903 and numbered 3400–3449. The cars had two pairs of double drop-bottom type dump doors 2-foot 8-inch by 3-foot 5-inch that closely followed those on early wooden ore cars. ACF designed these as a shortened version of a coal hopper replacing the coal hopper bays with 4 drop doors. The car body had two U-section ribs located above the bolsters and four equally spaced T-section ribs all riveted to the carbody. Unique on these all

Starting in 1899 Duluth & Iron Range began to purchase the first steel ore cars. This second group, 3400–3449 built by American Car & Foundry in 1903 was a single order for 100 cars placed in October 1902. After the order was placed it was split with 50 cars for D&IR and 50 for DM&N. This ACF builders photo shows that early steel ore car mimicked a shortened coal car. (American Car & Foundry Collection, John W. Barriger III National Railroad Library at UMSL)

steel cars was a wooden end sill reinforced with a U-channel. Some of these cars were sold to Malvern Gravel Co. in Arkansas in the 1930s. During 1910 the D&IR placed these cars in Class D.

These cars were quickly followed by 450 50-ton 22-foot ore cars 3450–3899 from Pressed Steel Car Co. in 1903 with 680 cu. ft. capacity. The cars had two pairs of double drop-bottom doors with each having an opening of 3-foot 2-inch by 3-foot 6-inch, creating a larger opening than the 1899-built cars. These doors became the standard ore cars doors until longitudinal doors were developed in 1909.

All were delivered with Barber arch-bar trucks, Westinghouse brakes and drawbar attachments, McCord journal boxes and Chicago and Tower couplers. The cars had six L-section side ribs with 6 half height diagonal L-section ribs. Sides only came half way down the side instead of full height sides of the 1899 order. This allowed easier access to door operating mechanisms. The K-brake equipment was relocated to the side just below the side slope sheet to allow easier maintenance. The cars did retain the fabricated fishbelly side sill from the 1899 order. These were the first cars with door opening and closing mechanisms controlled by brake wheel-type wheels located along the side adjacent to sill steps at a height allowing easy movement by a standing ore dock employee. During 1910 the D&IR placed these cars in Class D.

Pressed Steel Car Company

Class OS2-10

Pressed Steel Hopper Ore Car

Rated Capacity, 100,000 Pounds

Specifications 38-A

Diagram Tracing 8890

General Dimensions	
Height from top of rail to top of brake mast	9 feet 6½ inches
Height from top of rail to top of body	9 feet
Length over end sills	22 feet
Length inside of body	20 feet 4 inches
Length of drop doors in clear	6 feet 8 inches
Width over side stakes	8 feet 6 inches
Width inside of body	7 feet 11½ inches
Width of drop doors in clear	3 feet 6 inches

General Information	
Cubic contents, level	692 cubic feet
Cubic contents, with 10 inches average heap	827 cubic feet
Dead weight of body	17,097 pounds
Dead weight of trucks	15,203 pounds
Total dead weight	32,300 pounds
Maximum load	120,000 pounds
Ratio of paying freight to total weight of car when loaded	78.8 per cent.

Cars may be made of Pressed Steel or Structural Material, and equipped with any style of Trucks or other Specialties, as may be desired.

General Offices—Farmers Bank Building, Pittsburg

New York Office, 24 Broad Street, Blair & Co. Building; Chicago Office, 277 Dearborn Street, Fisher Building; London Office, 20 Broad Street House, New Broad Street, E. C.

Works—Allegheny and McKees Rocks, Pa.

Pressed Steel Car Co. ad for second group of Class D 50 ton steel ore cars 3450–3899, 450 cars built in 1903. Built to modified steel hopper car designs from the era, the diagonal bracing would be eliminated on future cars. (D.P. Holbrook collection)

The evolution in 50-ton steel ore car design is illustrated in this view at Pioneer Mine Shaft B, Ely, Minnesota, about 1915. From left to right, are two of 100 ACF-built 1910 cars from 11400–11499 series, the third and seventh car are two of 450 Pressed Steel Car Co.-built 1903 cars from 3450–3899 series and the fourth through sixth car are three of 750 Standard Steel Car Co.-built 1910 cars from 10500–11249 series. (University of Minnesota Duluth, Kathryn A Martin Library Archives and Special Collections, S3742 Box 7 Folder 20398)

Increased tonnage and a deteriorating wooden ore car fleet caused the D&IR to order additional 50-ton steel ore cars. An order was initially placed in April 1905 for 300 cars; increased in May to 500 cars from Standard Steel Car Co. Numbered 3900–4399, they were 22-foot, 710 cu. ft capacity cars with two pairs of double drop-bottom doors each having an opening of 3-foot 2-inches by 3-foot 6-inches. The carbody was modified from earlier designs and had two U-section ribs located above the bolsters. The remaining rib pattern consisted of two L and two U-section ribs equally spaced between the end ribs. The car had, what would later become standard on all future ore cars, a half-height side sheet carbody with slope sheets intersecting the sides to create a tub.

Between end sills and end slope sheets there was a diagonal L-section rib for additional support. Handbrake-type wheels to operate doors were located adjacent to sill steps on both sides of the car. K-brake equipment was located below the side sill between the trucks allowing for easy maintenance. These cars were built with Barber arch-bar trucks, Westinghouse brakes and drawbar attachments, and a mix of Climax and Chicago couplers. During 1910 the D&IR placed these cars in Class D. Beginning in 1931 a number was added to this class making them Class D3.

After the first order of ore cars having small coal hopper doors was delivered, the design for bottom-dumping doors was changed to four transverse doors operating as pairs for unloading. Suitable for their time, they did not allow the large chunks of Vermillion Range iron ore to dump easily. Main improvement of these cars was the door operating wheels located on the side of the car making it easier for men standing on the ground to operate the doors. Series 3900–4399 cars were built by Standard Steel Car Co. in 1905. Some would later be modified with longitudinal doors. Note the C and L stock ribs. (Standard Steel Car Co., D.K. Retterer collection)

Standard Steel Car Co. builders photo of D&IR 12750 built in 1916 shows off the evolving design changes. Basic construction remains the same but the lower half of the carbody has been modified to use less steel reducing the tare weight of the car and allowing easier maintenance of mechanical and airbrake components. (Standard Steel Car Co., D.P. Holbrook collection)

Further improvements in carbody design were made when 700 cars in 9000–9699 were built by Pressed Steel Car Co. in 1907. These cars mimicked the Standard Steel Car Co. design but had a different C and L stock rib arrangement, also, the bottom edge of the carbody side sheet is slightly longer on these cars. (Pressed Steel Car Co., D.K. Retterer collection)

Late in 1905 D&IR returned to Pressed Steel Car for 500 additional 50-ton 22-foot ore cars 4400–4899 with 710 cu. ft. capacity with two pairs of double doors each having an opening of 3-foot 6-inches by 3-foot. Pressed Steel modified the Standard Steel Car order 3900–4399 by changing the rib pattern to two U-section ribs located above the bolsters and four equally spaced L-section ribs between the U-section ribs.

The cars retained the diagonal L-section ribs attached between side sills near the ends and the slope sheets. These acted as slope sheet supports to supplement the two slope sheet supports attached to the top of the draft gear. All were delivered during March 1906 equipped with: Buffalo Car Wheel Works wheels, Simplex bolsters, Westinghouse brakes, Pressed Steel Car brake beams, Westinghouse drawbar attachments, Pittsburg Steel Spring Co. springs, McCord journal boxes, Susemihi side bearings and Franklin dust guards. During 1910 the D&IR placed these cars in Class D and in 1931 reclassed the cars as Class D4.

Satisfied with the 1905 order, the next group of 700 50-ton 22-foot ore cars 9000–9699 were ordered from Pressed Steel Car in 1907 with 710 cu. ft. capacity. These were identical to the 1906 Pressed Steel Cars and in 1910 were also placed in Class D. Beginning in 1931 a number was added to this class letter, making them Class D4.

SUMMERS TEST CAR

Summers Car Co. supplied a 50-ton demonstrator ore car, number 101, in 1909. Its arrival on the iron range would revolutionize future shipments of ore. Unique at the time, it was the first to have longitudinal dump doors consisting of two doors made from pressed steel shapes that extended over a portion of the inner truck wheels. The ends of the doors were flanged upwards which allowed them to open downward without obstructing the wheels. Side and end slope sheets had an inclination of 50 degrees with the doors, when open having the same slope. Doors were 5-foot 6-inch by 8-foot 6-inch, so large,

The first orders for cars with the new type doors were 800 Summers-type 50-ton 22-foot ore cars, 9700–10499 ordered from Standard Steel Car Company's Hammond, Indiana, plant in 1909 with 650 cu. ft. capacity for $842,517.67. The Summers Steel Car Co. patents were issued April 13, 1909. (Standard Steel Car. Co., D.K. Retterer collection)

that bridging of ore in the car was eliminated. Four heavy hinges attached to the doors were connected to 2½-inch steel shafts by chains, actuated by worm gears that rotated on a shaft connected to a cross shaft. This allowed the doors to be placed in any position from closed to full open without any additional mechanisms. Older iron ore cars required from six to ten men to unload the car. The new car could be operated by one man and unloaded within two minutes. D&IR Appropriation 486, of July 7, 1915, authorized the purchase of 101 for $300.00. The car had 820 cu. ft. capacity and was equipped with arch-bar trucks. Summers would go on to modify the design and produce many additional "Summers" ore cars for both the D&IR and the DM&N. After retirement of Rakowsky car 3049 in 1914, this car was renumbered second 3049 late in 1915. It was retired in 1928.

MORE 50-TON STEEL CARS

Innovation was needed. The standard 2 paired drop-bottom double doors were very slow dumping and this was causing congestion on the docks while cars were being dumped. Longitudinal dump doors for ore cars had been developed by two of the major builders. Standard Steel Car, with their Summers-type car, and American Car & Foundry with their Clark-type car. The first orders for cars with the new type doors were 800 Summers-type 50 ton 22-foot ore cars 9700–10499 ordered from Standard Steel Car in 1909 for $842,517.67. The new cars had a single door opening of 5-foot 6-inches by 8-foot 2-inches and rode on arch-bar trucks. The carbody was a single riveted plate steel side sheet with wide end platforms, with no corner or end posts, having an interior length of 17-foot 1-inch. End

Carbody modifications would follow quickly in 1910 when an additional 750 50-ton 22-foot cars in 10500-11249 series were ordered from Standard Steel Car. Note the changes in ladders, staff brake wheel mounting and the addition of small holes in the side sheets for maintenance. The left side of car had four small openings but the right side had five. It is unclear as to why they were different. (Standard Steel Car Co., D.K. Retterer Collection)

platforms had ladders on both sides leading up to a 4-foot wide poking platform. These would be the only cars built with the wide poking platforms. Cars also had the vertical staff brake wheel mounted at the extreme height of the carbody. The new door type caused these cars to be placed in Class E in 1910. A change in 1931 reclassified these cars as Class E6. Besides using the Class D and Class E for the steel ore car fleet, the D&IR often referred to the Class D cars (transverse doors) as "slow dumping cars" and the Class E cars (longitudinal doors) as "fast dumping cars" based on the their operating abilities. All of these cars were retired per AFE 3412 June 18, 1934.

Even before the new cars saw one year of service the D&IR had determined that they would order 1,000 50-ton ore cars with the new type of doors. An appropriation was approved for $102,433.47 with the order split between three car builders.

Quickly, the D&IR ordered 750 additional 50-ton 22-foot Summers-type cars 10500–11249 from Standard Steel Car in 1910 with 650 cu. ft. capacity. The cars had a single door opening of 5-foot 7-inch by 6-foot 6½-inch and rode on arch bar trucks. They continued the basic carbody style from the 1909 order with some modifications. Side sheets had four round openings on the L-side of the car and five round openings on the R-side for access to brake components. The car still did not have any end or corner posts, but the carbody top had angled metal corners to meet the slope sheet, allowing an interior carbody length of 18-foot 7-inches.

Sill steps were set in from ends to just below the end of side sheets to allow grab iron to be attached to side sheet for employees to get on or off the car. End ladders were modified with only one ladder at each end going up to a 3-foot wide poking platform. The vertical brake wheel staff was lowered to a small wooden end platform allowing easier access. These were placed in Class E in 1910. Beginning in 1931 this class became Class E7. The

Until 1910, ore car construction consisted of a carbody with exterior post construction. American Car & Foundry built 100 Clark-type 50 ton ore cars, D&IR 11400–11499 in 1910 with specially designed longitudinal rolling doors forming a hopper bottom. (American Car & Foundry Collection, John W. Barriger III National Railroad Library at UMSL)

D&IR/DM&N merger caused these cars to be renumbered to 40500–41249.

Wanting to sample additional builder's cars with longitudinal doors, D&IR ordered 150 50-ton 22-foot ore cars 11250–11399 from Pressed Steel Car in 1910 with 715 cu. ft. capacity. These were built at the Western Steel Car and Foundry plant in Chicago. Cars had a single bottom door opening of 5-inch 7-inch by 6-foot 6½-inches. These cars would be the first development of a standard ore car design with plate steel sides open below the side slope sheet attachment points, having side sill mounted bracing to the slope sheets.

Ladders were created by having corner posts with ladder rungs extending to the side sheets with sill steps inset to half way between the end axles and bolster. Side sills ended at the corner posts and angled end sills extended to the draft gear. These would also be first cars to have the end slope sheets angle from the bottom door opening to the corner posts, allowing easier unloading of the car. These cars also returned to the wide poking platforms. These were placed in Class E in 1910. Beginning in 1931 the classification became Class E6. The D&IR/DM&N merger caused these cars to be renumbered to 41250–41399.

The next order was placed with American Car and Foundry in 1910 for 100 Clark type 50-ton 22-foot ore cars 11400–11499 built under Lot 5809A with 650 cu. ft. capacity. Built to a new design with flat riveted plate steel sides and sloped corner posts supporting the end slope sheets, all bracing was internal to the carbody design. The cars had a special design of longitudinal rolling doors forming the hopper bottom. These were placed in Class E in 1910. A special design of longitudinal rolling doors 6-foot 6½-inches by 5-foot 7-inches formed the hopper bottom. AFE 1259 of March 11, 1924, AFE 1266 of March 27, 1924, and AFE 1273 of April 6, 1924, were approved for the application of extended sides and ends

Duluth resident and mechanical engineer, Gustav A. Rakowsky, took notice of the Summers and Clark designed ore cars and designed and applied for patents for his Rakowsky 50-ton automatic unloading ore car. This car was unique in having two halves that were attached to a frame body and dumped like two opposing dump truck bodies. Demonstrator car 100 was built by the "Automatic Unloading Car Co., a division of Northwestern Steam Boiler Co." at Duluth in 1908. Doors were operated by air which allowed unloading in ten seconds. D&IR and DM&N would test the car during 1909 and 1910 and then each acquire five cars built to a modified design in 1911. (Minnesota Historical Society collection)

to the 98 remaining cars. Sides and ends were extended 10-inches with $^{3}/_{16}$ plate steel and 3 x 3 angle iron for $7439.01. The modifications were removed per AFE 1884 begun on June 11, 1926, and completed on July 7, 1926. These cars quickly had bulging side sheet problems that caused all the cars to be retired by 1929.

Deliveries of 1,800 50-ton steel ore cars between 1909 and 1910 allowed the retirement of over 700 wooden ore cars. Remaining wood cars were now mostly in coal and gravel service and all wooden ore cars were retired by 1918.

RAKOWSKY ORE CARS

At the D&IR Board of Directors meeting on March 11, 1911, the purchase of five Rakowsky steel ore cars was approved on D&IR Appropriation 326 for $5,390.00. These cars were built by Automatic Car Unloading Co., a division of Northwestern Steam Boiler Co. of Duluth, and numbered 3045–3049. These cars were 22-foot 1-inch 50-ton ore cars with a capacity of 580 cu. ft. All cars were received and placed in service by November 1911. These cars were unique in their clamshell car bodies with each half rotating on pinions to allow dumping of the cars. The rigors of iron ore service caused these cars to suffer door operating problems, with 3049 being retired in 1914. The remaining four cars, 3045–3048 were held on roster out of service from 1914 to 1916. All four cars were written off accounting records in 1916.

MORE STEEL ORE CARS

Increasing ore business caused the D&IR Board of Directors to approve Appropriation

Standard Steel Car Co. builders photo of D&IR 11500 built in 1913 shows off the shortened carbody and extended draft gear. Ore car construction evolved from the first steel cars being built from foreshortened coal hopper designs, to rib-sided cars with L- and U-section supports, to plate steel riveted sides as depicted here, and then to a final design with plate steel covering only the upper half of the carbody. (Standard Steel Car Co. photo, DP Holbrook collection)

405 on July 11, 1912 for $1,150,000 to purchase an additional 1,000 steel 50-ton ore cars during 1913.

Standard Steel Car received the order for 800 Class E Summers-type 50-ton 22-foot ore cars 11500–12299 to be constructed at their Hammond, Indiana, plant with 695 cu. ft. capacity. The cars had a single door opening of 5-foot 6-inch by 8-foot 2-inch and rode on arch-bar trucks. Sides were constructed of riveted plate steel down to the side slope sheets, with the bottom side of carbody being supported by two large and two small riveted steel plates attached to the side sheets. Beginning in 1931 this class became Class E8. The D&IR/DM&N merger caused these cars to be renumbered to 41500–42299.

DM&IR 41641, one of 800 D&IR 11500–12299 50 ton Class E8 ore cars built by Standard Steel Car Co. in 1913. Cars were renumbered to 41500–42299 series after D&IR and DM&N merger. (D.P. Holbrook collection)

American Car & Foundry received the order for 200 Class E National-type 50-ton 22-foot ore cars 12300–12499 with 685 cu. ft. capacity riding on arch-bar trucks. The cars were unique in having stamped-steel carbody sides and side slope sheets that were riveted to the end slope sheets. Sides were supported by five U-section ribs. Corner posts were inset from the ends and a vertical staff brakewheel was mounted on the ends. These cars had a single longitudinal door opening 7-foot 1-inch by 5-foot 5-inches. Beginning in 1931 this group of cars became Class E9. The D&IR/DM&N

merger caused these cars to be renumbered to 42300–42499.

Appropriation 532 was approved at the June 13, 1916 Board of Directors meeting for the purchase of 750 50-ton 22-foot ore cars for $937,500. This order was again split between ACF and Standard Steel Car.

American Car & Foundry built 250 Class E National-type 50-ton ore cars 12500–12749 which were identical to the ACF 1913-built cars. Beginning in 1931 this group of cars received a change to Class E10. The D&IR/DM&N merger caused these cars to be renumbered to 42500–42749.

Standard Steel Car Co. built 500 Class E 50-ton Summers-type 22-foot ore cars 12750–

Class E8 ore car 42178, one of 800 D&IR 11500–12299 built by Standard Steel Car in 1913 is shown at Sellers Mine in June 1939. Note the Duluth, Missabe & Iron Range monogram without the Safety First slogan, used between 1937 and 1942. (University of Minnesota Duluth, Kathryn A. Martin Library Archives and Special Collections, Collection S2386 Box 27 Folder 5 detail)

The first 50 all-steel ore cars were supplied by American Car & Foundry in 1903, but D&IR would not return to ACF for additional ore cars until ordering the Clark-type ore cars in 1910. Not satisfied with that design, ACF improved on it with exterior post construction, hoping that it would solve a bulging side problem. D&IR ordered 200 Class E cars, 12300–12499 in 1913 based on this new design. These would be the final ore cars painted with the full "Duluth & Iron Range" name. Future orders would arrive with only D&IR reporting marks. (American Car & Foundry Collection, John W. Barriger III National Railroad Library at UMSL)

13249 with 695 cu. ft. capacity. The 1913 Standard Steel Car Co. car design was modified by having additional access openings in side slope sheets, and all were equipped with T-section Andrews trucks. Placed in Class E when purchased, the reclassifications of 1931 placing them in Class E11. The D&IR/DM&N merger caused these cars to be renumbered to 42750–43249.

Rib side support of ore car bodies returned to D&IR when 200 cars 12300–12499 were ordered from American Car & Foundry in 1913. Built to a National design, the carbody was supported by five U-sections riveted to the carbody. Renumbered to DM&IR 42300–42499 series, 42421 is shown at Proctor July 17, 1957. (Bob's Photo collection)

The teens would find the D&IR ordering multiple types of cars from American Car & Foundry. D&IR 12748 was part of 250 Class E ore cars 12500–12749 built in 1916. These were virtually identical to the 200 cars 12300–12499 built in 1913, with the main difference being Andrews T-section trucks instead of arch bars. (American Car & Foundry Collection, John W. Barriger III National Railroad Library at UMSL)

D&IR SLOW AND FAST DUMPING ORE CARS

Extremely happy with the longitudinal dump doors on newer 50-ton ore cars, D&IR began to pursue in 1928, a way to modify transverse door cars with longitudinal doors.

One hundred one of the 1905 cars 3900–4397 and 99 of the 1906-1907 built cars 4432–9618 were stenciled "work train cars for track department service only" per a July 13, 1928 letter from R.R. Moore, superintedent motive power, D&IR Two Harbors, and were *not* to be used for moving iron ore. All were transferred to the MofW department in December 1928. Per letter of June 6, 1929, from Thomas Owens, vice president and superintendent, these cars were to be returned to ore service, but stenciling was to remain on the cars. AFE 3123 approved February 4, 1931 transferred all 200 cars back to revenue service.

(On August 1, 1928 Mr. B.B. Thomas is-

Overhead view of D&IR 163 (actually DM&N 163) at Soudan Mine, August 12, 2014. Note the corner slope sheets attached to the end, side and side slope sheets. As built this car did not have corner slope sheets, and end slope sheet intersected the side sheets. This car matches D&IR 9000–9699 Class D 50-ton ore cars. (D.P. Holbrook)

sued a lettering advising that 500 of the 1905 and 1906-07 built cars were being sold to Duluth Iron & Metal. At that time it was advised that 341 cars should be retained and used exclusively for Soudan, Mary-Ellen and Siphon mine ore, account these mines being very slow-moving mines and the ores loaded at these mines were better loaded in these cars and this would not tie up fast-dumping cars in this service. The Siphon Mine was expected to close during 1929 and the Mary-Ellen was expected to close very soon. This would leave sufficient slow-dumping cars to protect the movement of Soudan ore.

Evaluation of the 1351 ore cars being considered for retirement in 1929 showed that 1261 were of the early slow-dumping variety.

Cars with longitudinal doors were considered a fast-dumping ore car which dumped contents almost instantaneously and only required one man to release the doors. One minute and one man were employed to clear a fast-dumping car of its contents.

Slow-dumping ore cars took an average of 4 men, 15 minutes each to perform the same service occasioned partly by having to climb the car and push out some of the contents. This was a result of the dump opening in the floor of the car being of much smaller dimensions, and also it had a transom bar across it, which acted as a bridge to hold the load from falling out of the car. The angle of the sides of the hopper were much flatter than in the fast-dumping type, which checked the velocity of the fall of the ore, whereas the type of car that was being retained had no transom, two doors, and a much larger opening. The side slope sheets of the fast-dumping cars were also much steeper.

Slow-dumping hoppers were badly corroded and worn, wood end buffers were decayed, draft channels and body bolsters needed to be reinforced, and the trap doors were badly worn. Cars were deteriorating rapidly.

The cars purchased in 1903 and 1905 were not convertible into fast-dumping type with longitudinal doors. But those built in 1906 and 1907 could be converted. The first two cars converted were 4656 and 9295 under AFE 1664 approved on May 19, 1925 for $1,322.99. Work was begun on September 5, 1925, and completed September 23. Besides replacing the doors, the end slope sheets were improved from 42 degrees to 48 degrees, making the cars easier to dump. Another car, 4596, was completed on November 10, 1925. Thirty-five additional cars were converted at Two Harbors during the 1927-1928 winter shopping season. These cars, 4400–4419, 9156, 9348, 9426,

Slow Dumping Ore Cars 1929			
Class-Built-Builder	**Number series**	**To be Retired**	**To Remain in service**
Class D Built 1903 ACF	3400–3449	50	
Class D Built 1903 PSC	3450–3899	434	
Class D Built 1905 SSC	3900–4399	487	
Class D Built 1906-07 PSC can be converted to fast dump	4400–4899 9000–9699	290	906
Fast Dumping Ore Cars 1929			
Class-Built-Builder	**Number series**	**To be Retired**	**To Remain in service**
Class E Built 1909 Summers	9700–10499		
3049		797	
Class E Built 1910 Summers	10500–11249		895
Class E Built 1910 ACF	11400–11499	90	
Class E Built 1913 SSC	11500–12299		800
Class E Built 1913 ACF	12300–12499		200
Class E Built 1916 ACF	12500–12749		250
Class E Built 1916 SSC	12750–13249		500
Totals		1351	4348

9566, and 9636 were rebuilt with longitudinal doors and 10-inch side extensions. AFE 1613 July 27, 1927 and AFE 1643 September 6, 1927 provided for the rebuilding of these cars.

There were 906 cars of this series retained for purposes other than hauling of ore, such as coal, gravel and sand, where the dumping apparatus did not materially affect the service of the car. It was furthermore contemplated, when the occasion arose and it was considered expedient, to convert these cars into fast-dumping type, and at that time increase the capacity by from two to three tons per car, strengthen the body, at an estimated cost of between $500 and $600 per car.

The 90 cars, 11400–11499, of 1910 construction remaining to be disposed of, were of the improved type of car, but they were so weak in construction that the body bulges in the center caused the entire car series to be pulled out of service.

It was thought that a very limited number of these cars could be disposed of at a figure somewhere between three and four hundred dollars. Once these retirements took place it left in service 4,348 cars, which was sufficient to take care of all requirements in addition to coal, gravel, and sand. All remaining cars were then of the fast-dumping type excepting 906 slow-dumping cars which were reserved for coal, gravel, and sand, with the possibility of converting them to a fast-dumping type when it was deemed expedient.

Over 320 of the 50-ton slow-dumping cars survived into the 1970s in work service. The four transverse doors that caused iron ore to slow dump because of their openings also made them a preferred car for dumping ballast. The dumping mechanism was operated via a chain connected to a shaft that was turned by a hand wheel on the side of the car. Each door had its own mechanism with a ratchet and pawl locking device that allowed the doors to be opened gradually to control the flow of ballast from the car.

D&IR 246 33-foot 4-inch coal car at Duluth circa 1905. The car was originally built as part of 15 coal cars 236–250 by Northwestern Car & Manufacturing in 1884. Renumbered to 2000–2014 series in 1887, the car would again be renumbered to 236–250 in 1895. Beginning in 1901 and completed by 1904 all were rebuilt to 36-foot cars with wooden sides atop a standard D&IR flat car as shown here. (Shorpy Photo Archives)

D&IR COAL CARS

At the same time as 300 wood ore cars were ordered in 1884, 15 coal cars 236–250 were built by Northwestern Car & Manufacturing. They were 33-foot 4-inch side-dump coal cars with 3-foot 4-inch inside height and 910 cu. ft. capacity. Sides were constructed of four 10-inch planks with 13 oak side stakes installed in stake pockets to support the sides. D&IR listed these as "coal, side dump." The cars were renumbered in 1887 to 2000–2014 and renumbered again in 1895 to 236–250. All were rebuilt to 36-foot cars with solid wooden sides between 1901 and 1904 at Two Harbors. Class H would be applied to these cars in 1911. Steel center sills were applied to car 248 in 1919. It appears that this may have been the only car rebuilt with steel center sills. All were off roster by 1934.

D&IR STEEL GONDOLAS

Prior to 1907, D&IR relied on a roster of solid-bottom wooden gondolas, most rebuilt from flat cars with added sides. Heavy use caused these to require constant repairs.

Increasing shipments of coal for mining companies, commercial customers and D&IR company coal created a shortage of gondolas. D&IR ordered three groups of 50-ton 41-foot steel drop-bottom gondolas in 1907 and 1908 from Pressed Steel Car Co. with 1,593 cu. ft. capacity. All were identical steel construction with seven side and three end T-section ribs and rode on arch bar trucks. The cars were equipped with 16 flat steel drop-bottom doors with three hinges per door along the longitudinal center sill and opened to the outside.

The doors were operated by a simple shaft with two chains attached to each door. Winding this mechanism caused the door to open or close. Center sills were of triangular cross section allowing the car to be virtually self-clearing making them suitable for hauling ore, sand, gravel or dirt. Iowa Central, Newburg & South Shore, and other carriers owned identical cars. Once old-growth timber had played out, the cars were used for hauling pulpwood, initially loaded longitudinal and in later years horizontally.

The first group 221–235 was built in 1907. These were numbered directly below wooden gondolas 236–250. All 15 cars were delivered between November and December 1907. At the time of the merger these were renumbered to 4221–4235 series. All were off the roster by 1971

The second group of steel gondolas, 201–220 were built in 1908. All twenty cars were delivered in April, 1908. At the time of the merger these were renumbered to 4201–4220 series. All were off the roster by 1971.

The last group of steel gondolas were built in 1908 and numbered 251–260. All ten were delivered in April, 1908. These were numbered directly above wooden gondolas, 236–250. At the time of the merger these cars were renumbered 4251–4260. All were off the roster by 1971. All three groups were placed in Class H beginning in 1911 when class letters began to be assigned. Car builder's order records from this time do not show D&IR ordering this last group of gondolas and it is thought these may have been originally built for Roebling Steel in 1906 who did not take delivery, and cars were purchased by D&IR in 1908

Pressed Steel Car Co. ad for Class H 50 ton 41-foot steel drop-bottom gondolas. D&IR would buy three groups, 201–220, 221–235 and 251–260 in 1907 and 1908. Fifty additional drop-bottom gondolas, 261–310 would arrived from Western Steel Car & Foundry in 1917. (D.P. Holbrook collection)

Appropriation 630 was approved at the Board of Director meeting on June 11, 1917, for $129,000 to purchase 50 Class H 50-ton 41-foot 8-inch steel gondolas 261–310 built by Western Steel Car & Foundry with 1,593 cu. ft. capacity. All 50 cars were delivered between February and March of 1918. They were identical to the Pressed Steel Car Co. cars built in 1907 and 1908. The cars were placed

Two orders were placed in 1907 for 45 steel drop-bottom gondolas from Pressed Steel Car Co. which would allow them to be loaded with coal, aggregates, pulpwood and lumber. Class H DM&IR 4211 is at Proctor, July 17, 1957. Having 1,532 cu. ft. capacity these cars allowed for a two for three replacement of flat car/gondola conversions and the drop bottoms made unloading less labor intensive. (Bob's Photo collection)

Maroon and gold paint and lettering on gondolas was adopted in 1958. Scotchlite monograms, reporting marks and numbers would replace the stencil versions beginning in late 1959. DMIR 4202 and 4282 are at Tower, Minnesota, in the summer of 1960. (Jim Maki Sr.)

An order for 50 additional steel drop-bottom cars was placed with Western Steel Car & Foundry (a Pressed Steel Car subsidiary created in 1902) in 1917 which were identical to the Class H cars built in 1907 and 1908. Early cars were delivered with arch-bar trucks but the 1917 order received Andrews U-section trucks which were replaced in later years. Class H1 DM&IR 4293 is at Proctor in 1957. (Bob's Photo collectioin)

DM&IR 4290 at Virginia, Minnesota, in the spring of 1962 is former D&IR 290, one of 50 Class H 261–310 acquired in 1918. Repainted into new-image maroon and gold paint at Proctor in September 1960. This car shows how DM&IR applied Scotchlite monograms, reporting marks and numbers to steel placards before application to the carbody. (Jim Maki Sr.)

in Class H in 1911 and in 1931 they received Class H1. At the time of the merger these cars were renumbered 4261–4310. All were off the roster by 1971.

D&IR FLAT CAR GONDOLA CONVERSIONS

By 1900, D&IR had equipped an unknown number of flat cars with gondola sides. Cars were used in general service and hauling coal. Company records are unclear as to the total number of cars converted and it appears that gondola sides were applied and removed from flat cars as needed.

During 1914 under Appropriation 459 for $4,125, flat cars in series 6100–6124 had wood gondola sides and side dump doors applied at the Two Harbors shops. After rebuilding, cars had an inside height of 3-foot 4-inches, an inside length of 34-foot 2-inches with 940 cu. ft. capacity. Sides were constructed of three 9-inch planks, one 13-inch top plank and 9 oak side stakes installed in the flatcar stake pockets. Originally these cars were built by Western Steel Car & Foundry in 1907 as 30-ton 40-foot flat cars. Class G was applied to these cars in 1914.

The side dump doors were replaced with solid sides in 1931. When rebuilt the sides were replaced with four 10-inch planks, 11 oak side stakes and corners reinforced with L-section steel. They retained the same numbers after the merger. All were off the roster by 1971.

An additional 55 cars in series 6125–6179 had wood gondola sides applied in 1926 at Two Harbors. Originally these cars were built by Western Steel Car & Foundry in 1907 as 30-ton 40-foot flat cars. Sides and ends consisted of four 10-inch planks with oak side stakes installed in stake pockets, creating a gondola with inside length of 34-foot 2-inches, inside height of 3-foot 4-inches and 934 cu. ft. capac-

Many flat cars were rebuilt with wooden gondola sides and stakes. DM&IR 6122, converted from flat car to gondola in 1914, is at Iron Junction in spring, 1961. (Jim Maki Sr.)

Two groups of Class I flat cars, 6100–6124 and 6125–6179 were received from Western Steel Car & Foundry in 1907. The first group was rebuilt to gondolas with dump doors at Two Harbors in 1914. The second group was rebuilt at Two Harbors with just gondola sides in 1926. All were placed in Class G1 when rebuilt. DM&IR 6139, at Endion on July 19, 1957, was retired on June 20, 1966. (Bob's Photo collection)

ity. The sides and ends were then removed in 1923, except for 6144 which was rebuilt with steel sides and ends in 1922. Fifty two of these cars, 6125–6131, 6134–6153, 6155–6179 were rebuilt again to gondolas with sides and ends consisting of four 10-inch planks with oak stakes between March 11, 1926 and May 11, 1926. Car 6135 was rebuilt in 1930 with steel sides and ends. Steel gondola sides were applied to car 6144 in 1922 and 6135 in 1930. Class G was applied to these cars in 1911. They kept the same numbers after the merger. All were off the roster by 1971.

Seventy seven cars from the 6180–6279 series flat cars built by American Car & Foundry in 1916 were rebuilt with wood gondola sideboards at Two Harbors on arrival from the builder. Sideboards consisted of four 10-inch planks and 13 oak side stakes installed in stake pockets. The cars had inside length of 34-foot 7-inches, inside height of 3-foot 4-inches and 933 cu. ft. capacity. Fourteen of these cars had the wood sideboard replaced with steel sidesheets in 1918. They kept the same numbers after the merger. Class G was applied to these cars in 1911. Fourteen of these had the wood-

One of 100 Class I flatcars 6180–6279 built by American Car & Foundry in 1916. DM&IR 6211 was one of 77 converted to a Class G2 gondola when delivered and is shown at Duluth on July 19, 1957. D&IR had the ability to convert them from flat cars to gondolas and choose to purchase flat cars and convert them at Two Harbors to gondolas. (Mainline Photos, J.M. Gruber collection)

DM&IR 6137 a Class G1 gondola at Proctor, in the late 1950s. Note the DM&IR stencil monogram lacking the "Safety First" wrap-around slogan which was only used from 1937 until 1942. The car was retired August 1, 1968. (D.P. Holbrook collection)

en gondola sides and ends replaced with steel gondola sides and ends in 1918. See D&IR roster for details on specific car numbers. All were off the roster by 1972.

D&IR FLAT CARS

The first cars purchased by most railroads were flat cars for initial construction. D&IR ordered 30 20-ton 34-foot combination flat cars from Northwestern Car & Manufacturing in 1883. No records have been found, but it appears that these cars may have been numbered 41–70. Combination type freight cars were minimally popular from 1850 until 1900 when railroads were trying to purchase equipment that could be used for multiple purposes by simple conversions. Cars of this type could easily be converted to box cars, stock cars, drop-bottom box cars and gondolas. From 1884 to 1887, 12 box cars and four stock cars were converted from these cars. Reference is made in 1893 D&IR diagrams to flat car 52 being converted to snowplow 52 and an early 1884 photo shows box car 68. When equipped with air brakes these cars were renumbered to odd numbers only 2201–2259 in 1888 and 1889. Because of the possibility of converting box cars back to flat cars, 30 numbers were allocated. Between 1883 and 1895, the odd numbers 2213–2259 were rebuilt to MofW support cars and box cars. When equipped with automatic couplers in 1895 and 1896, the 2201–2211 were renumbered to odd numbers 5001–5011 and rebuilt to 36-foot cars.

By 1887 the D&IR shops at Two Harbors was building equipment. Seventy-five 20-ton 34-foot flat cars 2261–2409, odd numbers only, were built during 1887. When equipped with automatic couplers in 1895 and 1896, the 2261–2359 were renumbered odd numbers 5053–5145. Cars 2361–2409 were renumbered odd numbers 5013–5051.

During 1888, D&IR ordered 200 30-ton 34-foot flat cars 2411–2809, odd numbers only, from LaFayette Car Co. When equipped with automatic couplers in 1895 and 1896, the 2411–2809 were renumbered odd numbers 5147–5541. Only 198 numbers were allocated accounted two cars being retired. Cars were equipped with Thielson trucks and attachments to allow movement of handbrake from B-end of car to side of car on the B-end.

Loading pulpwood and cut timber before pulpwood racks involved having a crosswise timber at the each end of the flat cars and loading the end piles parallel with the sides with side stakes in the stake pockets. The middle of the car was then loaded crosswise as depicted here on Class I flat car D&IR 5385, one of 200 5147–5541 odd number only built in 1888 by LaFayette Car Works. Logging chains would be installed on 5385 in 1927 and pulpwood racks in 1929. It would be retired in December 1933. (University of Minnesota Duluth, Kathryn A. Martin Library Archives and Special Collections, Collection S3742 Box 15 Folder 25)

Eleven years later, in 1899, D&IR ordered 50 30-ton 36-foot flat cars 5543–5641 odd numbers only from Illinois Car & Equipment Co. All were placed in Class I in 1911. The D&IR 1929 diagram book lists 40 of these cars, odd numbers 5563–5641, as being built by D&IR Two Harbors in 1899, but no additional supporting documents have been found.

Between 1890 and 1900 all 34-foot flat cars would be rebuilt to 36-foot cars at Two Harbors. All flat cars prior to the 1907 order once rebuilt had the same general appearance of truss-rod underframe with nine stake pockets per side and with the exception of the 1888 order from LaFayette, all rode on arch-bar trucks. Many

D&IR 5577, one of 50 Class I flatcars 5543–5641 built by Illinois Car & Equipment in 1899. (University of Minnesota Duluth, Kathryn A. Martin Library Archives and Special Collections, Collection S3742 Box 15 Folder 25)

of these early cars were rebuilt with steel center sills and changed from 25-ton to 30-ton cars between 1910 and 1920

Continued development of the logging industry and increased shipping of finished product caused the D&IR to place an order in 1902 with Western Steel Car & Foundry. These 50 30-ton 36-foot flat cars 5643–5741, odd numbers only, were delivered in 1903. The specifications included Westinghouse brakes, Tower couplers, Bryan tandem draft rigging, McCord journal boxes, National Railway Specialty Co. journal bearings, and arch-bar trucks. All were built with attachments to allow movement of handbrake staff and brake wheel from end of car to side of car on the B-end.

Shop capacity was available at Two Harbors, and D&IR built 50 30-ton 36-foot flat cars during 1905 and 1906. Forty-two of these cars, 5743–5825, odd numbers only, were built in 1905 and the remaining eight cars, 5827–5841, odd numbers only, built in 1906.

Later lacking shop capacity at Two Harbors, the D&IR ordered in October 1906, 25 Class I 30-ton 40-foot flat cars 6100–6124 from Western Steel Car & Foundry, all delivered in 1907. The cars were built with steel underframes, fishbelly side sills, had 13 stake pockets per side and rode on arch-bar trucks. All were converted to gondolas in 1914. An additional 55 identical cars were added to this order in 1906. Numbered 6125–6179, they were placed in Class I and delivered in 1907. Cars 6132, 6133 and 6154 were destroyed in 1913. The remaining 52 cars were converted with sides and ends to coal cars in 1916. Arch-bar trucks were replaced with Andrews U-section trucks during 1937 to 1939. The first car completed was 6140 on June 11, 1937.

The next group of flat cars were 100 Class I 30-ton 40-foot flat cars 6180–6279, ordered from American Car and Foundry on Lot 7934 on September 22, 1915, at a cost of $1,100 per car. All cars were delivered during 1916, had steel underframes, fishbelly side sills, 13 stake pockets per side and rode on arch-bar trucks. Immediately on delivery, 77 of these cars had wooden gondola sides and ends added at Two Harbors. Arch-bar trucks were replaced with Andrews U-section trucks during 1937 to 1939. All were retired by 1972

D&IR cars traveled widely. Examples being car 5411 being destroyed on Belt Railway of Chicago in Chicago during June 1921 and 5451 destroyed on Chicago Great Western in November 1921.

Appropriation 630 was approved at the June 11, 1917, Board of Director meeting for

After two successive orders for steel flatcars from Western Steel Car & Foundry in 1907, D&IR turned to American Car & Foundry in 1915 for additional steel flatcars. One hundred Class I 30-ton cars in 6180-6279 series were delivered in 1916. Seventy seven, before being placed in service, were equipped with sides and ends at D&IR Two Harbors. Additional cars would be rebuilt to gondolas and rack flats in later years. (American Car & Foundry Collection John W. Barriger III National Railroad Library at UMSL)

Satisfied with the first order of 100 steel flat cars built by American Car & Foundry in 1916, D&IR returned to ACF and ordered 25 additional Class I flat cars, 6280–6304 in 1917. Unlike the first order delivered on arch-bar trucks, these arrived with Andrews T-section trucks. (American Car & Foundry Collection, John W. Barriger III National Railroad Library at UMSL)

$43,125 for the purchase of 25 Class I 30-ton 40-foot flat cars 6280–6304 from American Car and Foundry on Lot 8406. The cars were identical to 6180–6279 series. All were delivered by December 1917, except 6301 which was delivered in January 1918. After the DM&IR merger these cars retained their numbers but were changed from Class I to Class K. All were retired by 1972.

A total of 264 flat cars in series 5013–6099 were on the roster on August 9, 1932, 222 of all-wood construction, and 42 with steel underframes. All wooden cars would not be permitted in interchange effective January 1, 1934, per ARA Rule 3, paragraph U2. A decision was made not to rebuild any of the 222 remaining wood-construction cars since most were in on-line service only. The 264 cars also did not meet ARA Rule 3, paragraph F2 that required flat cars to have stake pockets a minimum of 4- x 5-inches, and F1 requiring stake pockets to be spaced between 2- and 4-feet. Class I 6180–6304 (48 cars), Class V 2001–2379 (65 cars) and Class Y 2170–2228(18 cars) also did not meet the requirements of para. F2 or F1.

In 1963 AFE 9465 was approved on April 5, 1963, calling for AAR-mandated lading strap anchors to be applied to 6280–6304 and 12 Class V1 2500 series flat cars. By 1965 most remaining 40-foot flat cars had been equipped with pulpwood racks, with remaining plain flats being used for company service.

D&IR PULPWOOD RACK FLAT CONVERSIONS

By 1920 much of the old-growth white

pine timber in Northern Minnesota had been harvested and lumber harvesting had shifted from cut logs to smaller trees with shorter growing cycles. Initially these were hemlock and spruce which quickly gave way to poplar and aspen type trees. These were harvested for the wood fiber—pulpwood—they contained which was used for paper making. Unlike white pine logs for lumber production, which were easily hauled on logging flats and chain-equipped standard flat cars, the smaller pulpwood required flat cars and gondolas. Flat cars were used for hauling pulpwood that was cut into pieces 100 inches long. Some pieces were as short as 50 inches. Temporary bulkheads began to be applied to flat cars as early as 1915.

D&IR, DM&N and DM&IR responded to the need for pulpwood cars requested by paper companies by converting large numbers of older flat cars with pulpwood racks. Initially these were temporary installations of loose timber, but by 1910 the first cars were converted with milled lumber and beginning in 1930s, permanent steel pulpwood racks were installed. DMIR 5717 is shown at Two Harbors, in 1959. Note the outside diagonal is a steel C-channel but the remainder of the bulkhead is of wood construction. The far end of the car is equipped with a fabricated sheet metal snowplow for use in plowing yard tracks at Two Harbors. (Bassgen photo, D.P. Holbrook collection)

Prior to this time pulpwood was stacked parallel to the side of the car with vertical stakes placed in the end side stake pockets. The piles acted like bulkheads and allowed the remainder of the pulpwood load to be loaded crosswise. These loads were susceptible to shifting and pulpwood falling off the cars.

Appropriation 608 was approved at the June 11, 1917 Board of Directors meeting for the installation of pulpwood racks on 100 flat cars at a cost of $3,151. These conversions began during the winter of 1917 with the application of wood end racks to 5000-series flat cars. A total of 20 were converted during 1918. During various years from 1918 to 1965 the shops at Two Harbors continued to modify flat cars into rack flats. Log shipments had continued to decline during the 1920s, but the need for additional rack flats for loading pulpwood had increased.

AFE 3427 approved May 7, 1934, for 10 cars and AFE 3469 for 40 cars approved January 2, 1934, provided for the conversion of 50 Class G gondolas between numbers 6216–6278 to rack flats. Gondola sides and ends were

Permanent wood and steel pulpwood racks are shown on D&IR 6222 at Endion, Minnesota, in the early 1940s. It was one of 40 cars converted with pulpwood racks in 1935. (University of Minnesota Duluth, Kathryn A. Martin Library Archives and Special Collections, Collection S3742 Box 16 Folder 46)

removed and permanent combination wood and steel pulpwood end racks placed on these cars during winter 1934 and 1935. The sample rack flat, 6275 was released from Proctor rip track on July 20, 1934. All work was done on Proctor rip tracks with AFE 3427 completed between Sept 1934 and December 1934. AFE 3469 was completed between February and June 1935. Racks consisted of five vertical steel pieces creating a frame with header and footer of plate steel attached to deck with diagonal C-section channels attached to the bulkhead and side sill of car. These were the first rack flats equipped with permanent racks.

Declining needs for box cars and need for additional rack flats caused D&IR to start rebuilding Class F box cars into rack flats. AFE IR3506 was approved on April 11, 1935, and box cars 5348 and 5384 were rebuilt to rack flats 6305 and 6306 on May 21, 1935. The economic downturn and the arrival of World War II saw the rebuild program discontinued.

Twenty Class K2 flat cars were rebuilt with steel pulpwood racks per AFE 9867 approved on December 23, 1964. The remaining five cars were converted to rack flats per AFE 9900 on Feb 26, 1965. See roster notes for more details.

D&IR APRON FLAT CONVERSIONS

Prior to 1918, the D&IR began applying aprons to flat cars but did not start recording this information on annual equipment lists until 1919. This appears to be as a result of ICC Valuations of railroads requiring more detailed information on equipment owned. Starting in 1919, D&IR annual equipment lists kept great details on flat cars equipped with sideboards, large stake pockets, logging chains and pulpwood racks. Equipment lists discontinued keeping track of this information in 1923, resumed tracking in 1926 and then discontinued completely in 1930 except for cars equipped with pulpwood racks. Aprons consisted of wood sides, but no ends, applied to flat cars. Only one car, 5555, was converted in 1918, apparently as a test. Starting in 1919 additional cars were rebuilt; 21 in 1919 and 29 in 1920.

Many of these cars had large stake pockets and logging chains; some also were equipped with pulpwood racks. These conversions took place during a time when old-growth timber was playing out and pulpwood loading was increasing. Having this variety of equipment modified allowed D&IR to use cars in whatever service was required without making yearly modifications to each car.

D&IR COAL FLAT CONVERIONS

During 1916, flat car 5505 was converted to a temporary coal car by the addition of sideboards and ends. The car found service from 1916 until 1919 when the temporary sideboards and ends were removed and the flat car converted to a rack flat.

D&IR LOGGING EQUIPMENT

Besides the development of iron ore mining, Northern Minnesota had vast stands of old-growth timber. Logging was the number two industry and the hauling of logs, pulpwood and finished lumber added significantly to the D&IR bottom line. By the summer of 1901, D&IR rostered a total of 321 regular flat cars and 50 bunker flats equipped for log loading. Based on winter 1900-1901 log loadings, D&IR would not have enough equipment to fulfill the needs of the logging companies for the winter of 1901-1902. To fulfill this need D&IR leased 25 flat cars from the Chicago Great Western (CGW) for use during December 1901 through the end of March 1902.

BUNKER FLATS

The first cars purpose-built for logging were 50 30-ton 36-foot flat cars built at D&IR Two Harbors shop in 1900. These were numbered 6000–6049 and placed in Class I. The cars were built in 1900 without trucks and used arch-bar trucks from wooden ore cars during the winter month. D&IR called these cars "Bunker Flats." Quickly after construction as wood ore cars were retired, permanent arch-bar trucks were applied.

Fifty additional 30-ton 36-foot flat cars were built at Two Harbors during 1910. Numbered 6050–6099 and placed in Class I they were equipped with arch-bar trucks. Both groups of cars were identical to flatcars 5563–5641 built by D&IR in 1899, but all 100 cars were equipped with logging chains for hauling the larger logs that the skeleton log flats could not be used for.

LOG CARS

Seeing the traffic possibilities, D&IR started constructing a fleet of eventually 320 log flats at Two Harbors shops in 1901. All of these cars were constructed based on Russell designs using Russell castings. Numbered from 7000–7319, the cars were all identical. The 75 cars in 7000–7074 series were rebuilt in 1901 from 751–1020 series retired Haskell & Barker 25-ton wood ore cars. The next 50 cars in 7075–7124 series were rebuilt in December 1902 from additional retired wood ore cars. During December 1902 nine ore cars, 817, 934, 992, 1174, 1474, 1523, 1617, 1753, 2528 had their ore car sides removed and cars were used as log flats. All nine of these cars were retired during the summer of 1903.

Expanded logging caused seven cars in 7125–7131 to be built in 1905, 28 cars in 7132–7159 in 1906 and 160 cars in the 7160–7319 series in 1910. Cars 7160–7219 were constructed on Appropriations 288 for $4,872 for 20 cars and Appropriation 289 for $9,744 for 40 cars which were approved at the July 18, 1910, Board of Directors meeting. The remaining 100 cars were constructed under Appropriation 301 for 47,242, which also included 50 skeleton flat cars at the March 11, 1911, Board of Directors meeting. All were 20-ton 20-foot cars, built at a time when the hauling of logs to saw mills was a traffic factor of considerable importance.

Cars 7125–7219 were built at the D&IR Two Harbors shops from second-hand materials on hand. Each car consisted of a three-piece wood center sill with log bunks located between the bolster and ends of car. The log bunks each had an inward facing stake pocket attached to the bolster. Russell-designed cars did not have the stake pockets and it appears this was a D&IR design. The cars had with K-brakes

Before iron ore could be dug out of the ground the vast timberlands of Northern Minnesota had to be cleared of old-growth timber. D&IR built 320 Class K skeleton log flats 7000–7319 between 1901 and 1910 and all would be retired by 1928. Numbers were placed on the ends of logging bunks, 7152 on the left and 7283 on the right. (University of Minnesota Duluth, Kathryn A. Martin Library Archives and Special Collections, Collection S3742 Box 15 Folder 25)

and D&IR Standard 5-foot 4-inch rigid arch-bar trucks. Grab irons varied over the years as Safety Appliance Laws were revised.

Logging contributed a significant amount of traffic to D&IR yearly totals. During 1900 D&IR hauled 711,708 tons of sawed logs, a total of 12.34 percent of total traffic. Sawed logs were second to iron ore, which contributed 4.8 million tons or 83.23 percent of total traffic. By 1922 this had declined to 267,476 tons of sawed logs and pulpwood, and only 3.65 percent of total traffic. Heavy iron ore traffic in 1901 caused D&IR to embargo the movement of logs north of Two Harbors between May 1st and November 1st.

These cars could only be used for one thing, the hauling of forest products, mostly sawed logs, during the winter months. The apex of sawed log loading was 1915 with a steady decline until 1925 when the last of the lumber mills shut down. In order to get some additional use out of these cars during 1916, 100 cars were rebuilt at Two Harbors with additional side and center sill reinforcement which was constructed of 60 lb. retired rail. Rack ends were constructed so pulpwood could be loaded transversely across the deck of the car. Rack ends consisted of four vertical posts made of 4 x 6-inch lumber, 9-feet tall with three cross pieces on the inside of 2 x 10-inch pine or fir, with an oak 3 x 4-inch cross-piece of at the bottom which was fastened to the deck of the car. Across the back or outside of the rack was a piece of 60 lb. rail about 2/3 up from which two 1¼-inch steel rod braces were fastened through the car bunks. D&IR declared, with the addition of the 60 lb. rail that these were now "steel underframe" equipped. During the last year of operation in 1928 a total of 110 cars were still equipped with rack ends.

Federal inspectors determined in 1928 that to comply with safety regulations end ladders had to be installed. This cost was determined to be between $3000 and $4000 to comply with the required modifications. The depreciated value of these cars on March 1, 1928 was $2,430.64. It was not likely that the scrap value would be more than $60 per car. By 1928 the logging traffic for which the cars had been designed had all but vanished and D&IR had no other use for these cars. Three of the cars, 7071, 7192 and 7209, were sold FOB Duluth, MN to D.M. McCoy Lumber Co.in Rexton, MI at a sale price of $150 per car. No other buyers were found and with sufficient flat and rack cars on hand in 1928 it was determined that these cars could be retired. Most of the remaining cars were sold to Duluth Iron & Metal and off the roster by the end of 1928.

D&IR constructed 320 skeleton log flats from Russell designs at Two Harbors car shop between 1901 and 1910, Declining log loading of old-growth timber caused a little over 100 cars to be converted with pulpwood racks. D&IR 7260 is shown at an unknown location in the late teens or early 1920s. Note the car number applied to the ends of both cross members for identification. Just below this you can see a piece of 60-pound rail that was used beginning in 1916 to reinforce the underframes and support the pulpwood load. (D.P. Holbrook collection)

D&IR CABOOSES

Caboose information from 1884 to 1886 is sketchy. It appears the D&IR purchased nine 20-foot 4-wheel cabooses from Northwestern Car & Manufacturing numbered 21–29 in late 1884 or early 1885. References to some early equipment being "Stillwater" cars is supported

Duluth & Iron Range Cabooses

Number	Builder	Built	Sold or Retired	Length	Type	To DMIR	Notes
(1st)21	NC&M	1884	1892	20'8"	4W		1
(2nd)21	D&IR TH	1900	to DIR(3rd)21 1909/10	20'8"	4W		
(3rd)21	D&IR TH	1909/10	12/14/1933	31'	8W		
(1st)22	NC&M	1884	By 1887	20'8"	4W		
(2nd)22	D&IR TH	1888	to DIR(3rd)22 1909/10	20'8"	4W		
(3rd)22	D&IR TH	1909/10	12/14/1933	31'	8W		
(1st)23	NC&M	1884	By 1889	20'8"	4W		
(2nd)23	Wells & French	1890	to DIR(3rd)23 1909/10	20'8"	4W		
(3rd)23	D&IR TH	1909/10	8/10/1929	31'	8W		
(1st)24	NC&M	1884	By 1889	20'8"	4W		
(2nd)24	D&IR TH	1890	to DIR(3rd)24 1909/10	20'8"	4W		
(3rd)24	D&IR TH	1909/10	12/14/1933	31'	8W		
(1st)25	NC&M	1884	By 1888	20'8"	4W		
(2nd)25	D&IR TH	1888	By 1892	20'8"	4W		
(3rd)25	DMC	1892	to DIR(4th)25 1909/10	20'8"	4W		
(4th)25	D&IR TH	1909/10	Sold to CGW 1929	31'	8W		
(1st)26	NC&M	1884	By 1888	20'8"	4W		
(2nd)26	D&IR TH	1888	By 1890	20'8"	4W		
(3rd)26	Wells & French	1890	to DIR (4th)26 1909/10	20'8"	4W		
(4th)26	D&IR TH	1909/10	Sold to CGW 1929	31'	8W		
(1st)27	NC&M	1884	By 1887	20'8"	4W		
(2nd)27	D&IR TH	1888	to DIR(3rd)27 1909/10	20'8"	4W		
(3rd)27	D&IR TH	1909/10	Sold to CGW 1929	31'	8W		
(1st)28	NC&M	1884	By 1888	20'8"	4W		
(2nd)28	D&IR TH	1888	By 1892	20'8"	4W		
(3rd)28	DMC	1892	to DIR (4th)28 1909/10	20'8"	4W		
(4th)28	D&IR TH	1909/10	Sold to CGW 1929	31'	8W		
(1st)29	NC&M	1884	By 1886	20'8"	4W		
(2nd)29	D&IR TH	1886	to DIR (3rd)29 1909/10	22'8"	4W		2
(3rd)29	D&IR TH	1909/10	to DM&IR	31'	8W	C-129	3
(1st)30	NC&M	1887	to DIR 95 by 1900	43'8"	8W		4
(2nd)30	D&IR TH	1900	to DIR (3rd)30 1909/10	22'8"	4W		
(3rd)30	D&IR TH	1909/10	12/14/1933	31'	8W		
(1st)31	NC&M	1887	to DIR 96 by 1900	43'8"	8W		4
(2nd)31	D&IR TH	1900	Burned Hornby 02/13/1902 to DIR(3rd)31 1909/10	22'8"	4W		
(3rd)31	D&IR TH	1909/10	Sold to D&N 06/05/1929	31'	8W		
(1st)32	NC&M	1887	By 1889	43'8"	8W		5
(2nd)32	Wells & French	1890	to DIR(3rd)32 1909/10	20'8"	4W		6
(3rd)32	D&IR TH	1909/10	to DM&IR	31'	8W	C-132	3
(1st)33	NC&M	1887	to DIR 97 1896	43'8"	8W		5
(2nd)33	D&IR TH	1900	to DIR (3rd)33 1909/10	22'8"	4W		
(3rd)33	D&IR TH	1909/10	Sold to CGW 1929	31'	8W		
(1st)34	NC&M	1887	to DIR 98 by 1900	43'8"	8W		5
(2nd)34	D&IR TH	1900	to DIR (3rd)34 1909/10	22'8"	4W		
(3rd)34	D&IR TH	1909/10	Sold to CGW 1929	31'	8W		
(1st)35	D&IR TH	1888	to DIR (2nd)35 1909/10	20'8"	4W		
(2nd)35	D&IR TH	1909/10	to DM&IR	31'	8W	C-135	3
(1st)36	D&IR TH	1888	Burned Hornby 02/13/1902 to DIR (2nd)36 1902	22'8"	4W		6
(2nd)36	D&IR TH	1902	to DIR(3rd)36 1909/10	22'8	4W		

Number	Builder	Built	Sold or Retired	Length	Type	To DMIR	Notes
(3rd)36	D&IR TH	1909/10	to DM&IR	31'	8W	C-136	3
(1st)37	Wells & French	1888	Burned Hornby 02/13/1902 to DIR (2nd)37 1902	20'8"	4W		
(2nd)37	D&IR TH	1902	to DIR(3rd)37 1909/10	20'8"	4W		
(3rd)37	D&IR TH	1909/10	to DM&IR	31'	8W	C-137	7
(1st)38	D&IR TH	1888	to DIR (2nd)38 1909/10	20'8"	4W		
(2nd)38	D&IR TH	1909/10	to DM&IR	31'	8W	C-138	3
(1st)39	PCC Lot #1609	1889	to DIR(2nd)39 1909/10	20'8"	4W		
(2nd)39	D&IR TH	1910	Sold Duluth Iron & Metal Nov. 1935	31'	8W		8
(1st)40	PCC Lot #1609	1889	to DIR(2nd)40 1909/10	20'8"	4W		
(2nd)40	D&IR TH	1910	Sold Duluth Iron & Metal Nov. 1935	31'	8W		8
(1st)41	PCC Lot #1609	1889	to DIR(2nd)41 1909/10	20'8"	4W		
(2nd)41	D&IR TH	1909/10	Sold to CGW 1929	31'	8W		
(1st)42	PCC Lot #1609	1889	to DIR(2nd)42 by 1907	20'8"	4W		
(2nd)42	D&IR TH	1907	to DIR(3rd)42 1909/10	20'8"	4W		9
(3rd)42	D&IR TH	1909/10	Sold to CGW 1929	31'	8W		
(1st)43	PCC Lot #1609	1889	to DIR(2nd)43 1909/10	20'8"	4W		
(2nd)43	D&IR TH	1909/10	to DM&IR	31'	8W	C-143	10
(1st)44	PCC Lot #1609	1889	to DIR(2nd)44 1909/10	20'8"	4W		
(2nd)44	D&IR TH	1909/10	to DM&IR	31'	8W	C-144	10
(1st)45	DMC	1892	to DIR(2nd)45 1909/10	20'8"	4W		
(2nd)45	D&IR TH	1909/10	Sold to CGW 1929	31'	8W		
(1st)46	DMC	1892	to DIR(2nd)46 1909/10	20'8"	4W		
(2nd)46	D&IR TH	1909/10	12/14/1933	31'	8W		
(1st)47	DMC	1892	Wrecked 07/19/1901	20'8"	4W		
(2nd)47	D&IR TH	1902	to DIR(3rd)47 1909/10	20'8"	4W		
(3rd)47	D&IR TH	1909/10	to DM&IR	31'	8W	C-147	11
(1st)48	DMC	1892	to DIR (2nd)48 1909/10	20'8"	4W		
(2nd)48	D&IR TH	1909/10	Sold to CGW 1929	31'	8W		11
(1st)49	DMC	1893	to DIR (2nd)49 1909/10	20'8"	4W		
(2nd)49	D&IR TH	1909/10	12/14/1933	31'	8W		11
(1st)50	DMC	1893	to DIR(2nd)50 1909/10	20'8"	4W		
(2nd)50	D&IR TH	1909/10	12/14/1933	31'	8W		11
(1st)51	DMC	1893	to DIR 84 04/05/07	20'8"	4W		
(2nd)51	D&IR TH	1907	to DIR(3rd)51 1909/10	20'8"	4W		
(3rd)51	D&IR TH	1909/10	to DM&IR	31'	8W	C-151	12
(1st)52	DMC	1893	to DIR(2nd)52 1909/10	20'8"	4W		
(2nd)52	D&IR TH	1909/10	8/30/1934	31'	8W		13
(3rd)52	D&IR TH	1934	to DM&IR	31'	8W	C-152	14
(1st)53	DMC	1893	to DIR (2nd)53 1909/10	20'8"	4W		15
(2nd)53	D&IR TH	1909/10	to DM&IR	31'	8W	C-153	16
(1st)54	DMC	1893	to DIR (2nd)54 1909/10	20'8"	4W		
(2nd)54	D&IR TH	1909/10	to DM&IR	31'	8W	C-154	16
(1st)55	DMC	1893	to DIR (2nd)55 1909/10	20'8"	4W		
(2nd)55	D&IR TH	1909/10	12/14/1933	31'	8W		
(1st)56	DMC	1893	to DIR(2nd)56 2/14/07	20'8"	4W		
(2nd)56	D&IR TH	1907	to DIR(3rd)56 1909/10	20'8"	4W		
(3rd)56	D&IR TH	1909/10	to DM&IR	31'	8W	C-156	11
(1st)57	DMC	1893	to DIR (2nd)57 1909/10	20'8"	4W		
(2nd)57	D&IR TH	1909/10	to DM&IR	31'	8W	C-157	11
(1st)58	DMC	1893	to DIR(2nd)58 1909/10	20'8"	4W		
(2nd)58	D&IR TH	1909/10	to DM&IR	31'	8W	C-158	11
(1st)59	D&IR TH	1900	to DIR(2nd)59 1909/10	20'8"	4W		
(2nd)59	D&IR TH	1909/10	to DM&IR	31'	8W	C-159	11
(1st)60	D&IR TH	03/1903	to DIR 80 1906		8W		
(2nd)60	D&IR TH	2/17/1908	to DIR(3rd)60 1909/10	20'8"	4W		
(3rd)60	D&IR TH	1909/10	to DM&IR	31'	8W	C-160	11
(1st)61	D&IR TH	03/1903	to DIR 81 1906		8W		

Number	Builder	Built	Sold or Retired	Length	Type	To DMIR	Notes
(2nd)61	D&IR TH	2/17/1908	to DIR(3rd)61 1909/10	20'8"	4W		
(3rd)61	D&IR TH	1909/10	to DM&IR	31'	8W	C-161	11
(1st)62	D&IR TH	2/25/1905	to DIR(2nd)62 1909/10	20'8"	4W		17
(2nd)62	D&IR TH	1909/10	to DM&IR	31'	8W	C-162	11
(1st)63	D&IR TH	2/25/1905	to DIR(2nd)63 1909/10	20'8"	4W		18
(2nd)63	D&IR TH	1909/10	to DM&IR	31'	8W	C-163	11
(1st)64	D&IR TH	04/1906	to DIR(2nd)64 1909/10	20'8"	4W		
(2nd)64	D&IR TH	1909/10	to DM&IR	31'	8W	C-164	11
(1st)65	D&IR TH	04/1906	to DIR(2nd)65 1909/10	20'8"	4W		
(2nd)65	D&IR TH	1909/10	to DM&IR	31'	8W	C-165	11
(1st)66	D&IR TH	04/1906	to DIR(2nd)66 1909/10	20'8"	4W		
(2nd)66	D&IR TH	1909/10	to DM&IR	31'	8W	C-166	11
(1st)67	D&IR TH	04/1906	to DIR(2nd)67 1909/10	20'8"	4W		
(2nd)67	D&IR TH	1909/10	to DM&IR	31'	8W	C-167	11
(1st)68	D&IR TH	2/17/1908	to DIR(2nd)68 1909/10	20'8'	4W		
(2nd)68	D&IR TH	1909/10	to DM&IR	31'	8W	C-168	11
(1st)69	D&IR TH	2/17/1908	to DIR(2nd)69 1909/10	20'8"	4W		
(2nd)69	D&IR TH	1909/10	to DM&IR	31'	8W	C-169	11
(1st)70	D&IR TH	2/17/1908	to DIR(2nd)70 1909/10	20'8"	4W		
(2nd)70	D&IR TH	1909/10	to DM&IR	31'	8W	C-170	11
71	D&IR TH	1910	Retired 7/31/1931	31'	8W		19
72	D&IR TH	1910	to DM&IR	31'	8W	C-172	11
73	D&IR TH	1910	to DM&IR	31'	8W	C-173	11
74	D&IR TH	1910	to DM&IR	31'	8W	C-174	11
75	D&IR TH	1910	to DM&IR	31'	8W	C-175	11
76	D&IR TH	1910	to DM&IR	31'	8W	C-176	11
77	D&IR TH	1910	to DM&IR	31'	8W	C-177	11
78	D&IR TH	1910	Retired 1914	31'	8W		
79	D&IR TH	1910	to DM&IR	31'	8W	C-179	11
(1st)80	D&IR TH	1906	to DIR 88 May, 1910	31'	8W		20

by 1893 diagrams showing other builders' cabooses showing on D&IR diagrams as being "Stillwater Cabooses" referring to the design. These 20-foot 4-wheel cabooses would become a standard until 1905. Caboose 21 was used by D&IR President Charlemagne Tower, Jr. as the Tower, MN headquarters from 1884 until 1886. It was replaced by private car "B" in 1886.

The next group of cabooses, referred to as "Waycars" in diagrams, comprised five 43-foot 8-inches 8-wheel cars from Northwestern Car & Manufacturing numbered 30–34 purchased in 1887. The 1893 D&IR diagrams show 30 and 31 without a cupola and 33 and 34 with cupolas. It is unknown if they were built this way or if cupolas were added or removed after delivery. All were equipped with 8 bunks and 2 caboose stoves. Car 32 was retired prior to 1893. Cars 30, 31, 33, 34, were renumbered to 95–98 prior to 1900. This allowed 4-wheel and 8-wheel cabooses to be grouped together in continuous number series of each type.

Caboose 99 shows on roster as built in 1888 but the history of the car is unclear except that it was wrecked in 1903 per D&IR equipment lists.

From 1884 to 1892, when freight cars and cabooses were destroyed, the old number was reused. This caused the caboose roster between these years to become complicated with scattered old and new cars occupying a block of numbers.

By 1886 caboose 29 had been destroyed and a new caboose 29 was built at D&IR Two Harbors shops. Two Harbors shop constructed seven additional 20-foot standard 4-wheel cabooses in 1888. Numbers 22, 24, 27 were replacements to previously destroyed cars with the same numbers. Numbers 35–37 were new cars placed above the 8-wheel cars 30–34. Cabooses 29, 36, 37 are depicted in 1893 diagrams as not having cupolas. It is unknown if

Number	Builder	Built	Sold or Retired	Length	Type	To DMIR	Notes
(2nd)80	D&IR TH	1911	to DM&IR	31'	8W	C-180	10
(1st)81	D&IR TH	1906	to DIR(2nd)81 1920	31'	8W		21
(2nd)81	Rblt D&IR TH	1920	to DM&IR	40'	8W	C-181	22
(1st)82	D&IR TH	04/1906	to DIR(2nd)82 1910	31'	8W		
(2nd)82	D&IR TH	1910	to DM&IR	34'	8W	C-182	23
(1st)83	D&IR TH	04/1906	to DIR(2nd)83 1910	31'	8W		
(2nd)83	D&IR TH	1910	to DM&IR	34'	8W	C-183	23
(1st)84	D&IR TH	4/5/1907	to DIR(2nd)84 1920	40'	8W		24
(2nd)84	Rblt D&IR TH	1920	to DM&IR	40'	8W	C-184	25
(1st)85	D&IR TH	2/17/1908	to DIR(2nd)85 1920	31'	8W		26
(2nd)85	Rblt D&IR TH	1920	to DM&IR	40'	8W	C-185	27
(1st)86	D&IR TH	2/17/1908	to DIR(2nd)86 1919	31'	8W		
(2nd)86	D&IR TH	1920	to DIR (3rd) 86 1923	34'	8W		28
(3rd)86	Rblt D&IR TH	1923	to DM&IR	40'	8W	C-186	29
(1st)87	D&IR TH	4/21/1908	to DIR(2nd)87 1920	27'	8W		30
(2nd)87	Rblt D&IR TH	1920	to DM&IR	40'	8W	C-187	25
(1st)88	Rblt D&IR TH	1910	to DIR(2nd)88 1920	27'	8W		31
(2nd)88	Rblt D&IR TH	1920	to DM&IR	40'	8W	C-188	32
(1st)89	Rblt D&IR TH	1912	to DIR(2nd)89 1920	27'	8W		33
(2nd)89	Rblt D&IR TH	1920	to DM&IR	40'	8W	C-189	34
95	NC&M	1887	Rblt to DIR 89 07/11/12	43'3"	8W		35
96	NC&M	1887	Rblt to DIR 84 04/05/07	43'3	8W		36
97	NC&M	1887	Rblt to DIR 87 04/21/08	43'3"	8W		37
98	NC&M	1887	12/14/1933	43'3"	8W		38
99	NC&M	1888	Wrecked 1903	43'3"	8W		39
100	D&IR TH	1903	To MofW 100 10/1908	52'6"	8W		40
5232	D&IR TH	1923	To MofW W5232 6/15/44	37'3"	8W		41

1. Rebuilt to 51'1" tool car 21 in 1892 with no cupola. Rebuilt to wrecking train caboose 150 with cupola by 1910

2. No cupola

3. WUSR applied 1934

4. 8-bunks, 2 caboose stoves, no cupola, side door

5. 8-bunks, 2 caboose stoves, side door

6. Wrecked Two Harbors 10/1/02 repaired

7. Steel Center Sills applied 1927

8. Used as welders car. Taken over by Track Dept. 1918

9. Rebuilt 2/14/07 from (1st)42 - Rebuilt to 3rd 42 in 1909-10

10. Wood Underframe with steel rails added applied 1921

11. 6-bunks, Steel Underframe applied 1927

12. 6-bunks, Steel Underframe applied 1935

13. Remodeled to 4-wheel caboose and renumbed DIR 22 for display at Two Harbors, MN 8/30/34

14. 6-bunks, Steel Underframe when built

15. 6-bunks, Peteler Steel Center Sills 1935

16. 6-bunks, Steel Underframe applied 1934

17. Built 5/25/05 to replace 99 per D&IR list Jan 1905 - Rebuilt to 2nd 62 in 1909-10

18. Built 5/25/05 to replace 99 per D&IR list Jan 1905 - Rebuilt to 2nd 63 in 1909/10

19. 6-bunks, Steel Center Sills 1927, destroyed 6/13/31 Two Harbors

20. Former caboose 60 rebuilt to caboose 88 May 1910

21. Former caboose 61, rebuilt to 40' (2nd)81 at Two Harbors 1920

22. Steel Center Sills applied 1935

23. Wood Underframe with steel rails added applied 1921, end cupola

24. Rebuilt from 51 per D&IR list Jan 1906, then rebuilt to 40' (2nd) 84 1920 by adding six feet to carbody

25. 6-bunks, Wood Underframe with steel rails added applied 1921, Steel Underframe applied 4/10/36

26. Rebuilt to 40' (2nd)85 at Two Harbors 1920

27. Wood Underframe with steel rails added applied 1921, Steel Undeframe applied 1936

28. Burned Babbitt 5/11/23, Rebuilt to (3rd) 86

29. 6-bunks, side doors, Wood Underframe with steel rails added applied 1921, Steel Underframe applied 1936, Center Cupola, rebuilt from (2nd) 86

30. Rebuilt from 97 4/ 21/08, rebuilt to 40' (2nd) 87 at Two Harbors 1920

31. 6-bunks, Rebuilt from 80 May 1910, Rebuilt to 40' (2nd)88 at Two Harbors 1920

32. Steel Underframe applied 1936

33. 6-bunks, rebuilt from 95, rebuilt to 40' (2nd)89 at Two Harbors in 1920

34. Steel Underframe applied 1936

35. Former DIR 30 renumbered 95 in 1890. Rebuilt to 89 7/11/12

36. Former DIR 31 renumbered 96 by 1896. Rebuilt to 84 4/05/07

37. Former DIR 33 renumbered 97 by 1896. Rebuilt to 87 4/21/08

38. Former DIR 34 renumbered 98 in 1890. Rebuilt D&IR Two Harbors 1890. Passenger trucks.

39. Unknown former number. Wrecked per D&IR list Jan. 1904

40. Built March 1903. To MofW service as #100 in Oct. 1908

41. Rebuilt from boxcar 5232 in 1923 at Two Harbors as temporary caboose

Abbreviations:

DMC -	*Duluth Manufacturing Co. West Duluth, MN*
D&IR TH -	*Duluth & Iron Range Two Harbors, MN Shops*
NC&M -	*Northwestern Car & Manufacturing, Stillwater, MN*
PCC -	*Pullman Car Co.*
4W -	*4-wheel*
8W -	*8-wheel*
Rblt -	*rebuilt*
LS&M -	*Lake Superior & Mississippi Railroad, West Duluth, MN*
D&NM -	*Duluth & Northern Minnesota, Knife River, MN*
CGW -	*Chicago Great Western*
LSRM -	*Lake Superior Railroad Museum*
MNS -	*Minneapolis, Northfield & Southern*
EJ&E -	*Elgin, Joliet & Eastern*
D&N -	*Delaware & Northern*

the cupolas were removed of if the cars were built this way.

The next group of 20-foot 4-wheel cabooses were six cars ordered from Pullman in 1889 and numbered 39 to 44. All of these occupied a solid block and were not replacements for any destroyed cars.

D&IR placed an order in 1890 for four additional 20-foot 4-wheel cabooses numbered (2nd)23, (2nd)26, (2nd)32 and (2nd)37 for earlier cabooses with the same number that had been destroyed.

Displayed at the Two Harbors depot on July 18, 1965, is D&IR 22 rebuilt from 8-wheel caboose Second 52 on August 30, 1934. (D.P. Holbrook collection)

D&IR 50 fresh outside the car shops at Two Harbors in spring 1907. Rebuilt to an 8-wheel caboose in February 1911, it would be sold for scrap to Hyman-Michaels on May 28, 1975. (D.P. Holbrook collection)

After this order, D&IR placed two orders with Duluth Manufacturing Co. for 16 additional 20-foot 4-wheel cabooses. The 6 cars built in 1892 were, numbers (2nd)25 and (2nd)28 as replacements for earlier cabooses and 45 to 48 as new numbers. The 10 cars built in 1893 were numbered 49–58 occupying new numbers.

By 1900, D&IR shops at Two Harbors were building more and more equipment for the ever-expanding railroad. Two Harbors built a number of 20-foot 4-wheel cabooses beginning in 1900: numbers (2nd)30, (2nd) 33, (2nd) 34 and (2nd)59 in 1900, (2nd)36, (2nd)47 and 3 (2nd) 7 in 1902. Two new 4-wheel cabooses, 62 and 63, were added to the roster on May 25, 1905, to replace wrecked caboose 99. Nine more were added, 64–67, in April 1906 and (2nd)60, (2nd)61, 68–70 on February 17, 1908. This would conclude 4-wheel caboose construction at Two Harbors and all future cabooses would be 8-wheel.

During 1903, Two Harbors built 52-foot 6-inch long caboose 100 with wooden truss-rod carbody having an offset cupola and 8-foot wheelbase passenger car trucks. The car had a seating area at one end for mixed train service but had no side door for baggage. The car kept the number and was converted to Telephone & Telegraph Department service in 1908. Also added during March 1903 were two 8-wheel cabooses 60 and 61 which were renumbered to 80 and 81 in 1906

Rebuilding of 4-wheel cabooses to 27-foot 8-wheel cabooses at Two Harbors began in 1906 with 51 renumbered and rebuilt to 84.

The first newly constructed 27-foot 8-wheel cabooses were 82 and 83 in 1906. They would be the only D&IR or DM&N cabooses constructed with the cupola located at the very end of the carbody. The next two 8-wheel cabooses 85 and 86 were built in 1908. Built with the cupola off-set from the ends this would become the standard caboose design for future cabooses. Former long caboose 97 was

Early D&IR cabooses were 4-wheel 20-foot cars that during 1909 and 1910 were rebuilt to 31-foot 8-wheel cabooses retaining the same numbers. DM&IR caboose C-163 is former D&IR caboose second C-63 and is at Proctor on July 17, 1957. It had 6 bunks for sleeping and was equipped with a steel center sill in 1927. Unseen in this view, it was equipped with a Witte engine for electrical power for VHF radios in 1947. Note the vertical stanchion in the front center of the cupola to support the radio antenna. C-163 was dismantled on February 2, 1961. (Bob's Photo collection)

rebuilt to 27-foot 8-wheel caboose 87 on April 21, 1908. Caboose 80 was rebuilt and renumbered to 88 in 1910. Long caboose 95 was rebuilt to 27-foot 8-wheel caboose 89 in 1912

Appropriation 305 was approved by the D&IR Board of Directors at the July 18, 1910 board meeting for $32,942.50 to dismantle and rebuild 50 4-wheel cabooses to 8-wheel cabooses at Two Harbors. Also approved at the same meeting was Appropriation 234 for $11,860 for the construction of ten new 31-foot cabooses numbered 71–80. Cars 71–79 were built in 1910 and 80 in 1911. Draft sill failures on the 50 rebuilt 4-wheel cabooses caused D&IR to approve Appropriation 670 at the June 10, 1916 Board of Directors meeting for $7,600 to strengthen draft sills.

Beginning August 26, 1920, seven 27-foot cabooses 81, 84–89 were rebuilt to 40-foot cabooses at Two Harbors on AFE 367 with project completed on December 26, 1920, at a cost of $12,691.54. Cars had wood sheathed carbody, off-set cupola and 5-foot 4-inch arch-bar trucks. All seven were rebuilt with a side

First D&IR 84 is pictured in 1910 at what is thought to be the Eveleth depot. It was rebuilt from 4-wheel caboose 51 on April 5, 1907 when the side door was added. Carbody is painted a dark red with yellow lettering. (Minnesota Historical Society)

door and small amount of floor space inside for handling LCL on local trains.

Appropriation 1035 was approved at the July 1923 Board of Director meeting for $2,035 for the construction of one new 36-foot caboose. Second 86 had burned at Babbitt on May 11, 1923, and was rebuilt to Third 86 under this appropriation.

A caboose shortage in 1923 caused at least one box car, 5232, to be rebuilt into a temporary caboose. On June 15, 1944, it was converted to bath and shower car W5232. It is possible other temporary cabooses were rebuilt, but no records have been found to support this.

At the time of the merger D&IR rostered 44 8-wheel wood-sheathed cabooses.

DM&IR C-129, former D&IR C-29 is at Proctor on March 31, 1963. Note the steel underframe which was applied under AFE 4329 during late 1939. This was part of a project to rebuild 20 27-foot cabooses with steel underframes. Once completed, only five cabooses remained with wooden underframes. (D, Repetsky, John C. La Rue, Jr. collection)

Steel center sills were installed on DM&IR C-160, former D&IR C-60 in 1927 seen at Two Harbors during the mid-1940s. Sold to Laona and Northern on April 30, 1965 this car survives today. (W.C .Olsen collection)

DM&IR C-177, former D&IR 77 at Virginia, Minnesota, June 5, 1959, with a wreck-damaged end platform. Note the black lettering and roof indicating it is assigned to the Iron Range Division of DM&IR. (Joe Douda, D.P. Holbrook collection)

The view of DM&IR C-186 shows the access doors for the Witte engine and exhaust stack for powering radios installed in 1948. The engine and radios were removed from C-186 in 1951, but the exterior modifications were left in place. Note the vertical rod adjacent to the exhaust stack that was the radio antenna. It was converted to work car W-187 on January 29, 1975, and then later donated to Lake Superior and Mississippi tourist operation at West Duluth. (Bruce Meyer)

PASSENGER CARS

Once industrial development of the Vermillion iron range started the transportation of people became a necessity. The first two passenger cars were ordered from Barney & Smith in 1884. Combination car 2 was delivered in 1884, was 58-foot 8-inches long, had a wooden carbody with open platforms, and rode on two 7-foot wheelbase 2-axle trucks. Trucks were classified as D&IR Type No. 1. Car had a 23-foot baggage area with a centered side door and a 33-foot 6-inch passenger section having nine windows per side with seating for 32 people and one bathroom. Day Coach 1 was delivered in 1885 with the same basic details.

D&IR Passenger Cars

Kind	Trucks	Nos	Length	Builder	Built	Retired/Disposition
Day Coach	4 axle	1	58'8"	Barney & Smith	1885	By 1916
Comb, Baggage and Passenger (open ends)	4 axle	2	58'8"	Barney & Smith	1884	Sold to Bowe-Burke Mining Co 09/02/33 - Retired AFE 3348 09/08/33
Business Car (open platform both ends)	4 axle	(1st)3	56'6"	D&IR T.H	1887	Rebuilt 1895 - Renumbered from 3 to named car Vermillion in 1888
Comb, Baggage and Passenger (open ends)	4 axle	(2nd)3	58'8"	CA&StL	Unk	Purchased from C&A 1900 - Former C&A 34 - Retired AFE 3370 12/14/33
Day Coach (open ends)	4 axle	4	61'	Ohio Falls Car Co	1887	Sold 1917
Day Coach (open ends)	4 axle	(1st)5	61'	Ohio Falls Car Co	1887	Rebuilt by Hicks Loco & Car Works 1903 into Combination car (2nd)5
Comb, Baggage and Passenger (open ends)	4 axle	(2nd)5	60'7"	Ohio Falls Car Co	1887	Retired AFE 3370 12/14/33
Comb, Baggage and Passenger (open ends)	4 axle	6	58'6"	Ohio Falls Car Co	1887	Out of service Nov 1914
Mail/Baggage	4 axle	7	62'	ACF	1908	Retired AFE 3370 12/14/33
Mail/Baggage	4 axle	(1st)8	62'	ACF	1908	Renumbered to Valuation car X7 in 1914
Mail/Baggage	4 axle	X7	62'	ACF	1908	Converted to tool car 801 1929
Baggage/RPO/Express all-steel	6 axle	(2nd)8	70'8"	Barney & Smith	1914	Scrapped ????
Baggage/RPO/Express	4 axle	(1st)9	58'8"	Pullman	Unk	Purchased from C&IC Railway 1888 - Changed to Valuation Car X9 in 1914 then changed to work service car 185 in July 1920
Baggage/RPO/Express	4 axle	X9	58'8"	Pullman	Unk	Valuation car - Retired ????
Baggage/RPO/Express all-steel	6 axle	(2nd)9	70'8"	Barney & Smith	1914	Converted to W9 5/13/65 AFE 9744 9/24/65
Baggage/RPO/Express	4 axle	(1st)10	58'8"	Pullman	Unk	Purchased from C&IC Railway 1888. Changed to X10 in 1914
Baggage/RPO/Express	4 axle	X10	58'8"	Pullman	Unk	Changed to Valuation car X10 in 1914 - Then to work service as X10 in 1931
Baggage/RPO/Express all-steel	6 axle	(2nd)10	70'8"	Barney & Smith	1914	Retired 7/15/61 - Converted to W10
Parlor Car (open ends)	4 axle	11	60'6"	Unknown	Unk	Purchased from C&IC Railway 1888 - Sold Sept 1918
Day Coach (open ends)	4 axle	12	59'3"	Unknown	Unk	Purchased from C&IC Railway 1888 - Rebuilt to wrecking supply car 193 in 1928
Day Coach (open ends)	4 axle	13	59'3"	Unknown	Unk	Purchased from C&IC Railway 1888 - Sold Sept. 1918
Parlor Car (open ends)	4 axle	14	60'6"	Unknown	Unk	Purchased from C&IC Railway 1888 - Sold Sept. 1918
Day Coach	4-axle	15	59'6"	Ohio Falls Car Co	1892	Sold Sept 1918
Day Coach	4-axle	(1st)16	61'8"	Ohio Falls Car Co	1892	Destroyed by fire 7/19/00
Day Coach	6-axle	(2nd)16	61'8"	Ohio Falls Car Co	1882	Rebuilt by Hicks Loco & Car Works in 1903 from ?? Out of service Nov. 1912
Coach	4-axle	17	57'9"	CA&StL	Unk	Purchased from C&A 1900 - Former C&A 43 - Rebuilt to Tool car 140 in 1916
Coach	6-axle	18	57'9"	CA&StL	Unk	Purchased from C&A 1900 - Former C&A 49 - Sold Sept. 1918
Coach (open ends)	4 axle	(1st)19	67'6"	Barney & Smith	1907	Converted to Motor Coach MC-1 in 1926
Combination, Baggage/Passenger (open ends)	4 axle	(2nd)19	67'6"	Rebuilt T.H. 1933	R1933	Converted from Motor Coach MC-1 07/31/33 to combination car (2nd) 19. Converted back to coach and work car W804 on 7/12/43. Donated to LSMT in 1974
Coach (open ends)	4 axle	20	67'6"	Barney & Smith	1907	To Telephone and Telegraph car 192 in 1927 - Then rebuilt to cradle car 196 in 1928
Day Coach (vestibule ends)	4 axle	21	66'10"	ACF	1908	Retired AFE 3370 12/14/33
Day Coach (vestibule ends)	4 axle	22	66'10"	ACF	1908	Retired AFE 3370 12/14/33

A special passenger train at Two Harbors in June 1901. The head passenger car is D&IR 3, a combination baggage-smoker built for Chicago, Alton & St.Louis, acquired by D&IR in 1900 and retired on December 14, 1933. The following coach is thought to be D&IR 16 acquired from Ohio Falls Car Co. in 1882 and rebuilt by Hicks Locomotive & Car Co. in 1903. (University of Minnesota Duluth, Kathryn A. Martin Library Archives and Special Collections)

Kind	Trucks	Nos	Length	Builder	Built	Retired/Disposition
Day Coach (vestibule ends)	4 axle	23	66'10"	ACF	1908	Retired AFE 3370 12/14/33
Day Coach (vestibule ends) all steel	6 axle	24	70'8"	ACF	1912	Conv to Combo Bag/Coach AFE 3482 2/4/35 11/05/34. Renumbered W-24 AFE 6284 05/28/48 - Radio Telephone added 1947. Installed compartments #1 and #2 in 1947, Compartment #3 added in 1948 - Sold LSMT 08/03/03
Café/Observation	6 axle	25	65'6"	Unknown	Unk	Acquired from HL&CW in 1904 - Rebuilt by HL&CW in 1913 - Renumbered to Telephone and Telegraph Car 194 in 1928
Parlor Car	6 axle	26	72'10"	Barney & Smith	1907	Sold to F.H. Keyes 4/10/34 -Retired AFE 3426 06/5/34
Parlor Car	6 axle	27	72'10"	ACF	1908	Sold to Bowe-Burke Mining Co 8/9/33 -Retired AFE 3342 08/11/33
Day Coach (vestibule ends) all steel	6 axle	28	70'8"	ACF	1912	Remodeled 1939 with 12 seat solarium section added - To W28 5/12/65 1965. Donated to LS&M in mid-1980's then to D&NM(WGN) in 1994
Day Coach (vestibule ends) all steel	6 axle	29	70'8"	ACF	1912	Remodeled 1939 with 12 seat solarium section added - to work service W29 5/17/65
Day Coach (vestibule ends) all steel	6 axle	30	70'8"	Pullman	1918	To work service W30 AFE 9286 12/31/62. Then donated to Minn. Transportation Museum St. Paul, MN in 1983
Day Coach (vestibule ends) all steel	6 axle	31	70'8"	Pullman	1918	Sold to Thunder Bay Recreation, Marquette, MI in 1962
Day Coach (vestibule ends) all steel	6 axle	32	70'8"	Pullman	1918	to work service W32 AFE 9286 12/31/62
Day Coach (vestibule ends) all steel	6 axle	33	70'8"	Pullman	1918	Donated to LSMT in 1976
Day Coach (vestibule ends) all steel	6 axle	34	70'8"	Pullman	1918	to work service W34 4/27/65 AFE 9744 9/24/65 Then sold to WGN in 1997
Business	4 axle	Vermillion	55'	D&IR T.H	1887	Former business car (1st) 3 - Rebuilt 1895. SCS added 1912 - Sold AFE 3100 01/02/31
Business (closed end/open platform)	6 axle	Soudan	61'3"	Pullman	1889	Purchased from DM&N March, 1917 - To work service 1920 and renumbered 184
Business (vestibule end/open platform)	4 axle	Minnesota	60'	Barney & Smith	1895	Remodeled in 1909 - SCS added 1916 - Purchased from OIM on Oct 12, 1921. Converted to wrecking car W802 6/29/33 per AFE 3291 10/11/32
Business (open platform both ends)	4-axle	A	61'3"	Pullman Car Co.	1889	Purchased by Minnesota Iron Co. Renamed "Soudan" in unknown year
Business	4-axle	B	60'	Barney & Smith	1895	Changed from "B" to named car Minnesota in December 1901. Car was owned by Minnesota Iron Co but used by D&IR

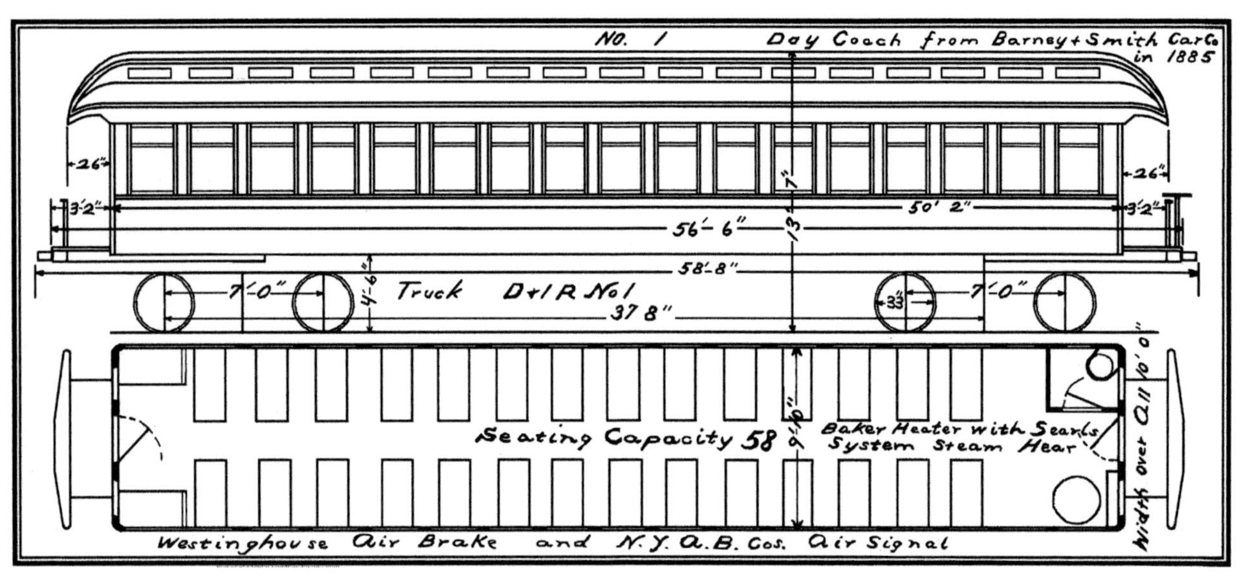

D&IR 1 day coach diagram. (D.P. Holbrook collection)

Car had a 56-foot 6-inch passenger area having 17 windows per side with seating for 58 people. Car 1 was off the roster by 1916 and car 2 in 1933.

Business car 3 was built at Two Harbors shop in 1887. The car was 56-foot 6-inches long, had a wooden carbody with open platforms, and rode on two 8-foot wheelbase 2-axle trucks, classified as D&IR Type No. 2. The car was named *Vermillion* in 1888 and assigned to D&IR President Tower. Tower did not like the paint on the car and had it repainted a "dirty white" in 1892. Rebuilt at Two Harbors in 1895, steel center sills were applied in 1912. The car was sold per AFE 310 January 2, 1931.

Orders were placed in 1887 with Ohio Falls Car Co. for two day coaches 4 and 5 and one combination car 6. All had wooden carbodies with open platforms on both ends and rode on two 8-foot wheelbase D&IR No. 2 2-axle trucks. Coaches had 17 windows per side with a seating capacity of 60 and one bathroom. The combination car had a 12-foot baggage section with a side door on each side and a 38-foot 2-inch passenger section seating 46 people and one bathroom. Coach 5 would be sent to Hicks Locomotive & Car Works in 1903 and rebuilt into a combination car. All were off the roster by 1933.

Between 1887 and 1888 it appears that

This view at Two Harbors circa 1910 shows from left to right: Baggage-Mail Express car either 7 or 8 built by American Car & Foundry in 1908, Combination Baggage-Smoker 5 built by Ohio Falls Car Co. in 1887 retired in 1914, Day Coach 1 built by Barney & Smith in 1885 and retired in 1915. (University of Minnesota Duluth, Kathryn A. Martin Library Archives and Special Collections)

passenger car numbers 7 and 8 were not used. No records have been found to determine if in fact there was a 7 and 8 purchased during 1887 or 1888. Numbers 7 and 8 would be used in 1908 for baggage cars.

The main line from Two Harbors to Duluth was completed on December 20, 1886. Duluth was becoming a center of commerce and neighborhoods were expanding eastward along Lake Superior. D&IR introduced passenger service from downtown Duluth to Lakeside, a suburb of Duluth, in 1887. This created a shortage of passenger cars. Needing additional passenger equipment quickly, D&IR purchased Baggage/Mail/Express cars 9–10, Day coaches 12–13 and Parlor cars 11 and 14 second-hand from Chicago & Indiana Coal Railway (C&IC) in 1888. No builder data has been found for these cars.

Baggage/Mail/Express cars 9–10 were 60 feet long. The cars had a 17-foot baggage section and 17-foot express section located on ends of car with a 4-foot wide side door on each side of each section. Centered on the car was a 16-foot mail section with a single window and a 26-inch wide side door. Day coaches 12–13 were 59-foot 3-inches long. The cars had a 51-foot passenger section seating 48 people and two bathrooms. Parlor cars 11 and 14 were 60-foot 6-inches long. All had wooden carbodies, open platforms on both ends and rode on two 8-foot wheelbase Pullman No. 5 2-axle trucks. Cars 9 and 10 would be placed in service for ICC valuations of railroads and renumbered X9 and X10 in 1914. Both were retired shortly after this service was completed.

An additional business car was ordered in 1889 from Pullman Car Co. A letter "A" was chosen to identify the car when delivered, account car being purchased and owned by Minnesota Iron Co., making it easier to distinguish between D&IR owned and numbered passenger cars. The car was 61-foot 3-inches long with wood carbody, open platforms and rode on two 9-foot 6-inch wheelbase 3-axle trucks. The non-observation end of the car had the platform enclosed at an unknown date. By 1900 D&IR had discontinued keeping track of this car on yearly equipment lists. The car was renamed *Soudan* but it is unclear when this took place. It lost its name and was converted to Engineer Car D&IR 184 in 1920 and then purchased by D&IR from Minnesota Steel Co. in November 1921.

A second business car "B" was purchased by Minnesota Iron Co., from Barney & Smith in 1895. It was changed from car "B" to named car *Minnesota* in December 1901. It was 60 feet long with a wood carbody, open platforms and rode on two 8-foot wheel base 2-axle trucks. The car was remodeled in 1909 at Two Harbors and steel underframe was applied in 1916. Sometime during its career the non-observation end acquired a vestibule type end. The name was removed and number 122 was placed on car from 1918 to 1920 when D&IR was under USRA control. Records show this carried as USRA 122 on equipment lists. Predecessor to Minnesota Iron Co., Oliver Iron Mining, sold the car to D&IR on October 12, 1921. It was approved to be converted to wrecking car W802 on AFE 3291 October 11, 1932, with work completed on June 29, 1933. Shop forces at Two Harbors neglected to paint out the word *Minnesota* on the car and a letter was issued July 12, 1933 for shop forces to correct.

Two new coaches, 15 and 16, were built by Ohio Falls Car Co. in 1892. Both were 61-foot wooden cars with 17 windows per side, rode on two Pullman No. 5 2-axle trucks, seating capacity of 64 and one washroom. Car 16 was destroyed by fire on July 19, 1900. Two Harbors shops constructed a new 16 by late 1903. Both were off roster by 1918.

Needing passenger cars quickly to fulfill increasing demands, D&IR found equipment on the used market and purchased one combine and two coaches from the Chicago & Alton Railroad in 1900. All three cars were open-platform wooden carbody cars riding on two 2-axle trucks built by Chicago & Alton

Business car *Minnesota* is shown with Parlor car 27 in 1922. Originally built by Barney & Smith in 1895, the car was numbered "B" and named *Minnesota* in December 1901. It was owned by Minnesota Iron Co. but used by D&IR. Remodeled in 1909 it acquired steel center sills in 1916. Minnesota Iron Co. became the Oliver Iron Mining division of U.S. Steel and OIM sold car to D&IR on October 12, 1921. It was converted to wrecking car W802 on June 29, 1933. (University of Minnesota Duluth, Kathryn A. Martin Library Archives and Special Collections, Collection S3742 Box 35 Folder 43)

on an unknown date. C&A combination car 34 was renumbered D&IR 3. On arrival from C&A it was rebuilt at Two Harbors and had seating capacity reduced to 24 people and baggage section enlarged to 29-foot 4-inches and a single bathroom. Diagrams from the time specify that this was a "Combination Baggage/Smoker" car indicating even at the turn of the century patrons wanted a separate area for smokers. C&A coaches 43 and 49 were renumbered D&IR 17–18. No diagrams or photos of these cars have been found. All were off roster by 1933.

Not much is known about Observation/Café car 25. It was 65-foot 6-inches long with a wooden carbody and first appears on D&IR equipment lists in 1905. Notes in diagrams indicate it was rebuilt by Hicks Locomotive and Car Works in 1913 and then renumbered to Telephone and Telegraph car 194 in 1928.

In 1907, age was catching up with the earliest passenger cars. A decision was made to order two coaches from Barney & Smith for delivery in 1907. Coaches 19–20 were 67-foot 6-inch open platform wooden carbody cars riding on two 8-foot wheelbase 2-axle trucks.

By the mid-1920s D&IR and DM&N were looking for cheaper alternatives to service their passengers. Full crews were required for steam-powered passenger trains and a significant cost savings could be had if passenger trains were replaced with self-propelled motor cars. Coach 19 was converted to motor coach MC1 in April 1926. The carbody was modified to a 28-foot 7-inch baggage compartment with a passenger compartment for 36 people. The car was powered by two 14H 104-horsepower 6-cylinder Red Seal Continental gasoline engines. Motor cars proved unsuccessful and the car had engines removed and was converted to combine (2nd) 19 in 1933. Car 20 was rebuilt to Telephone & Telegraph car 192 in 1927 and

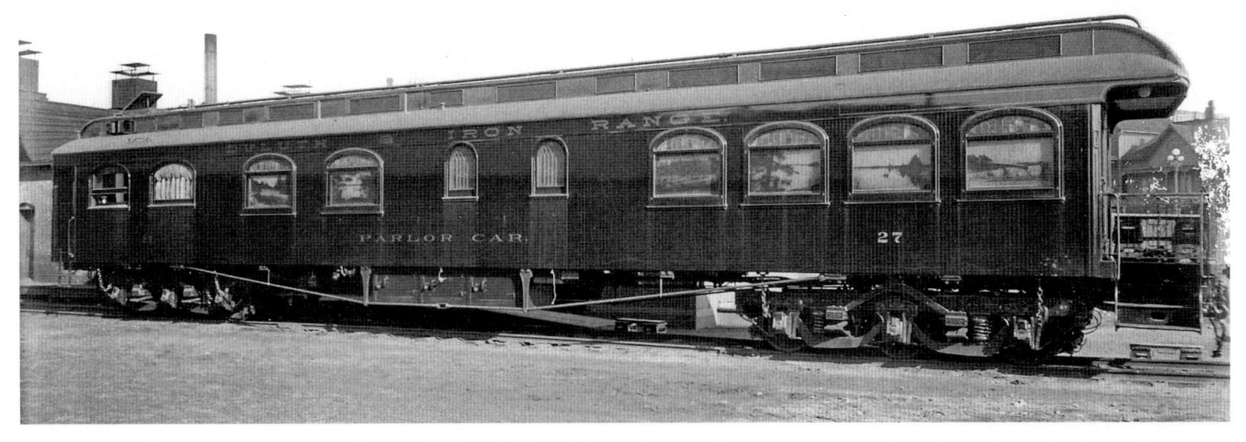

Parlor car 27 was constructed by AC&F in 1928. Note the various scenes depicted on the window glass. It was rebuilt in 1921 with displays and exhibits of the Superior National Forest including a miniature forest, stuffed animals, birds, fish and other items. Car 27, along with D&IR business car *Minnesota* toured from Ely, Minnestoa, to St. Paul, Omaha, and Kansas City, then returned to Ely during April and May 1922 promoting the tourism potential of the area. (University of Minnesota Duluth, Kathryn A. Martin Library Archives and Special Collections, Collection S3742 Box 16 Folder 50)

further rebuilt to cradle car 196 in 1928.

Extension of the D&IR main line from Biwabik to Virginia took place in 1908. Anticipating additional ridership, D&IR placed an order for six passenger cars from American Car & Foundry in 1907. Two baggage/mail cars, 7–8, three coaches 21–23 and one Observation/Café 27 were ordered October 25, 1907 and delivered in 1908. All cars rode on two Commonwealth 8-foot wheelbase 2-axle trucks and were of wooden construction.

Baggage/Mail cars 7-8 delivered on April 2, 1908, were 62 feet long, having a 20-foot mail compartment and a 39-foot baggage room both with single side doors on each side and the mail section having two windows per side. Between 1915 and 1920 the ICC undertook the large project of documenting almost every aspect of U.S. railroads in an effort to determine their net worth. These valuations, as they were called, created a need for various types of equipment, cabooses and passenger cars mostly, to house the individuals doing the valuation. D&IR elected to provide three different baggage cars for this purpose. Car 8 was renumbered X7 in 1914 and assigned to the "Valuations." Car X7 was retired December 1929, 7 was retired December 14, 1933 per AFE 3370.

Coaches 21–23 delivered in April, 1908 were 68-foot 5-inches long. Car 21 had a seating capacity of 68 and cars 22–23 seating for 70. All three were equipped with two washrooms and 23 was unique in having automatic ventilators. All three were sold in December 1933.

Observation/Café 27 was built by ACF, but an identical twin, 26, was ordered from Barney and Smith in 1908. Both were 72-foot 10-inch cars riding on two Commonwealth 10-foot wheelbase 3-axle trucks. They had a 19-foot 10-inch observation room that seated 14 people, a smoking room that seated 6, a café section that seated 12, two washrooms and a large pantry and kitchen located on the vestibule end. Both would be retired by 1934.

Car 27 was selected during the winter of 1921-22 to promote tourism to the Superior National Forest in the area around Ely, MN. Conceived by Vice President H. Johnson during 1921 the car was outfitted with exhibits inside the car that included a miniature forest, a canoe, snowshoes, and mounted specimens of fishes, birds and animals. Departing Ely, MN on April 21, 1922 with business car *Minnesota* for support personnel, the car was opened

On a fan trip in the early 1960s. DM&IR 29 is shown at Cloquet, Minnesota. Built by American Car & Foundry as part of a two-car order, cars 28 and 29 were both rebuilt with 12-seat solarium section in 1939. The 29 was placed in work service as W29 in 1965. (W.C .Olsen)

Car W24 was originally built as coach 24 by ACF in 1912 it was converted to a combine in 1935. A secondhand 4-kW 40-volt generator was installed in 1944 for electricity. Renumbered to W24 May 28, 1948, it became the companion support car for use with company business car *Northland*. Superior, Wisconsin, September 23, 1989. (D.P. Holbrook)

Two all-steel Baggage-RPO-Express cars D&IR 9 and 10 were ordered from Barney & Smith in 1914 to comply with an order from Congress on March 4, 1911 to replace all wooden RPO cars by July 1, 1916. Placed in MofW service as W9 September 24, 1965. (Owen Leander, John C. La Rue, Jr. collection)

Baggage-RPO 9 is at Endion, Minnesota, during the late 1950s. (D.P. Holbrook collection)

Replacing wooden passenger equipment, D&IR ordered five all-steel coaches 30–34 from Pullman in 1918. 32 would be placed in MofW service as W32 in 1962. (Owen Leander, John C. La Rue, Jr. collection)

to visitors in Tower, Aurora, Biwabik, Eveleth, Virginia and Two Harbors before arriving at the CMSTPM&O depot in Duluth.

It departed Duluth on CMSTPM&O and was displayed at St.Paul and Minneapolis on April 27 and April 28. Departing Minneapolis on MN&S for Northfield on April 29 it was interchanged to Chicago Great Western who handled car from Northfield to Omaha arriving May 10. CB&Q handled it to Kansas City on May 12 where it was displayed on May 13 through 15. The car departed Kansas City on CGW for Minneapolis and was back on the D&IR on May 26. Thousands of people toured the car and once back at Two Harbors the displays were removed and it was returned to regular passenger service.

Seeking to modernize an aging roster of wooden passenger cars, D&IR ordered three all-steel coaches with enclosed vestibules, 24, 28–29 from ACF in 1912. Appropriation 365 was approved at the D&IR Board of Directors meeting on July 11, 1912 for $45,000 for the purchase. They were 70-foot 8-inch long cars with seating capacity of 79, a 4-person men's smoking room with bathroom and a large women's dressing room and bathroom. Car 24 was rebuilt at Two Harbors to combination car 24 on February 4, 1935. Secondhand 4-kW 40-volt generating equipment was installed per AFE 5216 February 3, 1944 with work completed at Two Harbors on June 13, 1944. WPB purchasing restrictions at the time did not allow for new electrical equipment to be installed but no approval was necessary for secondhand purchases. The car was renumbered to work service W-24 on May 28, 1948, serving as a support car for business car *Northland*. It was sold to Lake Superior Museum of Transportation on November 3, 2003. Cars 28–29 were remodeled in 1939 by adding a 12-seat observation section. Both would be retired by the mid 1980s.

Congress would determine the next D&IR purchase of passenger cars. RPO clerks, early in 1911, called a strike for safer working conditions. This was precipitated by the large number of accidents involving wooden RPO cars. Congress reacted on March 4, 1911, and mandated that all future RPO cars must be

DM&IR 33, former D&IR 33, one of five coaches purchased from Pullman in 1918, is preserved at Lake Superior Museum of Transportation. Superior, Wisconsin. September 23, 1989. (D.P. Holbrook)

of all-steel construction and also required all wooden RPO cars be retired by July 1, 1916. To accomplish this, D&IR Board of Directors approved Appropriation 467 in July 1914 for $34,800 to purchase three combination baggage/mail/express cars from Barney and Smith. Delivered in 1914 as (2nd) 8, (2nd) 9 and (2nd) 10, all were of steel construction and were 73-foot 8-inches long and rode on two 10-foot 6-inch wheelbase 3-axle trucks.

These cars had a modern 3-foot long RPO section with desks and sorting table with the remainder of the car for baggage and express. The RPO section had a single door and three windows on each side, with the baggage/express section having a single door and two windows per side. This allowed baggage/mail cars (1st) 9 and (1st) 10 acquired from the C&IC in 1888 to be retired. Car 7 was kept on roster but not used in RPO service. It is interesting to note that car (2nd) 8 was leased to sister road DM&N during the winter of 1928/29. All would last until the end of passenger service on DM&IR on July 15, 1961.

The last passenger cars ordered by D&IR were five coaches, 30–34, built by Pullman in 1918. These were 78-foot 8-inch enclosed vestibule cars with two single and ten paired windows per side riding on two 10-foot 6-inch wheelbase 3-axle trucks. Even-numbered cars had a seating capacity of 82 and odd-number cars could seat 84. All would last until the end of passenger service on DM&IR on July 15, 1961 with some being used in MofW service and all off the roster by 1997. Car 30 is at the Minnesota Transportation, Museum Jackson Street roundhouse, 32 was donated to the Lake Superior and Mississippi tourist railroad at West Duluth and then to private owners Duluth and Northern Minnesota (D&NM) in 1996. Car 33 was donated to Lake Superior Museum of Transportation at Duluth and 34 is at the Wisconsin Great Northern railroad at Spooner, Wisconsin. ■

This overhead view of Duluth, Missabe & Northern yards at Virginia, Minnesota, during 1912 shows a mix of early 50-ton ore cars including Class U2, U3, U4 and U7 versions. Note the ore cars with company coal spotted at the coaling tower. (D.P. Holbrook collection)

DM&N No	CL	Description	Total	O.L	C.F	Lbs	Builder	Built	Trucks	Notes
7700–7708	T	Wood Ore Car	9	22′	454	70,000	Rebuilt from other classes	R1900	Arch Bar	42
(2nd)7009–7033	J1	Refrigerator Car	25	36′	1533	60,000	ACF Lot #4148	1906	Arch Bar	43
(2nd)7034–7058	J2	Refrigerator Car	25	36′	1916	50,000	Standard Steel Car Co	1908	Arch Bar	44
(2nd)7059–7083	J3	Refrigerator Car	25	36′	2030	50,000	Peteler Car Co	1910	Arch Bar	45
8000–8004	U	Steel Ore Car	5	22′	668	100,000	Schoen Pressed Steel Car	1899	Pressed Steel	46
8005–8354	U1	Steel Ore Car	350	22′	680	100,000	Pressed Steel Car	1903	Notes	47
8355–8404	U2	Steel Ore Car	50	22′	682	100,000	ACF Lot #2517	1903	Player Arch Bar	48
8405–9704	U3	Steel Ore Car	300	22′	689	100,000	Standard Steel Car	1905	Arch Bar	
9705–9954	U3	Steel Ore Car	250	22′	689	100,000	Standard Steel Car	1906	Arch Bar	
9955–10554	U3	Steel Ore Car	600	22′	689	100,000	Standard Steel Car	1907	Arch Bar	
10555–11704	U4	Steel Ore Car	1150	22′	689	100,000	Pressed Steel Car	1907	Arch Bar & A. L-section	
12000–12099	U5	Steel Ore Car	100	22′	650	100,000	ACF Lot #5809B	1910	Arch Bar	49
14000–14149	U6	Steel Ore Car	150	22′	669	100,000	Western Steel Car & Foundry	1910	Arch Bar	50
14500–15499	U9	Steel Ore Car	1000	22′1″	669	100,000	Western Steel Car & Foundry	1913	A. T-section	
16000–16749	U7	Steel Ore Car	750	22′	667	100,000	Standard Steel Car Co	1910	Arch Bar	51
18000–18004	U8	Steel Ore Car	5	22′1″	700	100,000	NSBM	1911	Arch Bar	52
19000–19999	U10	Steel Ore Car	1000	22′1″	704	100,000	Western Steel Car & Foundry	1914	A. T-section	
20000		Steel Ore Car	1	22′	654	100,000	Ralston Steel Car Co	A1916	A. T-section	53
20001–21000	U11	Steel Ore Car	1000	22′	704	100,000	Western Steel Car & Foundry	1916	A. T-section	
21001–21025	U12	Steel Ore Car	25	21′7″	935	190,000	ACF Lot #9907	1925	A. U-section	54
21026	U12	Steel Ore Car	1	21′5″	935	190,000	ACF Lot #9532	A1929	A. U-section	55
21101–21125	U13	Steel Ore Car	25	21′6″	1046	190,000	Pullman Car Co. Lot #5384	1925	Pflager	
21126	U13	Steel Ore Car	1	21′6	965	190,000	GACC	1931	Pflager	56
21999		Steel Ore Car	1	21′6″	965	190,000	GACC	A1930	Pflager	57
22000		Steel Ore Car	1	21′5	935	180,000	ACF Lot #9532	A1929	A. U-section	58
22001–22125	U14	Steel Ore Car	125	21′5″	935	182,000	GACC	1928	Dalmen 2-Level	
23001–23125	U15	Steel Ore Car	125	21′5	943	182,000	Standard Steel Car Co	1928	Dalmen 2-Level	
23171–23175	U16	Steel Ore Car	5	21′6″	1080	190,000	Standard Steel Car Co	1925	Pflager	59
23176–23177	U16	Steel Ore Car	2	21′6″	1055	190,000	Standard Steel Car Co	1925	Verona	60
24000–24799	U17	Steel Ore Car	800	21′6″	1000	166,600	PS Lot #5554	1937	ASF	61
24800–24999	U17	Steel Ore Car	200	21′6″	1000	166,600	PS Lot #5554	1937	National B1	62
25000–25499	U18	Steel Ore Car	500	21′6″	1000	166,000	GACC	1937	ASF	

Abbreviations:

R = Rebuilt
A = Acquired
A = Andrews trucks (L, T or U section)
C.F. = Cubic Feet
CL = Class
O.L. = Outside Length
ASF = American Steel Foundries
ACF = American Car & Foundry
GACC = General American Car Co
PS = Pullman Standard
ARL= Armour Refrigerator Lines
NSBM = Northwestern Steam Boiler & Mfg Co
OIM = Oliver Iron Mining
STDX = Standard Steel Car Co

Notes:

1. Shown in 1895 diagram book. Class X or Phelan Flat(21 cars) used for boarding cars and spreaders. All off roster by 1901

2. Shown in 1895 diagram book. No class shown. Trucks removed from ore cars during winter and placed under these log flats for service during the winter months. All sold in 1899

3. Renumbered to 3073 and 3074 during early summer 1905. Trucks used from retired wood ore cars

4. 218 rebuilt between 1893 and 1895 with different bracing and side supports(same cu. ft. and capy). Cars 202 and 364 rebuilt to Class T ore cars in 1900 - 300 equipped with Fox Steel Trucks. Ordered May 1892 - Delivered in late 1892

5. All off roster by 1904. All built as iron ore cars (partial iron construction), All but 388 converted to wooden by May 1895.

6. 945 rebuilt between 1893 and 1895 with different bracing, side supports, height and truck spacing(same cu. ft. and capy) Equipped at this time with Diamond trucks. See drawing Cars 684, 783, 1080, 1164 rebuilt to Class T in 1900. 870-879 equipped with Fox Steel Trucks

7. Built 3 thru 7-1895 Shipped 6 through 07-1895 1329 rebuilt to Class T ore car between 1895 and 1901

8. Built 12-1895 through 6-1896 Cars 1422 and 1941 rebuilt to Class T ore car in 1900 Car 1422 was first car rebuilt as a trial car and differed having a cu. ft. ,465.9 Broken into two groups 1401–1800 and 1801–2000, shipped 4=5-96, 5-96. Originally ordered 400 cars, an additional 200 cars added to order in Feb 28, 1896 per Railroad Gazette.

9. By 1901 following cars still 34 feet: 2006, 2009, 2010, 2012–2014, 2017, 2018, 2022, 2023, 2025, 2029, 2030, 2032, 2041, 2042 and 2050–2169. By 1902 remaining 34-feet cars were rebuilt to 36-feet cars and changed to Class V from Class W

10. By 1901 following cars rebuilt to 36 feet: 2001–2005, 2007, 2008, 2011, 2015, 2016, 2019–2021, 2024, 2026–2028, 2031, 2033, 2034–2040, 2043–2049. 2000 renumbered to 2229 in 1896

11. Bought second-hand from Munising Railway Co in 1898. Former Munising Railway 100–165 series cars

12. Bought second hand from the North West Railway in 1900

13. Former Class W flat car 2000. Rebuilt to 36 feet in 1896

14. Built from retired 200–1200 series wood ore cars

15. Steel centersill applied 1910. Letter G added as prefix to these cars to identify them as gondolas, starting 1919 and ending 1933. DM&N

flat cars converted to gondolas in 1919. All on roster from 1919 until 1932 when converted back to flatcars and rack flats: 2007, 2009, 2012, 2014, 2018, 2022, 2024, 2027, 2030, 2034–2036, 2041, 2043, 2048, 2050, 2053, 2057, 2058, 2060, 2064–2066, 2068, 2070, 2072, 2076, 2086, 2088–2091, 2095, 2098, 2099, 2101, 2105–2107, 2110, 2115, 2118, 2124–2126, 2130, 2131, 2133, 2134, 2137, 2146, 2147, 2150, 2153, 2157, 2160, 2164, 2168, 2170, 2188, 2222–2226, 2228, 2229, 2232–2236, 2238, 2240–2243, 2246, 2249, 2250, 2255–2260, 2265–2268, 2270, 2274, 2275, 2277, 2279, 2285, 2296, 2298, 2303, 2304, 2313, 2329, 2340, 2342, 2343, 2345, 2355, 2363, 2368, 2370–2372, 2374–2376. DM&N flat cars converted to gondolas in 1932: 2006, 2008, 2011, 2013, 2017, 2021, 2023, 2026, 2034, 2042, 2047, 2052, 2059, 2063–2065, 2067, 2075, 2082, 2086, 2087, 2089, 2090, 2095, 2098, 2101, 2104–2106, 2110, 2117, 2123, 2125, 2129, 2130, 2132, 2136, 2137, 2145, 2146, 2149, 2152, 2159, 2163, 2167, 2169, 2221, 2223, 2225, 2228, 2233–2235, 2237, 2240, 2242, 2245, 2248, 2249, 2255–2259, 2264, 2266, 2267, 2273, 2274, 2276, 2279, 2295, 2297, 2302-2303, 2313, 2328, 2339, 2341, 2342, 2344, 2362, 2367, 2369, 2371, 2373–2375

16. DM&N flat cars converted to gondolas in 1919. All on roster from 1919 until 1932 when converted back to flat cars: 2501, 2505, 2515-2516, 2529, 2535, 2546, 2558, 2564-2565, 2576, 2580-2581, 2587, 2589, 2592, 2596, 2624-2625, 2629, 2635, 2650, 2567, 2571, 2685, 2691. DM&N flaat cars converted to gondolas in 1932: 2500, 2504, 2514-2515, 2528, 2534, 2545, 2557, 2563, 2564, 2575, 2579-2580, 2586, 2588, 2591, 2595, 2623-2624, 2628, 2634, 2649, 2656, 2670, 2684, 2690
17. Reclassed as Class O sometime between 1901 and 1904. Then re-classed to Class M in 1905
18. 3011, 3019, 3032, 3035 rebuilt to refrigerator cars during 1893 and 1895 and renumbered 5000–5003. 5003 off roster by 1896, 5000–5002 to box cars 3032, 3075, 3076 in Nov. 1905. 3018 destroyed 1901 and 3000 renumbered (2nd) 3018 1901
19. 2nd 3011 replaced 1st 3011 that was rebuilt to reefer 5000
20. 2nd 3019 replaced 1st 3019 that was rebuilt to reefer 5001
21. 2nd 3035 replaced 1st 3035 that was rebuilt to reefer 5004
22. Purchased second hand
23. Box car 3046 rebuilt to stock car S3046 1912 - Retired Oct 1920
24. Box car 3062 rebuilt to stock car S3062 1912 - Retired May 1920
25. Former reefer 5000 and 5001
26. Cars 3148 and 3149 rebuilt to stock cars 4002 & 4003 in 1906
27. Rebuilt with double doors in 1910 At Proctor for automobile serivce
28. 4000 Retired 8-16, 4001 retired 12-15
29. Rebuilt from box cars 3148, 3149 in 1906 3149 to box car ?? To stock 4003 5-35, to box car 3149 12-31-37. Classed as class K, then class P, at some point they quit changing classes
30. Rebuilt from box cars 3013, 3011, 3020, 3021, 3008, 3038, 3004, 3006 in 1907. Used in oil trade.
31. Rebuilt from box cars 3012, 3024, 3037, 3040, 30xx and 30xx in 1908. Used in oil trade
32. Ordered 11-15. Built 5-16 thru 6-16
33. Rebuilt from box cars 3011, 3019, in 1893. ilce boxes removed in 1905. Renumbered to 3075, 3076, as Class M box cars in 1905
34. Rebuilt from box cars 3032,3035 in 1895. 5003 Off roster 1896. 3032 back to Class M box car 3032, 1905
35. Purchased second hand. 5007 destroyed 12-07. 5008 destroyed 1-09
36. Ordered 11-18-1905 ACF Chicago plant
37. Steel underframe - Rebuilt from 25 traded-in wood ore cars: 6007, 6015, 6030, 6032, 6099, 6265, 6324, 6382, 6384, 6386, 6445, 6480, 6561, 6587, 7162, 7185, 7211, 7224, 7293, 7343, 7407, 7517, 7604, 7610, 7624
38. Built 1-1898 through 4-1898 Shipped 3=4-98
39. Ordered 4-1899
40. Ordered 11-1899 Built 5-6-1900
41. Ordered 1-1900. Built 6=7-1900
42. Rebuilt from ore cars 202, 364, 684, 783, 1080, 1164, 1329, 1422, 1941 and reclassed class T
43. Renumbered from 5009–5033 in 1931
44. Renumbered from 5034–5058 in 1931
45. Renumbered from 5058–5083 in 1931
46. Cars 8002 sold to State of Minnesota 1-08. 8001 and 8004 sold to NP 11-09, 8003 sold to NP 10-11, 8000 to work service 1-16
47. 175 cars Arch Bar, 175 cars Pressed Steel
48. 100 cars ordered by D&IR on 10-15-1902 - Order split to deliver 50 cars to DM&N and 50 cars to D&IR 3400-3449 ACF Detroit plant
49. Clark type ore car
50. Built by Pressed Steel Car at Western Steel Car & Foundry Chicago plant
51. 170 cars sold to OIM 3-16
52. Rakowsky ore cars(First 3 delivered 12/11, last car delivered 1/12)18004 retired 11-15, 18000-18003 retired 1/16
53. Car never given a DM&N class - Demonstrator car for Ralston 1915 to 1916 - Built 5-13-1915 - Acquired 1916
54. Blt ACF Chicago Plant. Blt 11/1924. Delivered in 1925
55. Former DM&N 22000
56. Former DM&N 21999
57. Renumbered to 21126 in late 1931. Former GATX 1000 sample ore car
58. Renumbered to 21026 in late 1931. Former ACFX 9001 90-ton sample ore car built 6-23-23 under Lot #9532
59. Purchased 1931(secondhand) Former STDX 23171–23175
60. Purchased 1931(secondhand) Former STDX 23176, 23177
61. Built at Michigan City plant
62. Built at Michigan City plant. Cor-Ten Steel

-continued from page 107

frigerator cars, 5000, 5001 in 1893 and 5002 in 1895, by installing "ice boxes" on the inside of the box cars. An additional car, 3035 was converted in 1895 and numbered 5003. Cars had a light weight of 29,400, 1,433 cu. ft and rode on 5-foot 6-inches archbar trucks. These cars retained the boxcar sliding doors and did not have rooftop ice hatches. By June 21, 1899, the sliding doors had been replaced with a smooth-fitting refrigerator type door, 5-foot wide by 6-foot high. Car 5003 was destroyed in 1896 and during November 1905 the remaining three cars had the ice boxes removed and were renumbered back to box cars: 5000 to 3075, 5002 to 3076 and 5002 to 3032. The cars were equipped with Chicago couplers, truss rods, double-sheathed carbody and ends, and end doors 2-foot wide by 2-foot 3-inches high on the B-end of the car. Cars were probably painted with a white or yellow car body, mineral red roof and black underframe and trucks. A June 21, 1899, diagram shows "Summer and Winter Car" lettering on the door, a banner to the left of the car door with the car number and the arched "Duluth, Missabe and Northern" to the right of the door.

The second group of refrigerator cars was

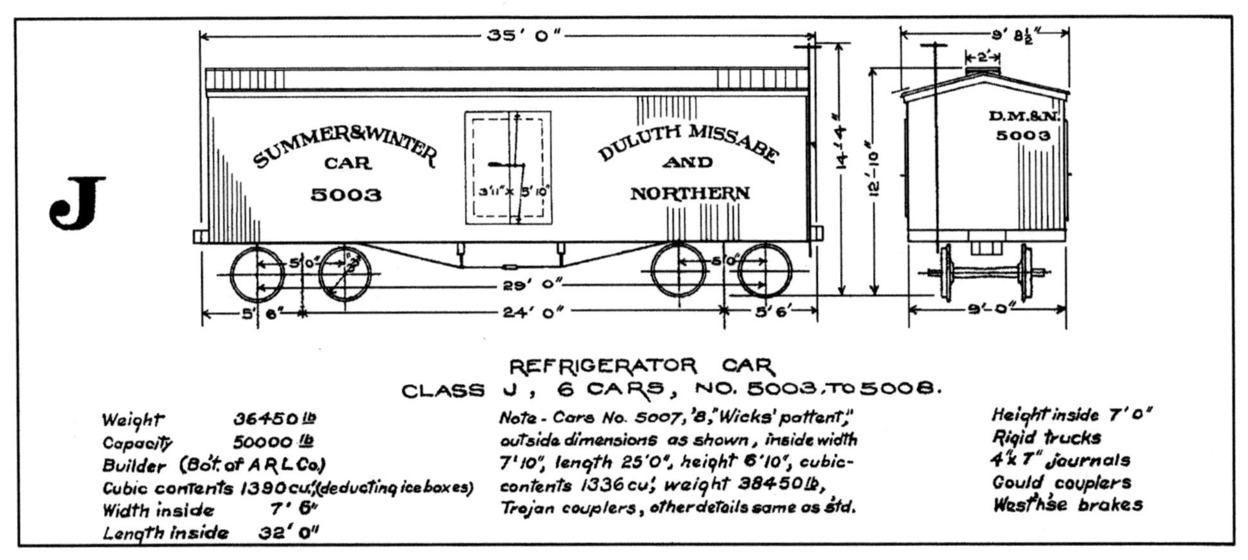

DM&N Class J 5003–5008 diagram. (D.P. Holbrook collection)

four Class J 25-ton 35-foot reefers 5003–5006 purchased second-hand from Armour Refrigerator Lines (ARL) with 1,390 cu. ft. capacity. The 5003 (2nd) was acquired early in1898 and 5004 to 5006 in May 1898. These cars were equipped with: Rigid trucks with 4 x 7-inch journals, Gould couplers, truss rods, double-sheathed carbody and ends. Details of icing equipment are not known for the 5003 to 5006. The cars had 3-foot 11-inches wide by 5-foot 10-inches high door openings. It is not known if these cars were painted white or yellow, but initial lettering style was a large arched Summer & Winter Car logo to left of door with number below and the arched "Duluth, Missabe and Northern" to the right of the door. The 5003 to 5005 were destroyed on April 7, 1906, 5006 destroyed on August 5, 1905.

Two additional Class J 25-ton 35-foot reefers 5007 and 5008 with 1,336 cu. ft capacity were purchased secondhand in December 1901 from F.M Pease, an equipment dealer, for $580.75 per car. These were originally built by ARL and equipped with "Wickes patent" ice tanks and rode on 5-foot wheelbase rigid arch bar trucks. The Wickes Refrigerator Car Co. received its first patents in 1880 and by 1898 a total of about 10,000 refrigerator cars had been built under these patents. The Wickes system incorporated two tanks at each end, 3-foot 7-inches long, and 2-foot 2-inches wide by 5-foot 2-inches deep that were built using an oak framework on the inside and nailed interwoven strips of galvanized iron 2 inches wide and 1 inch apart. 5007 was retired in 1908 and 5008 was off the roster by 1913.

By 1905 DM&N was satisfied that there was money to be made by hauling perishable traffic and an order for 25 Class J1 30-ton 36-foot refrigerator cars 5009–5033 was placed in 1905 with American Car and Foundry. All were delivered in 1906 at a cost of $1133.20 per car. Equipped with Bohn ice tanks, ice hatches were located in all four corners of the roof. They were equipped with Rigid arch-bar trucks with 4½ x 8-inch journals, Griffin wheels, Simplex bolsters, Westinghouse brakes, Pressed Steel brakebeams, Chicago couplers, Railway Steel Spring Co. springs, McCord journal boxes, wood underframe with truss rods, Miner drawbar attachments and plain side bearings. They were constructed with swinging doors, 3-foot 11-inches wide by 5-foot 11-inches high and had double-sheathed carbody and ends. As built, cars had a light weight of 39,700 and 1,533 cu. ft. Over the years the cubic capacity changed as cars were rebuilt with various refrigeration and heating systems. These cars were delivered with the stacked "Duluth, Missabe and

D&IR leased refrigerator cars before making a purchase (Chapter 2). DM&N converted four box cars with ice boxes in 1893 and 1895 and then purchased second-hand reefers from Armour Refrigerator Lines in 1898 and 1901. The first car builder-built reefers, 25 Class J1 5009–5033, were acquired in 1906 from ACF. This builder photo shows 5020 in either yellow or white with black lettering and hardware details. They were delivered with Bohn ice tanks. All but four were rebuilt in 1916 with Moore overhead ice tanks with three hatches centered on the car above the center door and on either side of the running board. (American Car & Foundry Collection, John W. Barriger III National Railroad Library at UMSL)

Northern" lettering to the left of the side door and the reporting number to the right of the door. It appears from photos that cars were painted either white or a light yellow with black lettering, hardware and underframe.

During 1916 all cars, except for 5009, 5011, 5013 and 5030, were rebuilt with Moore overhead ice tanks equipped with three hatches on each side of the roof centered over the doors. At the same time, a permanent under-slung heater was installed with a "stovepipe" stack on the roof. This allowed for the cars to be used during the winter months when the Minnesota weather would have frozen anything loaded in the cars. The heater maintained the interior temperature at an appropriate level for shipping of produce and meat.

In 1931, the 5009–5033 were renumbered to 7009–7033, keeping DM&N initials, in anticipation of the DM&IR merger. By 1934, only the 7018, 7022, 7026 and 7032 remained on the roster. Beginning on August 21, 1937, the remaining cars began to receive DM&IR initials. In 1938, the 7022 and 7032 were rebuilt with Bohn ice tanks with two ice hatches located in opposite right hand corners of the roof and flat plate steel ends. 7018 was retired in 1941 and 7026 placed in MofW service in 1942 and renumbered to W7026. The 7022 was retired in 1960 and 7032 in 1963.

Almost as soon as the 25 Class J1 reefers were delivered, the DM&N placed an order in 1907 with Standard Steel Car for an additional 25 Class J2 25-ton 36-foot refrigerator cars 5034–5058. All were delivered in 1908 at a cost of $1586.82 per car. Cars were equipped with: Rigid arch bar trucks with 4¼ x 8-inch journals, 4-foot wide by 9-foot 11⅝ inches high swinging doors, 10 feet steel channel side sills, steel fishbelly underframe, double-sheathed carbody and ends with an outside length of 38-foot 7-inches. Refrigeration was provided

Standard Steel Car Co. would build 25 Class J2 refrigerator cars DM&N 5034–5058 in 1908. They would be renumbered to DM&N 7034–7058 in 1931 to avoid conflicts with D&IR box cars in the 5000 series in anticipation of the merger of both carriers into DM&IR. Note roof and ends are in either yellow or white and only the lettering, hardware and underframe are painted black. 7055 was retired in 1954. (Standard Steel Car Co., D.K. Retterer collection)

Endion, Minnesota, was the location that refrigerator cars gathered on the DM&IR. The freight house here was the transload and consolidation point for all towns and cities on the DM&IR. Class J2 7037 is seen here during July 1962. (Joe Collias)

by Bohn ice tanks located at opposing right hand corners of the roof. As built, cars had a light weight of 45,900 pounds and 1916 cu. ft. Over the years the cubic capacity changed as cars were rebuilt with various refrigeration and heating systems.

During 1916 five cars, 5036, 5037, 5044, 5056 and 5057 were rebuilt with Moore overhead ice tanks equipped with 3 hatches on each side of the roof centered above the side doors. In 1931, the 5034–5058 were renumbered to 7034–7058, keeping their DM&N initials, in anticipation of the DM&IR merger. At the same time, a permanent underslung heater was installed with a stovepipe caboose stack on the roof. Beginning on August 21, 1937, the remaining cars began to receive DM&IR initials. Bohn ice tanks were re-installed as follows: 1936: 7036 and 7056, 1940: 7037, 1941: 7057 and 1942: 7044.

The wood ends deteriorated rapidly and Youngstown corrugated steel ends and steel roof were installed as follows: 1936: ten cars, 7034–7036, 7042, 7050–7054, 7056; 1938: two cars, 7043, 7048; 1940: two cars, 7037, 7038; 1941: four cars, 7045–7047, 7057; and 1942: one car, 7044.

Per AFE 3064 at a cost of $3,300 roofs and

Mechanical reefer? Never in revenue service as a mechanical, but 7049, retired and relettered W7049, had a "Trail-Aire" air conditioner and 50-gallon underslung gasoline tank installed to keep provisions for MofW workers cool. Proctor on March 31, 1963. (D. Repetsky, John C. La Rue, Jr. collection)

Class J3 7071 at Duluth in May 1955 Built by Peteler Car Co. as part of an order for 25 refrigerator cars 7059–7083 in 1910, 25 wood ore cars were traded in on this order. The cars are distinctive in having steel underframe and truss rods. Delivered with Moore overhead ice tanks, 7071 was rebuilt in 1939 with two Bohn ice tanks located at opposing corners of the car body. All Class J3 cars were retired by 1971. (W.C. Whittaker)

ends were requisitioned on December 13, 1935 and all cars were converted by July 28, 1936. Internal study had determined that the average life of wood ends was 17 to 20 years and roofs eight years, versus steel ends having an indefinite service life, and steel roofs a 15-year service life. Initially the roofs were ordered from Standard Railway Equipment and specified Hutchins outside metal roofs. Standard Railway Equipment recommended that Murphy Improved Pivoted Roof be substituted and this was approved on January 15, 1936. These roofs were supplied without hatch openings which DM&N cut and installed after roofs were replaced.

Six cars, 7039–7041, 7047, 7049, 7055 retained wooden ends until retirement. Some of the Youngstown end cars were again rebuilt with flat steel ends before retirement, including 7036 and 7047. At least one car, 7054, was equipped with 4/4 dreadnaught ends at a later date.

Car 7049 was chosen in 1948 per AFE 6314 approved June 30, 1948, to be rebuilt for work train service by installing a Trail-Aire air conditioner and a 50-gallon underslung gasoline tank, becoming the only DM&IR mechanical refrigerator car. Renumbered to W7049 at the same time, the car never saw revenue service as a mechanical reefer. All the Class J2 refrigera-

Class J3 7063 at Endion in 1962. Repainting revenue equipment in maroon with gold lettering began in 1959. This is the only known example of a refrigerator car in this paint scheme. (L. Miekiszak Photo, Burlington Route Historical Society collection)

tor cars were off the roster by 1968.

The last class of reefers purchased by the DM&N were 25 Class J3 25-ton 36-foot refrigerator cars 5059–5083 built by Peteler Car Co. in 1910 for $1187.50 per car. DM&N traded in 25 wood ore cars to Peteler Car Co. on this order. These cars were equipped with: Moore patent overhead ice tanks and underslung heater, 5-foot 4-inch Rigid arch-bar trucks, 4½ x 8 inch journals, 4-feet wide by 5-foot $11\frac{5}{8}$-inches high swinging doors, double-sheathed carbody and ends with 2030 cu. ft. capacity. A fabricated steel underframe with outside truss rods attached to the outside channel were used in an attempt to keep the car tight and square to avoid water from seeping into the insulation and causing damage to the car body. Over the years the cubic capacity changed as cars were rebuilt with various refrigeration and heating systems. These cars were delivered with stacked Duluth, Missabe and Northern lettering to the left of the side door and the car number to the right of the door. It appears from photos that cars were painted either white or a light yellow with black lettering, hardware and underframe.

In 1931, to avoid conflict with D&IR box cars in the 5000 series, 5059–5083 were renumbered to 7059–7083, keeping their DM&N initials, in anticipation of the DM&IR merger. By 1937, 7059, 7060, 7064, 7066–7068, 7072, 7074, 7076–7079, 7081, 7082, and 7083 (15 cars) had been retired. Beginning on August 21, 1937, the remaining 10 cars, 7061–7063, 7065, 7069, 7070, 7071, 7073, 7075, 7080 began to receive DMIR reporting marks.

Eight cars were rebuilt with Bohn ice tanks with two ice hatches located at right hand corners on either end of car and plate steel ends as follows: 1937, under AFE 3394, four cars 7061, 7063, 7069, 7070, completed on January 6, 1938, 1939: one car, 7071, 1941: two cars 7075, 7080 and 1942: one car 7065. Cars 7062 and 7073 retained their Moore ice tanks until being renumbered into MofW service in 1950 by placing the letter "W" in front of the number. All of the J3 class cars were off the roster by 1971.

On September 30, 1938, AFE 3792 for $270 was approved for the purchase and installation of "Frigiwarm" curtains on nine refrigerator cars; 7022, 7032, 7043, 7048, 7061, 7063, 7069, 7070 and 7071. These curtains allowed the icing of one end of the car only to allow merchandise to be loaded in one end and perishable traffic in the other. This project was completed on December 14, 1938

Declining LCL perishable movements and increased icing cost caused DM&IR in 1964 to look into dry ice for cooling needs on LCL refrigerator cars. AFE 9882 for $3,915 was approved on January 5, 1965, to modify 23 of the remaining refrigerator cars by installing new insulated curtains and dry ice racks. Seventeen cars would be modified on one end only and six with both ends. Thirteen of the cars were former DM&N cars. Modified on one end only were: 7037, 7043–7045, 7053, 7054, 7057, 7065, 7069, 7075, and 7080. Modified on both ends were 7061 and 7070. The remaining ten cars were former D&IR cars and are covered Under D&IR (Chapter 2). High Kold-Pak insulated quilts (curtains) for this project was supplied by CanPro Corp. of Fond Du Lac, Wisconsin. Rebuilding started on January 26, 1965, and all work was completed by June 18, 1965.

DM&N STOCK CARS

Two Class I 18-ton 37-foot 8-inch stock cars, 4000, 4001 with 2,147 cu. ft. capacity were purchased from Armour Refrigerator Lines in 1900 at a cost of $434.50 per car. Cars were equipped with Rigid arch bar trucks, Gould couplers and Westinghouse air brakes. It is unknown what year ARL built these cars. They were unique in having the door off-set to the right center of the car with an opening of 4- foot 11-inches wide by 7-foot 9-inches high. Cars had end doors on both end 2-foot wide by 2-foot 5-inches high. It is unknown what color these cars were painted. Diagrams indicate letterboards located just below the roof line, with Duluth, Missabe & Northern spelled out to the left of the door opening and the reporting number located to the right. They both were retired, 4001 in December 1915 and 4000 in August 1916.

Two Class K 40-ton 40-foot 8-inch stock cars 4002, 4003 with 2,730 cu. ft. capacities were converted from Class P box cars 3148, 3149 in 1906. Originally built by Western Steel Car & Foundry in 1906, they were rebuilt to stock cars at DM&N Proctor Shops in December 1906.

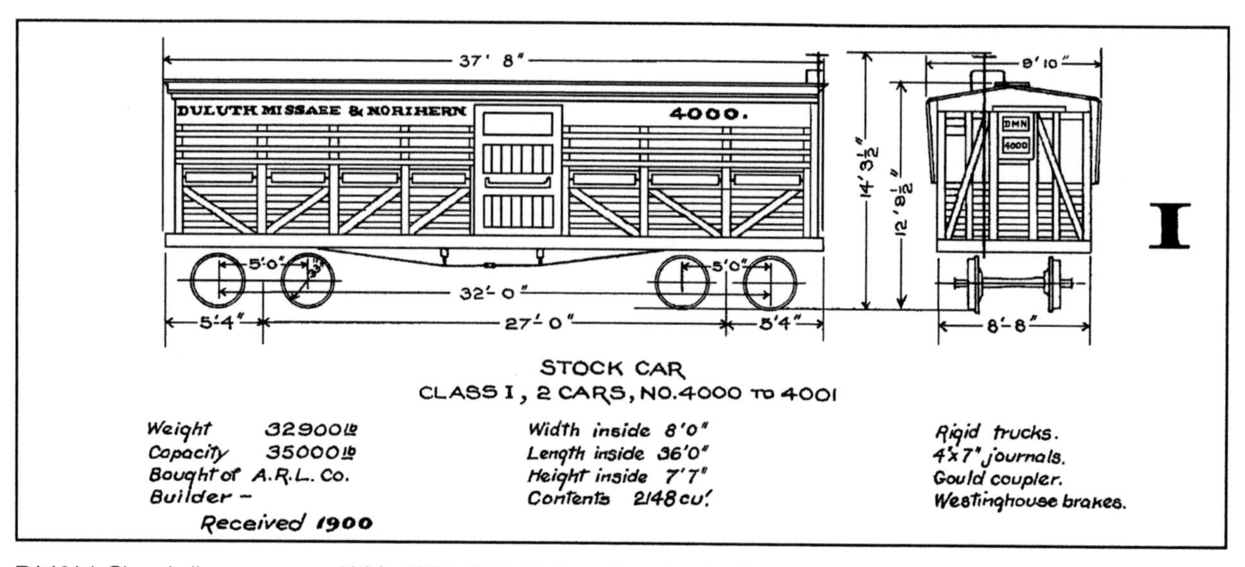

DM&N Class I diagram, cars 4000, 4001. (D.P. Holbrook collection)

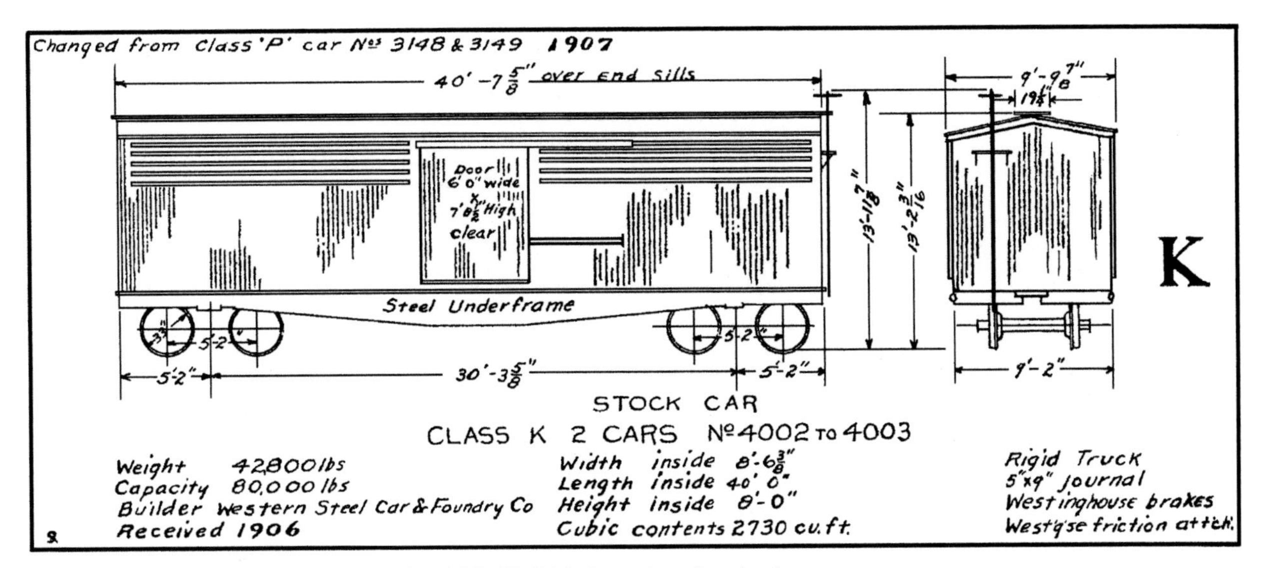

DM&N Class K diagram, cars 4002, 4003. (D.P. Holbrook collection)

DM&N records showed these as stock cars but the *ORER* carried these as box cars in oil service. They were equipped with Rigid arch bar trucks and Westinghouse air brakes. Constructed as double-sheathed box cars, the upper car sheathing was removed and slatting for ventilation was installed in this area. DM&N assigned Class K to these stock cars when initially rebuilt, then later Class P. Car 4002 was rebuilt back to a box car prior to 1935 and then converted back to stock car 4002 on May 23, 1935, and finally retired in 1970. Car 4003 was rebuilt back to box car 3149 sometime prior to 1934. It was again converted back to stock car 4003 on May 23, 1935, and then again back to box car 3149 on December 31, 1937.

DM&N rebuilt 25-ton double-sheathed boxcars 3046 in 1912 and 3062 in 1913 at the Proctor Shops to Class N 25-ton 34-foot stock cars with 1,840 cu. ft. capacity at a cost of $439.50 per car. To indicate they were stock cars, DM&N added a letter prefix of "S" and numbered them S3046 and S3062. Cars had 5-foot wide by 6-foot 3½-inches door opening with 5-feet wheelbase arch bar trucks. They were originally purchased in 1899 from Armour Refrigerator Lines, as part of a 20-car group of 34-foot box cars numbered 3043–3062. Constructed as double-sheathed box cars, part of the upper car sheathing was removed and slatting for ventilation was installed in this area. It is unknown what year ARL built these cars. Both were retired, S3062 in May 1920 and S3046 in October 1920.

DM&N BOX CARS

The first box cars ordered by DM&N were 36 Class L 25-ton 34-foot double-sheathed cars 3000–3035 with 1,705 cu. ft. capacity from Duluth Manufacturing Co. of Duluth in 1892 at a cost of $499.80 per car. Ten of these cars were built in 1892 with the remainder being built in 1893. Sometime during 1903–04 these cars were placed in Class O and then in 1905 they were placed in Class M along with cars in 3036–3042 series. They had 5 foot wide by 6 foot high door openings with vertically sheathed wood doors that opened to the left. End doors 2-foot wide by 2-foot 3-inches high were located on the A-end of car. In 1893 and 1895, three of these cars, 3011, 3019, 3032 were rebuilt to refrigerator cars 5000–5002. Refrigerator car 5002 was renumbered back to box car 3032 in 1905. Refrigerator cars 5000, 5001 were renumbered 3075 and 3076 in late 1905 above the then just-built automobile cars 3073, 3074. These cars appeared to have been delivered with a small version of the arched lettering to the right of the door. By 1905, the

DM&N 3006 is one of 36 25-ton 34-foot box cars 3000-3035 built by Duluth Manufacturing. Ten were built in 1892 with remainder completed in 1893. Placed in Class L, they would all be placed in Class O between 1901 and 1904 and finally Class M in 1905. 3006 was rebuilt to oil car 4011 in 1907 and retired during November, 1916. (Shorpy Photo Archives)

road number appeared to the left of the door opening and the stacked Duluth, Missabe & Northern lettering was located to the right. All were off the roster by 1925.

The second group of box cars ordered by DM&N was ten Class M 25-ton 34-foot double-sheathed cars with 1,705 cu. ft. capacity from Pullman Car Co. under Lot 2173 in 1896 at a cost of $542.95 per car. These cars were numbered, 3011(2nd), 3019(2nd), 3035(2nd) and 3036–3042. The backfilling of these car numbers to replace the rebuilt refrigerator cars in the 3000–3035 series was for accounting and equipment trust reasons. The cars had 5-foot wide by 6-foot 3½-inches wood door openings and rode on 5-foot 6-inch arch bar trucks. Doors opened to the left and were equipped with lumber-type end doors 2-foot wide by 2-foot 3-inches high on the A-end of the car only. Cars were delivered with the road name spelled out in an arch to the right of the door opening. Sometime between 1901 and 1905 the lettering was changed to a stacked road name to the right of the door. All were off the roster by 1925

Traffic continued to increase, but financial troubles were putting a strain on the budget. Looking to increase boxcar ownership, the

Ten Class M 34-foot boxcars were ordered from Pullman Car Co. on Lot 2173 in 1896.Three refrigerator cars had been rebuilt from 3000–3035 series box cars between 1893 and 1895 so three of the Pullman built cars assumed these numbers: 3011, 3019, 3035 with the remainder being placed in 3036–3042 series. Class lettering would not be painted on cars until 1899. (Pullman Palace Photographs, Archives Center, National Museum of American History, Smithsonian Institution)

The first 40-foot boxcars on DM&N were Class P cars 3100–3149 built by Western Steel Car and Foundry in 1906. They introduced a steel underframe to the DM&N roster with fishbelly side and center sills and I-beam crossmembers. DM&N 3144 is shown in yard at Endion, north of downtown Duluth on July 19, 1957. (Bob's Photo)

DM&N acquired 20 Class N 25-ton 34-foot double-sheathed box cars 3043–3062 with 1,840 cu. ft. capacity from Armour Refrigerator Lines in 1899 at a cost of $439.50 per car. Equipped with right-opening wood doors 5-foot wide by 6-foot 3½-inches high and truss-rod underframes, they rode on 5-foot wheelbase rigid arch bar trucks. Cars 3046 and 3062 were rebuilt to stock cars S3046 in 1912 and S3062 in 1913 at the DM&N Proctor Shops. The cars were delivered with the stacked Duluth, Missabe and Northern lettering to the left of the door opening and the road number to the right. All were off the roster by 1930

In 1901 the American Railway Association recommended a 36-foot boxcar design with 2,442 cu. ft. capacity, which was adopted by the MCB in 1904 as recommended practice. Many railroads and car builders built cars closely following this design. Under the U.S. Steel umbrella, an order was placed in November 1902 from Western Steel Car and Foundry for ten Class N1 30-ton 36-foot 10-inch double-sheathed box cars 3063–3072 with 2,448 cu. ft. capacity at a cost of $890.43 per car. D&IR received 25 identical cars 5174–5222 (even numbers) at the same time. All were delivered in 1903. Cars had wood doors covering a 5-foot 11¼-inches wide by 7-foot 6-inches high door opening, truss-rod underframes and rode on arch-bar trucks. They were delivered with the stacked road name to the left of the door opening and the road number to the right. All were off the roster by 1930.

Unlike sister D&IR who continued to buy and build 36-foot box cars, DM&N moved on to 40-foot box cars in 1905. An order was placed with Western Steel Car and Foundry for 50 Class P 40-ton 40-foot 7-inch double-sheathed steel-underframe box cars 3100–3149 with 2,730 cu. ft. capacity in October 1905 at a cost of $1,221.98 per car. All were delivered in 1906 with doors 6-foot wide, 8½-inches high.

Underframes were girder type, which was fabricated from fish-belly steel girders for center and side sills and structural shapes for cross-members. The cars were equipped with Griffin wheels, arch-bar trucks, Simplex bolsters, Westinghouse brakes, Pressed Steel Car brake beams, Chicago couplers, Railway Steel Spring Co. springs, McCord journal boxes, Westinghouse drawbar attachments and Susemihi side bearings. Delivered with double sheathed ends, these ends were replaced on most cars with Murphy 6/7 reverse corrugated steel ends by 1916. Ladders were unique on the new ends, having an S-shape in the vertical stiles to allow them to be attached to the new Murphy ends. In later years a weather guard was added above the top door rollers of the door track. Delivered with arch-bar trucks these were replaced with Andrews U-section trucks between 1937 and 1939. They were delivered with the stacked road name to the left

New maroon and gold paint and Scotchlite monogram, reporting marks and numbers were applied at Two Harbor shops in October 1960 on Class F box car DM&IR 3108. By this date over 100 cars had received the new image. The white signboard says "FOR COAL AND ROUGH FREIGHT ONLY." (Jim Maki Sr.)

of the door opening. At the time of the merger, 45 of these cars were still on the roster. All were retired by 1968.

Needing additional box cars for growing business, the DM&N ordered 100 Class P1 40-ton 42-foot 4-inch double-sheathed steel underframe box cars 3150–3249 with 2,755 cu. ft. capacity from Standard Steel Car Co. in 1907 at a cost of $1209.48 per car. Door openings were 6-foot wide by 7-foot 6-inches. The cars had a fishbelly steel underframe with 10-inch channel side sills. All were rebuilt with Murphy steel roofs and Murphy 7/7 reverse corrugated steel ends by 1916. Once equipped with the Murphy ends, these cars had unique 2-section end ladders that were split between the top and bottom corrugations on the ends. Five of these cars, 3150–3154, were rebuilt to automobile cars in 1910 at DM&N Proctor Shops. At the time of the merger, 92 of these

Unlike sister D&IR who stuck with 36-foot box cars, DM&N began purchasing 40-foot box cars with 50 Class P cars in 1906. The next order was for 100 Class P1 boxcars 3150–3249 from Standard Steel Car Co. in 1907. Fresh from Two Harbors shop during February 1943 3200 shows off its as-built steel underframe. Cars were delivered with wood ends which were replaced with Murphy 7/7 reverse corrugated ends by 1916. Unique to DM&IR, car puller tabs were only installed to the left end of each side above the truck. (W.C. Olsen)

DM&IR 3240 is an Endion July 19, 1957. Note the chalk mark X next to the light weight and the notation above it in chalk to reweigh, indicating cars needs a new light weight. (Bob's Photo collection)

Introduced in 1910, the large DM&N ball monogram was painted on box cars to replace the stacked Duluth, Missabe & Northern. Class P1 DM&N 3236 is at Two Harbors in June, 1938. Note the replacement 7/7 Murphy end. The horizontal hand rail along the edge of the roof is for tying off support ropes that were attached to wooden angles supporting walkways alongside the carbody sides for wheel barrows to move cement at construction projects such as bridges and ore docks. (Bob's Photo collection)

cars were still on the roster. Originally built with arch-bar trucks with 5-foot 4-inch axle spacing which were changed to 5-foot 6-inch Andrews U-section trucks between 1937 and 1939. Doors had modernized door hangers applied during late the 1940s and early 1950s. This involved either a reinforcing metal strip being applied near the top of the door, or a wider metal plate with the door hangers attached. The cars were delivered the stacked Duluth, Missabe & Northern lettering. All 50 cars were still on roster when merger took place with last of the cars retired in 1971.

A number of the Class P1 cars were equipped with a horizontal handrail at the edge of the roof during the 1920s. This was about the time bulk cement in railway construction work was becoming popular. Many railroads had devised angled brackets for hanging planks for wheelbarrows off the side of the roof that allowed the unloading of cement into wheelbarrows with men using the planks along the side of the car as a runway from the boxcar door opening to the construction site. Because of these cars having radial roofs, DM&N could not use angled brackets and it appears DM&N installed the handrails on the edge of a few of these cars to allow for a tie-off location for supporting these walkways.

Seven Class P1 box cars, 3158, 3181, 3190, 3195, 3224, 3229 and 3240, were reconditioned at Two Harbors in 1937 and assigned to Emeralite Surfacing Products who operated a quarry and rock crusher three miles west of Ely for loading emeralite used in the manufacturer of roofing materials. Emeralite is also called Ely greenstone.

Freight car construction had evolved by

All DM&N boxcars prior to 1923 were of double-sheathed construction. Ordered from ACF in 1923, 100 Class P2 single-sheathed box cars 3300–3399 would become the only single-sheathed boxcars on the roster. Most lasted until the early 1970s before being scrapped or placed in MofW service. Note the DM&N initials to the right end of the car and the road number to the left below the monogram. (American Car & Foundry Collection, John W. Barriger III National Railroad Library at UMSL)

After delivery the Class P2 box cars received sill steps below the door for ease of entry at the car door when in LCL service and also car puller tabs above the left truck on each side. 3350 is at Endion on July 18, 1957. (Bob's Photo collection)

1923 when the next group of 100 Class P2 40-ton 40-foot 9-inch steel underframe box cars 3300–3399 with 3,153 cu. ft. capacity were purchased from American Car & Foundry under Lot 9601. The purchase, was approved on AFE 677 on January 26, 1923, at a cost of $2679.33 per car. The first 34 cars were delivered in September 1923 with the remainder delivered in October and November. Door openings were 6-foot wide by 8-feet 7-inches. Constructed as single-sheathed cars, they were unique as all box cars purchased by DM&N prior to this order were of double-sheathed construction. Cars had fishbelly steel underframes, 7/8 Murphy ends and Youngstown steel doors. They were delivered with the DM&N ball monogram and car number stenciled towards the left side and the reporting marks towards the right side. All 100 cars were on the roster at the time of the merger with the last car being retired in 1973.

AFE 1329 was approved on March 4, 1926, approving the installation of sill steps under the side doors of 250 box cars. This allowed for easier access when cars were being used in LCL service. Cars rebuilt were one Class N1, 3064; 48 Class P 3100–3147; two Class P stock cars, 4002, 4003; 99 Class P1

Class P2 3336 at Proctor in June, 1972. Painted maroon during the 1960s, Scotchlite lettering for reporting marks and numbers were applied to metal placards and then applied to the car sides. (Ed Stoll)

box cars 3150–3177 and 3179–3249, and 100 Class P2 box cars, 3300–3399. The project was completed on January 15, 1927

DM&IR handled 620 cars of lime during 1964, generating $34,687 in gross revenue. Paper bulkheads were being used across the door openings on the Class P2 box cars in this service. Many times these were tearing and allowing spillage, which created a fire hazard when the lime was exposed to water. Application was made on AFE 9947 for $13,720 on May 13, 1965, to equip 13 Class P2 box cars with removable bulkheads constructed from 2 x 6 lumber and steel channels. This would eliminate the paper bulkheads and cars would be placed in assigned lime service. Cars 3309, 3312, 3314, 3315, 3322, 3331, 3341, 3357, 3374, 3380, 3388, 3396 and 3397 were converted between July 9, 1965 and September 22, 1965. Renewed plywood linings were applied at the same time. Two additional cars, 3312 and 3357, were approved on AFE 9946 for $585 on May 13, 1965, to have ¼-inch plywood linings installed for lime loading. Lime Service placards were added to lime service box cars starting November 25, 1963, at left of the door opening. These were eliminated on November 12, 1968. Most of these cars saw service hauling lime from Duluth docks to Saginaw for interchange to D&NE railroad and movement to Northwest Paper at Cloquet, Minnesota.

One of the more interesting sales was car 3368 sold on July 27, 1972 to Erie Mining at Hoyt Lakes, Minnesota. All were off the revenue roster by 1973.

DM&N VEHICLE CAR

As the iron range developed so did the infrastructure within the mining towns and locations. Soon highways appeared and the need to transport automobiles from Duluth to these communities created the need for specialized equipment to haul vehicles.

Two Class O 25-ton 40-foot double-sheathed vehicle cars, 100, 101, were built at DM&N Proctor Shops in 1905 with 3,250 cu. ft. capacities at a cost of $600 per car. Trucks and brake equipment were obtained from retired wood ore cars. Side door openings were 7-foot 6-inches wide by 8-foot high door openings that opened to the right. Cars were equipped with end loading doors measuring 7-foot wide by 7-foot 11½-inches high located on the B end of the car, with the high mounted brake wheel and staff mounted on the left edge. The underframe was wood with four truss rods, had Chicago couplers and 5-foot 6-inch wheelbase arch-bar trucks. During summer 1905 these cars were renumbered

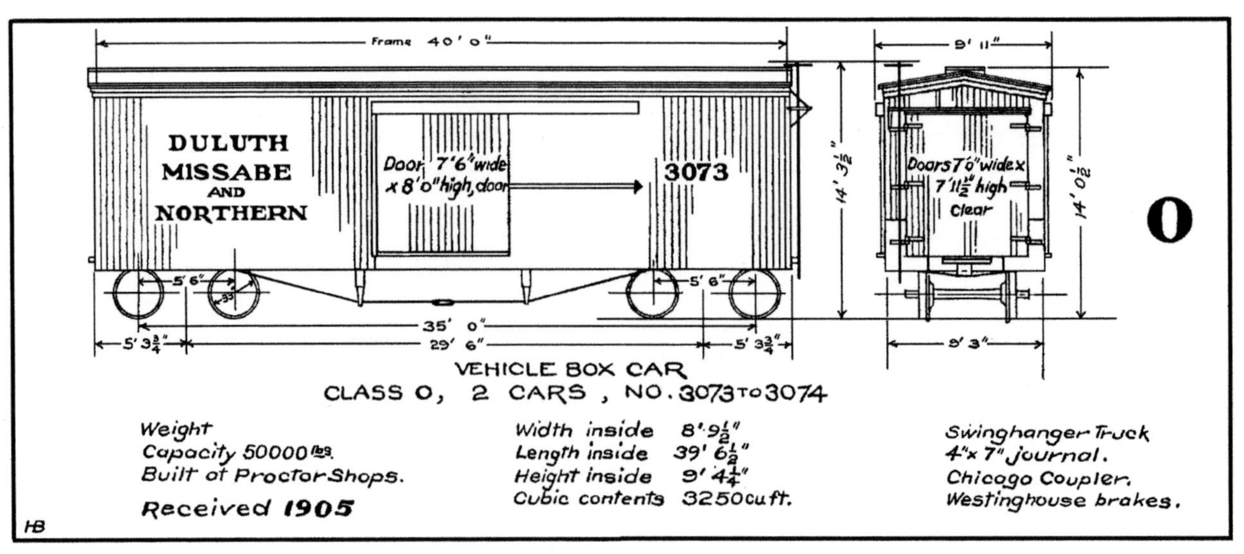

DM&N Class O diagram, cars 3073, 3074. (D.P. Holbrook collection)

to 3073, 3074. Cars were built with the stacked road name to the left of the door opening. It is unknown what color these cars were painted. Car 3073 was destroyed in wreck on PRR on November 15, 1920 and 3074 was destroyed on August 5, 1921.

The need for additional automobile cars caused the DM&N to select five double-sheathed cars from the 100 Class P1 cars built by Standard Steel Car in 1907 for rebuilding. Cars 3150–3154 were rebuilt at DM&N Proctor Shops in 1910 and equipped with 12-foot wide wood single sliding door which was centered with 3-foot to the left and 9-foot to the right of the center of the car. Class P1 was retained after rebuilding. This door opened to

Short-lived Class P1 DM&N 3150, shown in this builders photo by Standard Steel Car Co in 1907, was one of five cars, 3150–3154, rebuilt with double doors at Proctor in 1910 for automobile service. Note that the Class P1 stencil has yet to be applied to this car. The underframe is built up using C-channels for side and end sills and fishbelly center sills. Built before the application of the Safety Appliance Act, cars were delivered with no grab irons or sill steps on left end of car or ladders on ends. (Standard Steel Car, D.K. Retterer collection)

Once converted to automobile service Class P1 DM&N 3150 looked like this at Two Harbors on July 20, 1957. Monogram, reporting marks and number were placed to right of door to avoid being covered when doors were open. (Bob's Photo collection)

the left side of the door opening. During 1942 all five were rebuilt with two 6-foot wide by 7-foot 6-inch wood doors, with the right door centered on the carbody and the 6-foot left door located to the left of this door. The left door opened to the left side and the right side door to the right. Cars were rebuilt with Murphy steel underframes and Murphy 7/7 reverse corrugated steel ends by 1916. Originally built with arch-bar trucks with 5-foot 4-inch axle spacing, these were changed to 5-foot 6-inch U-section Andrews trucks between 1937 and 1939. All were delivered with the stacked Duluth, Missabe & Northern lettering. It is unknown if any of the cars were repainted with the large DM&N ball monogram to the left of the side door and stacked DM&N initial and numbers to right of the door. AFE 9484 was approved on April 25, 1963, to transfer 3150 and 3151 to work equipment service with the addition of prefix W to the numbers. All of the revenue cars were off the roster by 1967.

DM&N OIL CARS

After 1900, more and more oil was hauled by DM&N in barrels. Because of smell, leakage and the potential fire hazard when loaded in box cars it was determined that assigned service cars with adequate ventilation were needed. D&IR rebuilt box cars with small openings on the upper sides and ends to allow for ventilation. No photos or rebuilding documents have been located for the DM&N cars.

The first oil cars were rebuilt at DM&N Proctor Shops in 1907 from Class M 25-ton 34-foot boxcars. Numbered 4004–4011, eight box cars were chosen for rebuilding: 3013 (4004), 3011 (4005), 3020 (4006), 3021 (4007), 3008 (4008), 3038 (4009), 3004 (4010) and 3006 (4011).

Six additional cars were rebuilt in late 1908, 30xx (4012), 30xx (4013), 3012 (4014), 3024 (4015), 3037 (4016) and 3040(4017). All of these cars were stenciled "Oil Trade Service". It is unknown what additional modification were done to these cars, as DM&N diagrams only show these as notes on the diagrams for the Class M box cars which were built as double sheathed box cars. The 3000–3035 was originally built by Duluth Manufacturing Co. in 1892 and 1893, and the 3036–3042 was originally built by Pullman Car Co. as part of Lot 2173 in 1896. Both groups were built as 25-ton 34-foot box cars with 1,705 cu. ft. capacity, arch-bar trucks and Westinghouse air brakes. Cars 4007, 4009, 4014, 4015 and 4017 were rebuilt back to box cars during 1910 to 1912 and received their old box car numbers.

The last oil car, 4008, was destroyed on September 27, 1921, and retired per AFE 944 approved on March 11, 1924.

DM&N HART CONVERTIBLE GONDOLAS

DM&N purchased 59 Class Q 50-ton 41-foot 6-inches Hart Convertible gondolas 4026-4084 with 1,040 cu. ft. capacity in 1907 from Rogers Ballast Car Co. at a cost of $1,363.45 per car. Cars had the ability to be set up for self-clearing center dumping, and side dumping with the use of Lidgerwood powered plow. The cars ends could be dropped to allow the Lidgerwood plow to be pulled through multiple cars via cable, pushing the contents out through the side doors. By placing the cars in this number series it is apparent that they were intended for revenue service. It should be noted that a gravel spreader plow car numbered with non-revenue number R4085 was purchased from Rogers at the same time.

By 1923 only seven cars, 4035, 4038, 4044, 4052, 4053, 4075, 4079, still had bottom and side dump doors. Starting in 1938, DM&IR began converting all of the cars by removing the drop doors and drop ends and replacing them with solid bottoms and fixed ends. Arch-bar trucks were replaced by Andrews U-section trucks between 1937 and 1939. By 1943, 51 cars were true gondolas with only four cars, 4027, 4037, 4042 and 4059 having solid floors and ends but retaining the side doors.

Hart convertible gondolas were popular around the turn of the century, but unlike most railroads DM&N placed their 59 Class Q cars purchased in 1907 into a revenue number series 4026–4084. Cars had the ability to be self-clearing either by center or side dumping using a Lidgerwood-powered plow that was pulled through the cars. By 1943 51 cars had been converted to true gondolas. Outside truss underframes were rebuilt to traditional center sills on 4060–4070 by spring 1959. W4083 is shown at Proctor on July 17, 1957, and was one of two side dump gondolas left, the other being W4084. (Bob's Photo collection)

Hart cars converted to gondolas took on the appearance of a flat car with gondola sides added. Car 4037 is shown in the Missabe Jct. yard at Duluth on July 27, 1957. Note the tapered side posts held to the car by U-bolts. (Bob's Photo collection)

The outside truss underframes were rebuilt to traditional center sills on 4060–4070 by spring 1959. Car 4053 was destroyed on PRR in 1963 and retired per AFE 9674 on January 27, 1964, proof again that DM&IR cars did get off-line. All were off the revenue roster by 1971.

DM&N WOOD ORE CARS

The DM&N initially was built from Stoney Brook Jct., Minnesota to Mountain Iron. DM&N interchanged their ore with the Duluth & Winnipeg at Stoney Brook Jct. for movement by D&W from there to their Lake Superior docks at Allouez, Wisconsin. DM&N and D&W agreed to each supply 200 ore cars each to cover this movement. D&W never fulfilled this part of the contract. The worsening relationship with D&W caused the DM&N to extend their railroad into Duluth and build their own ore docks.

The first wood ore cars were ordered from Duluth Manufacturing in May 1892. Constructed in 1893, the 180 Class P 25-ton 22-foot ore cars 200-379 had 294 cu. ft. capacity. Sometime between 1893 and 1895 car 218 was rebuilt with different side bracing. The cars had a side truss built up with six vertical corner and side posts connecting the side sill (lower sill) and rail sill (upper sill) sections with diagonal Howe truss design bracing between the posts, with the middle section being spaced closer together. Unloading was made easy with 35-degree end and 70-degree side slope sheets extending to the top side rail of the car with two transverse mounted drop doors mounted in the center of the hopper bottom. They rode on 5-foot 4-inch wheelbase arch-bar trucks with 4 x 7-inch journals.

Before this order was delivered the DM&N placed an additional order in October 1892 with Duluth Manufacturing for an additional 575 cars, later increased to 801 Class O 25-ton 22-foot ore cars 400–1200 constructed during 1893 and 1894 with 295 cu. ft. capacity. Cars 800–879 were equipped with Fox trucks, all

Class P wood ore car 342 built in 1892 by Duluth Manufacturing Co. had the honor of being the first load of iron ore shipped on the DM&N. It is shown on display at Duluth Union Depot on October 18, 1892. (University of Minnesota Duluth, Kathryn A. Martin Library Archives and Special Collections, S2386 Box 29 Folder 4)

others with 5-foot 4-inch wheelbase arch-bar with 4x7-inch journals. These cars were an improved version of the Class P cars with 52-degree end and 70-degree side slope sheets extending to the top side rail of the car. The cars had a side truss built up with six vertical corner and side posts connecting the side sill and rail sill sections with Howe truss bracing between the second and third and four and fifth posts. The change in slope sheet angle caused the end slope sheets to be supported by the vertical support in sections one and five with a small flat platform area on either end of the car atop the top side rail.

Ore traffic continued to increase and an order was placed for 200 Class R 25-ton 22-foot ore cars 1201–1400 in 1895 with Pullman Car Co. and built under Lot #2102, with 300 cu. ft. capacity. These cars differed from previous designs and were only 7-foot tall from rail to top side rail. The side and end slope sheets were 52 degrees and only extended halfway up the carbody, with the remaining part of carbody vertical to the top side rail. Cars had side truss built up with seven vertical corner and side posts connecting the side sill and rail sill with no Howe truss type bracing. All rode on 5-foot 4-inch wheelbase arch-bar trucks with 4 x 7-inch journals.

An additional order was placed in May 1895 for 300 25-ton 22-foot ore cars to be built by Pullman Car Co. and 100 cars to be built by Duluth Manufacturing Co. This order was changed to 400 cars from Pullman and 200 cars from Duluth Manufacturing in June 1895. Early in 1896, Duluth Manufacturing was in financial trouble and the 200-car order was cancelled and the 200 cars were added to the Pullman Car Co. order in February 1896. Pullman built all 600 cars under Lot No. 2166 with 1401–1800 shipped during April and May 1896 and the 1801–2000 delivered in May 1896. Cars were all placed in Class S, had 305 cu. ft. capacity, cost $442.89 per car and rode on 5-foot 4-inch wheelbase arch-bar trucks with 4x7-inch journals. Car design was identical to the 1895 Pullman built Class R 1201–1400. These would be the last 25-ton capacity wood ore cars purchased by DM&N. An additional side raising timber was added in 1900 and 1901 on top of each side rail to increase the heaped capacity.

Continuing to be satisfied with Pullman built cars, DM&N ordered 400 Class T 35-ton 22-foot ore cars 6000–6399 from Pullman Palace Car Co. under Lot No. 2298 in late 1897 with 454 cu. ft. capacity. All cars were delivered by spring 1898 in time for the opening of the navigation season. Cars were equipped with Chicago couplers, Hinson drawbar attachments, Detroit springs, Westinghouse air

High air hoses had not been introduced in 1895. DM&N 1290 is shown in June 1895 at Pullman, Illinois, part of 200 25-ton cars in DM&N 1201–1400 series. This series would be placed in Class R by 1899. (Pullman Palace Photographs, Archives Center, National Museum of American History, Smithsonian Institution)

Financial complications with Duluth Manufacturing Co. and its subsequent plant closing, DM&N in 1895 ordered 300 wood ore cars from Pullman and 100 from DMC. The order was increased to 400 and 200 respectively but early in 1896 DMC went into bankruptcy and DM&N ordered all 600 cars from Pullman. DM&N 1401–2000 were built between December 1895 and June 1896. This series would be placed in Class S by 1899. (Pullman Palace Photographs, Archives Center, National Museum of American History, Smithsonian Institution)

Another 35-ton capacity car, DM&N 6209 was delivered from Pullman Car Co. in 1898 as part of 400 Class T cars 6000–6399. Standards were not established until 1900 for the side-raising timbers, with this group being delivered without them. Note the built date is not stenciled on this car but does appear on DM&N 7289. (Pullman Palace Photographs, Archives Center, National Museum of American History, Smithsonian Institution)

Larger 35-ton capacity wood ore cars would begin arriving in 1899 and continue until 1900 totaling an eventual 1,700 cars. DM&N would also rebuild a handful of older pre-1899 cars to this design at Proctor shops. DM&N 7289, built in 1900 as part of Class T 6900–7399 series, has the side-raising timbers above the top frame member. Steel ore cars would become standard in 1903. (Pullman Palace Photographs, Archives Center, National Museum of American History, Smithsonian Institution)

brakes, cost $439.87 per car and rode on 5-foot 4-inch arch-bar trucks with 4½ x 8-inch journals. Cars were taller than previous ore cars by almost a foot but had the same basic design as the Class S cars and reintroduced the Howe Truss type bracing between the vertical posts.

Pullman did not have the capacity to build the next order so DM&N ordered 500 Class T 35-ton 22-foot ore cars 6400–6899 from American Car & Foundry in April 1899 with 454 cu. ft. capacity. All were delivered late in 1899 and were identical to the Pullman Class T cars built in 1898.

DM&N ordered 500 Class T 35-ton 22-foot wood ore cars 6900–7399 from Pullman Palace Car Co. under Lot No. 2516 in November 1899 and built between May and June 1900 with 454 cu. ft. capacity. The order was increased to 800 Class T ore cars in January 1900 with these cars being built between June and July 1900. The additional 300 cars were identical in appearance and numbered 7400–7699. Cars were equipped with Griffin wheels, 5-foot 4-inch wheelbase arch-bar trucks with 4½ x 8-inch journals, Westinghouse automatic brakes, National hollow brakebeams, Chicago couplers, Hinson drawbar attachments, Detroit No. 46 springs and Barber roller side bearings. Both groups cost $675.87 per car.

During 1900 to 1903 two Class P cars, 202,

364; three Class O, 783, 1080, 1164; one Class R, 1329; and two Class S, 1422, 1941 cars were rebuilt to Class T ore cars. Class O car 684 was rebuilt to a Class T type ore car to replace car 6172 but was not renumbered. During 1905 all eight were renumbered 7700–7707 to keep them in the newest number series. The first car rebuilt was 1422. Unable to purchase new steel ore cars quick enough to fulfill the increasing ore shipments, DM&N rebuilt 1700 wooden ore cars, 6000–7699 from 1905 to 1907 allowing them to be used for another 10 years.

Cars built prior to 1900 had an additional 6 x 8-inch sideboard, called a side-raising timber, added to the top of the side rail of the car to increase capacity starting in 1900. This allowed a larger heap to be placed in the car with 186 cars completed in 1900, 397 cars in 1901 and 126 cars in 1902. This completed the modification to all cars in the 6000-7699 and 7700-7707 series.

Class O, P and Q cars were virtually identical. Bodies and framing were of wood construction with hopper bottom doors and truss rod supports. The four wooden doors had no mechanical elevating mechanism and were held closed by a crowsfeet bar. All were equipped with spring draft gear, link and pin couplers, cast wheels, wooden bolsters and air brakes. The link and pin couplers were quickly replaced by automatic couplers.

Class R and S cars were built with lower sides but were of the same capacity as Class O, P and Q cars. Framing was more substantial with more steel and cast iron used for bolsters, door hinge supports, center crossbeams, corner bands, and casting pockets for ends braces and stakes. All were delivered with automatic couplers.

Class T was a larger capacity car then Class R and S. Bottom doors were larger and more steel was used in their construction. These cars were at the limit of capacity for construction of wooden ore cars.

During 1902 the DM&N leased 350 wood ore cars to D&IR to assist in handling their ore traffic without purchasing any additional cars.

As wood ore cars began to be replaced some were sold off for secondhand use including: three cars to Zenith Furnace Co. of Duluth on July 13, 1903, ten to Minnesota Land & Construction Co. in May 1902, four to Powers and Simpson Co. in 1906, 15 to F.M. Pease in August 1902, 982 cars to F.M. Hicks of Chicago and 321 to Hicks Locomotive & Car Works in 1906, one car to Zenith Furnace Co. in 1907, six cars to South Dakota Central Railway in 1907, 20 cars to Bayfield Transfer Railway Co. in May 1907, ten to Mahoning Ore & Steel in 1908, and two cars to Utica Mine in 1909. DM&N dismantled 150 cars and used parts for constructing 50 flat cars, 2230–2279, in 1902-03 and 100 flatcars, 2280–2379, in 1904-05.

DM&N AFE 937 was approved on March 10, 1924 for the sale of wooden ore cars 6189, 6408 and 7373. All three of these had been placed in non-revenue company service in late 1919. This ended the wooden ore car era on DM&N.

DM&N IRON ORE CARS

Along with the first 180 wood ore cars ordered from Duluth Manufacturing, an order for 20 Class Q 25-ton 22-foot ore cars 380–399 of iron construction with 294 cu. ft. capacity was made in May 1892. They were delivered in 1893 with Diamond trucks. Cars had 35-degree end slope sheets and 70-degree side slope sheets when built, making them easy to unload.

During the 1850s gas-pipe iron freight cars were developed by Bernard J. LaMothe. He had designs for flat, box and refrigerator cars. Between the late 1860s and the late 1880s a number of companies built iron pipe constructed freight cars including the United States Tube Rolling Stock Co. The successor, Iron Car Co. took over Minnesota Car Co. works in Duluth in October 1889, renaming it Minnesota Iron Car Co. At that time the intent was to build another 1,500 of the iron cars

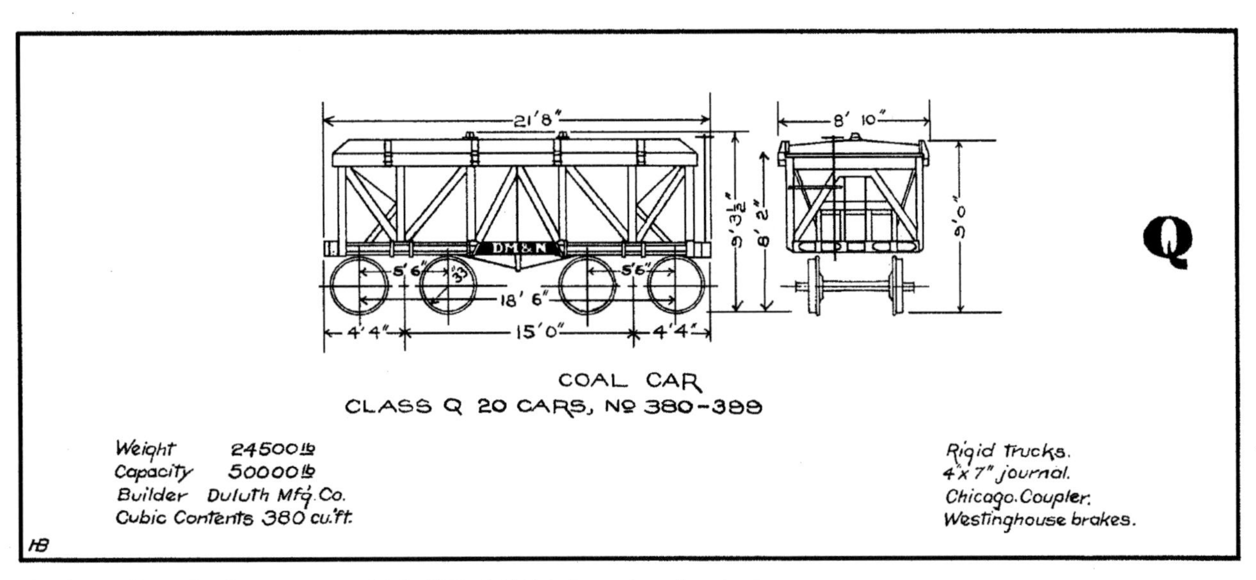

DM&N Class Q diagram, cars 380–399. (D.P. Holbrook collection)

at the Duluth plant primarily for private car companies. Short-lived, the Minnesota Iron Car Co. was sold to Duluth Manufacturing Co. in August 1891.

It appears that Duluth Manufacturing may have used the Minnesota Iron Car Co. designs for an ore car and produced the 20 DM&N 25-ton iron ore cars.

All of the cars, except 388, were converted to wooden construction by May 1895. Prior to January 1900, cars 387 and 388 were rebuilt to Class O ore cars. By 1901, 387 and 388 had been retired and all 18 remaining cars placed in coal service. The last eight cars were sold in June 1904.

DM&N STEEL ORE CARS

The first steel ore cars, five Class U 50-ton 22-foot cars 8000–8004 with 668 cu. ft. capacity were purchased from Schoen Pressed Steel Car in 1899 at a cost of $902.40 per car. The small order was placed to allow DM&N to determine if these "new" steel cars could stand up to the rigors of iron ore service and were consider experimental. The cars were a radical design change with the complete hopper and underframe constructed of steel. Design of these cars mimicked Pressed Steel Co. coal hoppers from the era, with 8-foot of the middle section of car removed. Hopper doors were wood and equipped with a door trapping mechanism for operating the doors. They had 6 T-section side ribs and pressed steel fishbelly side sills, center sills and slope sheet bracing. Unique for ore cars, they had two hopper bottom doors much like coal cars from the era, but these were narrower dumping only between the rails. Each door was only 2-foot 10½-inches square. D&IR purchased 50 identical cars numbered 3050–3399. Most of these cars would have a short service life on DM&N. 8002 was sold to the State of Minnesota and converted to a state scale test car in January 1908. Cars 8001 and 8004 were sold to NP in November 1909 and 8003 to NP in October 1911. Car 8000 was placed in work service in January 1916.

Considering the performance of the first five steel cars, modifications were made to the design and the DM&N placed their first large order for steel ore cars. A total of 350 Class U1 50-ton 22-foot ore cars 8005–8354 with 680 cu. ft. capacity were ordered from Pressed Steel Car in 1903 at a cost of $1107.34 each. D&IR ordered 450 identical cars numbered 3450–3899. They were built wider and lower than the Class U cars. Framing was heavier with two pairs of double drop bottom steel doors with an opening of 3-foot 2-inches by 3-foot 6-inches, creating a larger opening than the

The first partial side sheet 50-ton ore cars were 300 Class U3 cars 8405–9704 built by Standard Steel Car in 1905. Transverse mounted pairs of drop doors were used to bottom dump the cars and brake wheel type wheels were used to operate the doors from the side of the car. Built with single sill steps and a vertical grab iron, cars would be equipped with side by side sill steps on each end to make it easier for trainmen to ride the car. (Standard Steel Car Co., D.K. Retterer collection)

1899-built cars. These doors became the standard ore cars doors until longitudinal doors were developed in 1909. They were equipped with MCB couplers with friction draft gear and standard journals 5½ x 10-inches, larger than the Class U cars. Trucks were split evenly between Standard Steel and Pressed Steel arch-bar trucks. They had six L-section side ribs with 6 half-height diagonal L-section ribs. Side sheets only came halfway down the side instead of full-height sides of the 1899 order. This allowed easier access to door operating mechanisms. The K-brake equipment was relocated to the side just below the side slope sheet to allow easier maintenance. The cars did retain the pressed steel fishbelly side sill from the 1899 order. These were the first cars with door opening and closing mecha-

The usefulness of the Class U3/U4 and D3/D4 ore cars is depicted in this ballast dumping scene on the Missabe Division during the summer of 1950. The MofW employees are using the dumping mechanism wheels on the side of the cars to control the ballast flow out of the doors. Note the white dot with black line in the upper left corner of the first car. On the black line and not visible is the word"weld" indicating that this car had welded sloped sheet extension inside the car to prevent leaking of taconite pellets. Many of these cars were used for shipping taconite pellets from Reserve Mining's Silver Bay, Minnesota, plant before they had completed their docks at Silver Bay. (D.P. Holbrook collection)

The 600 Class U3 50-ton ore cars delivered by Standard Steel Car Co. in 1907 introduced a modification to the carbody replacing the two center L-section ribs with C-channels. This was done to strengthen the side sheets and prevent bulging of the carbody. (Standard Steel Car Co., D.K. Retterer collection)

nisms controlled by brake wheel type wheels located along the side adjacent to sill steps at a height allowing easy movement by a standing ore dock employee. They had a light weight almost two tons heavier than Class U cars. Initially, when delivered, these cars could not be used in solid trains since the bridge over St. Louis River was not safe for such loads. It was rebuilt during winter 1903-04 and solid trains of the Class U1 cars began to be operated.

Appropriation 1831 was approved on July 7, 1927, to equip six cars: 8022, 8076, 8096, 8128, 8161 and 8177, with extended sides and steel covers for handling coal screenings from Duluth docks to the power plant at Proctor. Conversion began on February 1, 1927, and were completed on June 1, 1927. Another six cars, 8021, 8071, 8095, 8115, 8156 and 8157, were also converted under AFE 1643 between October 3, 1927, and March 16, 1928. It is unknown what color they were painted but they were delivered with the stacked DM&N name spelled on the carbody side. All cars retained their DM&N numbers after the merger with all retired by 1949.

Under U.S. Steel's corporate umbrella, the D&IR ordered 100 50-ton 22-foot ore cars with 682 cu. ft. capacity from American Car & Foundry under Lot 2517 on October 15, 1902, at a cost of $1107.34 per car. After the order was placed, delivery was split between DM&N and D&IR. Fifty of these cars were delivered to DM&N in 1903, placed in Class U2, and numbered 8355–8404. The other 50 went to D&IR and numbered 3400–3449. The cars had two pairs of double drop-bottom type dump doors 2-foot 8-inches by 3-foot 5-inches that closely followed those on early wooden ore cars. ACF designed these as a shortened version of a coal hopper replacing the coal hopper bays with four drop doors. The car body had two U-section ribs located above the bolsters and four equally spaced T-section ribs all riveted to the carbody. Unique on these all steel cars was a wooden end sill reinforced with a U-channel. All were off the roster by 1944. It is unknown what color cars were painted but they were delivered with the stacked DM&N name spelled out on the carbody side. All were off the roster by 1937.

Satisfied with their performance, DM&N began a serious effort in 1905 to replace the aging wooden ore car fleet by placing three orders with Standard Steel Car Co. The 1905 group was 1,300 cars numbered 8405–9704 costing $899.60 per car. The 1906 group was 250 cars numbered 9705–9954 costing $1052.57 per car. The 1907 group was 600 cars numbered 9955–10554 costing $1037.91 per car. All were Class U3 50-ton, 22-foot cars, with 689 cu. ft capacity. Bottom drop-door openings were 6-foot 8-inches long by 3-foot 6-inches wide.

The carbody was modified from earlier designs and had four rolled steel U-section ribs, two located above the bolsters. The remaining rib pattern consisted of two L and two U-section ribs equally spaced between the end ribs. Cars had what would become standard on most future ore cars, a half-height carbody with slope sheets intersecting the sides to create a tub. Between end sills and end slope sheets there was a diagonal L-section rib for additional support. Door operating hand brake type wheels were located adjacent to sill steps on both sides of the car. K-brake equipment was located below the side sill between trucks allowing for easy maintenance. Cars were equipped with New York Car Wheel Co.'s T.M. special 700-pound wheels, arch-bar trucks, Simplex bolsters, Westinghouse 10 x 12-inch cylinder brakes, SSC brake beams, Chicago and Climax couplers, Westinghouse

Former Class U4 DM&IR 10668 has been placed in work service denoted by the W prefix. The letter D is the stencil group denoting the year car is supposed to have air brakes tested. (Owen Leander, John C. La Rue, Jr. Collection)

Showing the chain-connected transverse paired drop doors, Class U4 50-ton ore car 11424 is at Endion on July 19, 1957. Note the double sill steps. The white circle with the word "weld" is located to the left and right of the U-channel ribs indicating the car had the extensions added to the slope sheets to prevent leakage of iron ore and taconite pellets. (Bob's Photo collection)

drawbar attachments, with friction draft gear, Railway Steel Spring Co. truck springs, McCord and Symington journal boxes, Camel journal bearings and Woods and Susemihi side bearings. Two 15-inch channels were used for center sills.

AFE 3031 was approved July 1, 1930, for equipping all Class U3 ore cars with additional end platforms, making a wider place for ore punchers to stand. These cars retained DM&N numbers after the merger; all retired by 1959.

Also during 1906, DM&N ordered 750 Class U4 50-ton 22-foot ore cars 100–849 with 689 cu. ft. capacity from Pressed Steel Car costing $1052.57 per car. They had half height side sheets with U-section ribs above the trucks and 4 L-section ribs equally spaced between these ribs to support the side sheets. K-brakes were relocated from directly below side sheets to below side sill at the same location. The standard 4 drop doors were operated by brake wheel type wheels located adjacent to sill steps. The cars were broken up into groups with various types of center plates, side bearings, couplers, journal boxes and side bearings. All were equipped with arch-bar trucks except 600–629 which had Andrews L-section trucks. Cars would retain their DM&N numbers after the merger with all being retired by 1959.

Happy with the initial 1906 order, DM&N ordered an additional 1150 Class U4 50-ton 22-foot ore cars 10555–11704 with 689 cu. ft. capacity from Pressed Steel Car in 1907 at a cost of $1053.28 per car. Cars were of identical construction as the 100–849 series built in 1906. Cars were equipped with arch-bar trucks except 10555–10629 and 10655–10679 which were equipped with Andrews L-section trucks. All were equipped with Westinghouse draft gear except: 10555–10604, Sessions draft gear; 10605–10629, McCord draft gear; and 10630–10654, Waugh draft gear. Various types of couplers, journal boxes and wheels were used on these cars, the most interesting being P.R.R. wheels on car 11652 and four Griffin and four P.R.R. wheels on car 11653. The

D&IR and DM&N both sampled ACF 50-ton Clark ore cars. DM&N purchased 100 cars 12000–12099 in 1910. Built to a new design with flat riveted plate steel sides and sloped corner posts, cars were equipped with a special design longitudinal rolling doors that formed the hopper bottom. Cars had bulging side sheet problems and all were retired by 1919. (American Car & Foundry Collection, John W. Barriger III National Railroad Library at UMSL)

standard 4 drop doors were operated by brake wheel type wheels located adjacent to sill steps. All future cars would eliminate the wheel and have a standard trapping shaft operated by wrench or trapping machine. They were painted with two coats of Illinois Steel Co. No. 845 black paint, including trucks. These cars would retain their DM&N numbers after the merger with all retired by 1959.

Correspondence during November 1928 indicated that many of the Class U1, U2, U3 and U4 cars were in need of heavy repairs. It was also noted that all four classes had small transverse doors and were difficult to unload.

Other builders started to offer their own versions of 50-ton ore cars and DM&N would sample cars from three builders in 1910. An order for 100 Class U5 Clark-type ore cars 12000–12099 was placed with American Car and Foundry under Lot 5809B in 1910 at a cost of $1111.59 per car. One hundred identical cars were built for D&IR on Lot 5809A in 1910 and numbered 11400–11499. Cars were 50-ton 22-foot cars with 660 cu. ft. capacity equipped with arch-bar trucks. Built to a new design with flat riveted plate steel sides and sloped corner posts supporting the end slope sheets, all bracing was internal to the carbody design. These cars would have a short service life on the DM&N. A special design of longitudinal rolling doors 6-foot 6½-inches by 5-foot 7-inches formed the hopper bottom. Starting with this order all future cars would have lance (poke hole) openings in the sides for steaming of the ore when it became frozen in the cars. Painted black, cars were delivered with the stacked DM&N spelled out above the road number on the sides. The interior framing and door mechanism were not strong enough to withstand the rigors of iron ore service which caused cars to develop bulging sides. These cars had a short service life of eight years.

The second 1910 order for 150 Class U6 50-ton 22-foot ore cars 14000–14149 with 689 cu. ft. capacity was placed with Western Steel Car & Foundry to be built at their Chicago plant at a cost of $1080.69 per car. They were equipped with arch-bar trucks. These were the first successful large longitudinal door ore cars with an opening of 6-foot 6½-inches by 5-foot 7-inches. Doors opened downward from pivots on the side sills. Intermediate side sills were eliminated and side sills became necessarily more complex including the connection between them and the draft sills. Corner posts were set back to the carbody; this created a center sill and draft gear that extended past the carbody. These were the first cars received from builders with rolled steel wheels. Painted

Class U7 DM&N 16715 one of 750 50-ton steel ore cars 16000–16749 built by Standard Steel Car in 1910 in as-delivered paint. Photo circa late teens. Note the folded over top, corner, and end sheets riveted on the inside of the carbody. (D.P. Holbrook collection)

black, the cars were received with the stacked DM&N spelled out and the car number stenciled below. Correspondence during February 1929 indicated that 75 of the 150 cars had buckled with two being destroyed. They required constant light repairs to keep them in service. Authorization was received on AFE 1917 on March 31, 1929, to begin to replace side sills and slope sheets to reinforce the cars. The cars would retain their DM&N numbers after the merger with all retired by 1958.

The third 1910 order for 750 Class U7 50-ton 22-foot ore cars 16000–16749 with 667 cu. ft. capacity was placed with Standard Steel Car Co. at a cost of $1086.80 per car. Equipped with arch-bar trucks they had bottom door openings of 5-foot 6-inches by 8-foot 2-inches. This group of cars integrated the side sill construction into an entire riveted plate steel carbody extending from the side sill to the top chord of the car. The cars had 5 small circular openings on the L-side and 4 small circular openings on the R-side of the car to access brake components and door mechanisms. Equipped with arch-bar trucks, 500 cars had Carmer pin lifters and 250 had Duplex Auto Uncoupling pin lifters. Cars were painted black and delivered with the stacked DM&N spelled out with the road number below this lettering. The opening of the Oliver Iron Mining Trout Lake Washing Plant at Coleraine, Minnesota, in 1909 caused rapidly expanding mining operations on the west end of the Mesabi iron range. Unable to keep up with demand for cars to haul iron ore from the pit to the washing plant, U.S. Steel, owner of both DM&N and OIM, directed that 170 of the Class U7 cars be sold to OIM in 1916 to quickly resolve the car shortage. OIM renumbered these cars into the 5000 series. All DM&N cars would retain their DM&N numbers after the merger with all retired by 1953.

Satisfied with the Western Steel Car and Foundry cars 14000–14149 built in 1910, DM&N placed an order in 1913 for 1,000 Class U9 50-ton 22-foot identical ore cars 14500–15499 with 669 cu. ft. capacity costing $1148.57 per car. These cars had Andrews T-section trucks which were an improvement over Class U6 arch-bar trucks. Bottom door size was increased to 6-foot 6½-inches by 5-foot 8½-inches. Side sills were riveted to subsills with four vertical supports above the side sill supporting the side slope sheets. Unique on these cars were two of the four vertical supports being closely spaced together in the middle of the car. They were painted black and delivered with the stacked DM&N spelled out with the road number below this lettering.

By 1929 almost half of these cars were experiencing the same buckling problems as Class U6. Authorization was received on AFE 1917 on March 31, 1929, to begin to replace side sills and slope sheets to reinforce the cars. These cars would retain their DM&N numbers after the merger, with all retired by 1965.

Superintendent of Motive Power and Cars, C.W. Seddon notified President W.A. McGonagle on November 20, 1926, of the condition of the DM&N ore car fleet. He noted that Class U1, U2, U3 and U4 cars should receive

Class U9 DM&IR 15162 is at Proctor on July 17, 1957 with the stencil group C having just been applied at Proctor car shop. Chalk notation on the side indicates that once load had been transferred, which has been accomplished, this car was to "Store Medium Heavy" indicating that it probably would not be returned to revenue service. (Mainline Photos, J.M. Gruber collection)

minimal maintenance and repairs because of the small transverse doors and the difficulty in unloading these cars. The U7 cars were slowly being disposed of. The U9 cars were experiencing buckling of frames on a regular basis and he suggested that heavy repairs should be scheduled. Seddon observed that the buckling was occurring where the side sills were cut away to allow for the gear wheel for the door dumping mechanism. He designed a repair that involved reinforcing the side sill by creating a three-quarter box section attached to the side sill channel. Strengthening the connection between the draft and side sills would be accomplished by attaching plates to prevent the spreading of the side sills. Average repair costs of $265 per car would increase by $150 with this rebuilding. He felt that these repairs would extend the service life of these cars by 15 to 20 years. This process was approved and between 1929 and 1934 most of the U9 cars received these modifications along with new side sheets and slope sheets.

The DM&N began testing a prototype Rakowsky car during the summer of 1908. This testing showed various weak points in the car design. The first of the redesigned cars were delivered to DM&N in 1909. The 50-ton 22-foot car had a capacity of 700 cu. ft. capacity and had a light weight of 33,000 lbs. The slope of

Overall view of Class U9 50-ton ore car DM&IR 14613 at Chisholm, Minnesota, July 9, 1984. Note the center patch and the left diagonal patch repairing the side sheets. (Bruce Meyer)

Gustav A. Rakowsky designed a unique 50-ton steel ore car in 1909 that was manufactured by the Automatic Unloading Car Co. division of Northwestern Steam Boiler Co. of Duluth. The initial design was modified from an outside steel frame to a sheet steel side as depicted in this photo of DM&N 18001, one of five purchased in 1909. (D.P. Holbrook collection)

the dumping surfaces was 52½ degrees. Testing was complete in 1910 and a decision was made to sample this type of car. DM&N ordered five of the Rakowsky-type cars per Appropriation 255 on July 17, 1910, for 5,390.00. Placed in Class U8, 18000–18004 were delivered in July 1911. Cars were lettered with simple DM&N reporting marks with periods after the initials. Heavy, unrestricted use of these cars showed numerous weaknesses; most notable was the inability of the clamshell carbody to close properly. Car 18004 was retired in November 1915 and the remaining 4 cars were retired in January 1916.

DM&N returned to Western Steel Car & Foundry in 1914 and placed an order for an additional 1000 Class U10 50-ton 22-foot ore cars 19000–19999 with 704 cu. ft. capacity from their Hegewisch, Illinois, plant at a cost of $1045.80 per car. They had Andrews trucks with a bottom door opening increased in size to 6-foot 4-inches by 6-foot 3½-inches. These were the final development in 50-ton ore cars. Side sills were made in a box section giving it greater strength. Construction was more substantial, sharp corners in the hopper were removed, body bracing more rigid, improved truck side frames, draft gear and stronger draft rigging. Corner posts were set back from ends of draft gear and carbody was half height plate steel with side and end slope sheets riveted to sides. Cars had integrated T-section corner posts above the truck bolsters with two additional side posts supporting the side slope sheets. They were painted black and delivered with the stacked DM&N lettering spelled out and car number located directly below this lettering. These cars would retain their DM&N numbers after the merger. Cars had a long service life, with the last being retired in 1965.

A single 50-ton 22-foot demonstrator ore car, with 654 cu. ft. capacity, was purchased by DM&N from Ralston Steel Car Co. in 1916. Numbered 20000, this car cost $1,283.00 and was never given a car class by DM&N. Built on May 13, 1915, it demonstrated on other carriers but was not acquired by DM&N until 1916. Carbody was constructed of a plate steel side sheet supported by four U-section ribs, all of riveted construction with the corner and end posts set back from the end of the draft gear. Side sills were composed of C-section channels turned inward to provide a flush surface for the side ribs. It was equipped with Carmer cut levers and rode on Andrews

Iron ore and cold temperatures did not mix well. To lengthen the ore shipping season steam was used to thaw the ore before dumping on the ore docks, causing DM&IR to roster and use steam until 1960. This process was called steaming ore. As many as 30 steam engines were in service during November to thaw the ore just at Proctor. Engines would be attached to steam lines and lances would be placed in lance openings on the side of ore cars to thaw the ore, which once thawed would quickly be taken to the docks for dumping. Infrared facilities were installed in 1960, thus ending the need for steam thawing of ore. Thawing also required an army of employees for placing the lances in and out of the cars. These scenes are at Proctor in the fall of 1950 showing Class U10 19551 and Class E11 42860 and sisters being thawed. (Two photos, W.C. Olsen)

T-section trucks. Painted black it was delivered with stacked DM&N lettering spelled out on the carbody with number below. It was retired in 1965.

Continuing to be satisfied with Western Steel Car & Foundry, the DM&N returned again in 1916 and placed an order for 1,000 Class U11 50-ton 22-foot ore cars 20001–21000 with 704 cu. ft. capacity at a cost of $1,217,624.41. Cars were a modified version of the Class U10 built in 1914 by having the tare weight reduced by modifying the side sheets, cutting them back to the inside slope sheets and creating a distinctive carbody. These cars would retain their DM&N numbers after the merger, with all retired by 1965.

By 1916, 50-ton steel ore cars had replaced all of the wooden ore cars. The few remaining wood cars were kept in coal and ballast service from 1916 to the fall of 1919 when all remaining wood cars were retired.

AFE 913 for 450 ore cars and AFE 914 for 50 ore cars were approved January 24, 1923, both for 50-ton cars. This order was

One thousand Class U10 50-ton steel ore cars 19000-19999 built by Western Steel Car & Foundry were delivered in 1914 differing from the Class U9 cars by having the side sill extend to the corner post. Additional modifications from the Class U9 cars were larger poking platforms on each end. (Western Steel Car & Foundry, D.K. Retterer collection)

Ralston Steel Car Co. built a single demonstrator 50-ton ore car in 1915. Acquired by DM&N in 1916 it was renumbered DM&N 20000. The demonstrator did not develop any orders and was retired in 1965. (D.P. Holbrook collection)

To reduce tare weight the side sheets were modified from a rectangle to a sloped side sheet on 1000 Class U11 50-ton ore cars purchased from Western Steel Car & Foundry in 1916. Good in theory, these cars were susceptible to bulging side sheets and future cars would return to the rectangular side sheets. DM&IR 20881 is at Proctor on July 17, 1957. (Bob's Photo collection)

DM&N placed orders in 1914 for Class U10 19000–19999 and 1916 for Class U11 20001–21000 from Western Steel Car and Foundry to replace the wood ore car fleet. Class U10 would have rectangular side sheets, but the Class U11 cars, in an effort to reduce the tare weight would have the side sheets cutout in line with the slope sheets. DM&IR 20714, Virginia, Minnesota, June 5, 1959. (Joe Douda)

cancelled when a decision was made to rebuild instead of purchase new and develop standards for a larger type ore car. Many of the 50-ton ore cars were rebuilt with new side sills and heavier gauge steel slope sheets between 1924 and 1947.

DEMONSTRATOR AND HIGHER CAPACITY ORE CARS

Many of the car builders had demonstrator 22-foot ore cars with capacities between 85 and 95 tons out on various railroads during the early 1920s, trying to attract car orders. American Car and Foundry and General American Car Co. built single sample ore cars, ACF in 1922 and GACC with an unknown built date. GACC and DM&N records are incomplete concerning the GACC demonstrator built date. Standard Steel Car would build seven different demonstrator ore cars to two different designs in 1925.

ACF demonstrator 90-ton ore car, ACFX 9001, built at Detroit on June 23, 1923, under lot number 9532, was tested on the Soo during late summer and early fall of 1923. It appears that this car was reconditioned by ACF under Lot number 1049 in 1929 as a 95-ton capacity ore car with 935 cu. ft. capacity. The car was purchased by DM&N for $2500 on AFE 2049 approved September 9, 1929. Delivered in November 1929, it was renumbered 22000, placed in Class U12 with a light weight of 49,300 lbs. Late in 1931 this car was renumbered to 21026 following identical cars 21001–21025. It was repainted black after delivery with DM&N ball monogram. It retained the DM&N number after merger and was retired by 1966.

Pullman built a 90-ton capacity sample Pullman-Commonwealth-type test ore car 5501 in 1922 on Lot number 5322. The car demonstrated on various carriers, but was later assigned to DM&N. It was not carried in inventory by DM&N, but was sold for scrap to Duluth Iron and Metal on August 22, 1934. It is unknown if car was retired becausse of damage, or design flaws.

The GACC demonstrator 95-ton ore car, GATX 1000, was purchased for $2500 on AFE 2323 approved September 10, 1930. It is unknown when GACC built this car. Delivered to DM&N in November 1930, car was renumbered 21999 and placed in Class U13. Car had 965 cu. ft. capacity and was equipped with Pflager 4-foot 2-inch wheelbase trucks, painted black and stenciled with the DM&N ball monogram. Late in 1931 this car was renumbered to 21126 after the 21101–21125 series and placed in Class U13.

During the late 1920s and early 1930s DM&N tested a variety of ore cars. ACFX 9001, built in 1923, was acquired by DM&N in 1929, numbered 22000 and then renumbered to 21026 in late 1931. (American Car & Foundry Collection, John W. Barriger III National Railroad Library at UMSL)

ACF and GACC both introduced single demonstrator ore cars in 1929 and 1930. Standard Steel Car Co. had previously released seven 85-ton capacity demonstrator cars with two car designs and two styles of trucks. Five cars, STDX 23171–23175 built with Pflager trucks and two cars STDX 23176 and 23177 with Verona trucks were built in 1925 for testing. Cars were returned to SSC in 1929. Shortages of ore cars caused DM&N to purchase these seven cars in February 1931 and renumber them DM&N 23171–23177. After 1960 all were assigned to company coal service. In 1964 23176 and 23177 would be sold to Keokuk Electro Minerals and during 1965 23171–23175 were sold to C&NW and renumbered 120090–120094. Note the distinctive Verona trucks on STDX 176. (Standard Steel Car Co., D.K. Retterer collection)

Three years after ACF, Pullman and General American supplied their 90- and 95-ton cars, Standard Steel Car Co. built seven 85-ton 22-foot sample steel ore cars in 1925. They were tested by various railroads including DM&N. DM&N returned these cars and they were stored at Hammond, Indiana, in 1929. Supt. Motive Power & Cars C.W. Seddon was sent to Hammond on January 24, 1931, to inspect the cars for possible purchase. Inspection determined that outside of being rusty all five cars could be returned to service with very little expense. Five had 1,080 cu. ft. capacity, equipped with Pflager cast steel trucks and numbered STDX 171–175. Two had 1,055 cu. ft. capacity, equipped with Verona cast steel trucks and numbered STDX 176, 177. The Verona trucks were unique at the time in having truck-mounted brake assemblies. All seven cars were purchased in February 1931 on AFE 2448 at a cost of $4,780.

They were repainted black with white stenciling and DM&N ball monogram at Hammond before movement. DM&N received them on April 9, 1931, from Omaha Ry. at South Itasca, Wisconsin. DM&N renumbered former STDX 171–175 to DM&N 23171–23175, STDX 176, 177 to DM&N 23176, 23177 and placed them in Class U16. Cast iron wheels on 23171–23175 were replaced with steel per AFE 2813 approved October 18, 1933. Capacity was increased to 100 tons in January 1938 and reduced to 70 tons in early 1940. After 1960, all five were assigned to company coal service between Duluth docks and Proctor. Retirement was approved on

Axle loadings had improved enough by the early 1920s that 50-ton loading for ore cars had become obsolete and 70- to 100-ton capacity cars began to be developed. Pullman, ACF and GACC all built demonstrator 90-ton capacity ore cars that toured the various iron ore railroads in the Midwest. ACF's demonstrator car was successful enough to cause DM&N to place an order for 25 Class U12 95-ton capacity ore cars 21001–21025 in 1924 with all delivered in 1925. Cars were unique in having plate steel sides supported by two plate steel ribs offset in the center of the car with sloping end and corner posts. (American Car & Foundry Collection, John W. Barriger III National Railroad Library at UMSL)

AFE 9505 on June 5, 1965. Cars 23176 and 23177 were sold on May 29, 1964 to Keokuk Electro Metals Co. and the remaining five cars were sold to C&NW on AFE 9952 August 25, 1965, numbered CNW 120090–120094 and continued in ore hauling service on CNW.

Quickly after the 1922 demonstrator cars were on DM&N property a decision was made in November 1924 to place small 25-car orders with ACF and Pullman for 95-ton ore cars under AFE 1030 approved August 2, 1924. This would allow ten 95-ton cars to replace 19 50-ton cars with a cost savings in repairs and maintenance by the reduction in the number of cars required for hauling ore.

Actual loaded capacity of 1,000 cu. ft. capacity cars was 1,244 cu. ft. with a 12-inch heap. Iron ore weighed 156.07 pounds per cubic foot. A car loaded with a full heap would have a total of 86 long tons of ore. A long ton is measured as 2,240 pounds. Mine loading practices showed that the majority of cars loaded were in the 70-ton range and during 1940

Interior view of Class U12 ore car. Note the end walkway is plate steel which would be replaced with open grid metal soon after delivery because of the slipping hazard created when wet. (American Car & Foundry Collection, John W. Barriger III National Railroad Library at UMSL)

DM&IR rerated the 80- to 95-ton ore cars to 70-ton capacity.

Delivered in May 1925, ACF built 25 Class U12 95-ton 22-foot ore cars 21001–21025 with 935 cu. ft. capacity in April 1925 for $3,300 per car. To obtain the 95-ton capacity, cars were taller than previous ore cars at 10-foot $2\frac{3}{8}$-inches and were wider at 10-foot 1½-inches. They were equipped with Andrews U-section trucks with 6½ x 12-inch journals. The cars were unique with plate steel sides with angled corner and end posts from the end sill to the top chord on the ends. They had two plate steel riveted side braces and instituted, what would become standard on all future cars, except for cast steel underframe cars, a wide 10-inches inverted C-channel side sill. A box cross section side sill was created by attaching the C-channel side sill to 13¾-inch plate top and bottom and another reversed C-channel, creating an end to end box section supporting the outside edge of the car. This box ran through the end of the bolster assembly to the end sill, creating a very rigid underframe. Empty-load brake equipment was used because the weight ratio between loaded and empty cars was greater than three to one. Steel used in side and slope sheet construction was increased from ¼ inch to $\frac{5}{16}$ inch thickness. Painted black, the cars would be first delivered with the new DM&N ball monogram. Capacity was increased to 100 tons in January 1938 and reduced to 70 tons in early 1940. Cars retained there DM&N numbers after the merger. One car was retired in 1964 and all remaining were retired in fall of 1965.

Delivered in June 1925, Pullman built 25 Class U13 95-ton 22-foot ore cars 21101–21125 with 1,046 cu. ft. capacity on lot number 5384 for $4500 per car. The AFE referred to the Pullman-built cars as Commonwealth type ore cars. Cars were unique in having cast underframes manufactured by Commonwealth Steel Co. with casting number 12987 cast in November 1924. Only other DM&N ore cars built with cast steel underframes were Class U14 22001–22125 and 21126. Side sheets were supported by three cast T-braces. These cars were one inch wider than the ACF 85-ton cars. They were delivered with Pflager trucks having 6½ x 12-inch journals and a 4-foot 2-inch wheelbase. Painted black and stenciled with the DM&N ball monogram, they retained DM&N numbers after the merger. Capacity was increased to 100 tons in January 1938 and reduced to 70 tons in early 1940. It was determined during 1930 that the Pflager trucks on these cars and Class U16 23171–23177 were all right to use for mainline movements, but they were banned from use between Missabe Jct. and Steelton because they were derailment-prone on bad track. By 1934, cars 21102 and 21108 were the only cars still equipped with Pflager trucks. All were retired by 1963.

While testing was still on going with the 85-ton cars, orders were placed in July 1925 for 125 cars from General American Car Co. and 125 cars from Standard Steel Car Co., all to be 81-ton 22-foot cars. This order was approved on AFE 1517 on January 25, 1927.

The GACC cars Class U14 22001–22125 were built at East Chicago in January 1928. They were unique in having cast underframes manufactured by Commonwealth Steel Co. with casting number 18746 cast in July, 1927. The only other DM&N ore cars built with cast steel underframes were 21101–21125 and 21126. Side sheets were supported by three cast T-braces. Cars had 940 cu. ft. capacity and were equipped with Dalman 2-level trucks having 6 x 11-inch journals. Cars 22001–22075 were received in January 1928 and 22076–22125 in February 1928. These were the first cars equipped with Enterprise Railway Equipment door operating mechanisms which would become standard on all future ore car purchases with the exception of 23171–23177. Cars were painted black graphite and stenciled with the DM&N ball monogram. Capacity was increased to 100 tons in January 1938 and reduced to 70 tons in early 1940. Cars retained

General American Car Co. also built 81-ton capacity cars. Sporting a General Steel Castings cast steel underframe, 125 Class U14 cars DM&N 22001–22125 were built in 1928 riding on Dalman two-level trucks. (General American Car Co., D.P. Holbrook collection)

Having 81-ton capacity, 125 Class U15 cars DM&N 23001–23125 were ordered from Standard Steel Car Co. in 1927 riding on Dalman two-level trucks. Class U9 and U25 cars were unique with the C-channels side sills only extending to the sill steps and not to the end corner posts. (Standard Steel Car, D.P. Holbrook collection)

their DM&N numbers after the merger with all retired by 1969.

The SSC cars Class U15 23001–23125 with 943 cu. ft. capacities were built in between December 1927 and February 1928. They looked much the same as the GACC cars, the exception being they were built with fabricated underframes. They were delivered with 5-foot wheelbase Dalman 2-level trucks. Cars were painted black and stenciled with the DM&N ball monogram. Capacity was increased to 100 tons in January 1938 and reduced to 70 tons in early 1940. Cars retained their DM&N number after the merger with all retired by 1969.

Standard Steel Car Co. would deliver 125 Class U15 81-ton ore cars 23001–23125 in 1928 riding on Dalman two-level trucks and fabricated underframes. General American Car Co. delivered 125 Class U14 81-ton ore cars 22001–22125 in 1925 with General Steel Castings Co. underframes. DM&N specified these designs to allow comparison of fabricated versus cast underframes. All ore cars purchased after 1925 received fabricated underframes because the cost comparison did not justify the cast steel underframes. (Standard Steel Car Co. photo, D.K. Retterer collection)

Built in 1928 as 81-ton capacity cars by Standard Steel Car, Class U15 23001–23125 were rerated as 70-ton cars, as shown at Roundout, Illinois, on April 10, 1967. The pulling tab below the side sill was not applied by DM&IR. (Owen Leander, D.P. Holbrook collection)

DM&N cars had been plagued for years with bulging top chords, causing extensive repairs to be needed to side sheets to keep cars in service. Class U14 and U15 cars delivered in 1928 addressed this problem by having the top edge of the side sheet bent inward and an additional 5 x 4½-inch steel L-section attached on the inside edge of the side sheet, creating a top chord that was reinforced on the outside edge with a number of T-section pieces that reinforced the top chord. This solution eliminated most of the side sheet bulging problems and was adapted on all future orders of ore cars.

Once the thawing sheds were built at Duluth and Two Harbors in 1960, it was determined that the Class U14 and Class U15 cars would not fit inside the sheds. All were placed in coal service in 1961. During early 1963 these cars were stenciled "For Coal and Limestone Service Only." As traffic requirements dictated, they were in and out of storage between 1963 and 1968, when they were all retired.

DM&N turned to PS and GACC in 1937 and ordered 1,500 83-ton 22-foot ore cars with 1000 cu. ft. capacity. This was the beginning of standardization of the DM&N

Pullman-Standard builder photo of Class U17 DM&N 24621. Note the lack of a stencil group which was not begun until 1952. (D.P. Holbrook collection)

and DM&IR ore car roster, with cars built to the same design allowing brake components, door mechanisms and grab iron to be interchangeably used. Cars introduced the angular end sills and standardization on the Enterprise door mechanism. Side sheets were attached to the outside of Z-section side posts located directly above the bolsters. The order was split with 1,000 Class U17 ore cars 24000–24999 ordered under AFE 3250 approved December 11, 1936, from Pullman-Standard on Lot number 5554. Eight hundred cars 24000–24799 came equipped with ASF 5-foot wheelbase trucks, the remaining 200, 24800–24999 were built from Cor-Ten steel and equipped with National 5-foot wheelbase B1 trucks.

The 500 Class U18 70-ton 22-foot ore cars 25000–25499, were ordered under AFE 3363 approved February 3, 1937, from GACC. These came equipped with ASF 5-foot wheelbase trucks. Both orders of cars were purchased on March 1, 1937, under a 15-year lease agreement between DM&N and Continental Illinois National Bank and Trust Compnay of Chicago at a cost of $3,229 per car.

The U17 and U18 cars were identical cars with the exception of trucks and light weight. Light weight for 24000–24799 was 43,400 lbs. with cars 24800–24999 being 41,000 lbs. because they were constructed from Cor-Ten

Class U17 24380 is at Proctor on June 16, 1959. This was the first standard-design 70-ton ore car built for DM&N with 1025 cars 23975–24999 built by Pullman-Standard in 1937 and 1938. Stencil group "A" in center of car means this car will need air brake testing during the winter shopping season between 1962 and 1963. After the merger, DM&IR would continue to buy this same design of car until 1957. (Bruce Meyer)

steel. All were stenciled, on delivery, with a capacity of 166,600 pounds, but records indicate that all were rerated to 140,000 pounds beginning in January 1938. Re-stenciling to this new capacity took a number of years.

Concerned with the mix of 50-ton, 70-ton

Starting in 1937 with Class U17 24000–24999 all future 70-ton ore cars on DM&N and DM&IR would be built to this standard design. Pulling tab in middle of side sill was not applied by DM&IR. Car is returning to DM&IR after being leased to EJ&E during winter 1966-67. Rondout, Illinois, April 10, 1967. (Owen Leander)

and 95-ton cars and the difference in braking capacity, DM&N approved AFE 3755 for $3,452 on January 2, 1937, to purchase and equip 82 85-ton large capacity ore cars with double brake shoes. Cars equipped were, three from Class U12 series 21001–21025 (21004, 21009 21025), four from Class U13 series 21101–21125 (21104, 21107, 21112, 21114), 18 from Class U15 series 23001–23125 (23004, 23006, 23011, 23032, 23043, 23044, 23047, 23056, 23061, 23069, 23074, 23076, 23083, 23093, 23101, 23112, 23119, 23123) and 62 from Class U14 series 22001–22125 (22001, 22003, 22004, 22007, 22009–22012, 22015, 22021–22023, 22025, 22026, 22028–22031, 22034, 22039, 22043, 22045–22049, 22051, 22056, 22057, 22059, 22063, 22065, 22069-22071, 22076, 22077, 22079, 22081, 22083, 22084, 22086–22088, 22090, 22092, 22094, 22097, 22099, 22101, 22105, 22107). This included the purchase of: double brake shoes, double brake shoe keys, double brake heads and brake hangers. Application took about nine months. This modification was removed in the late 1940s.

DM&N OPEN HOPPER CARS

Unlike sister road D&IR, the DM&N quickly embraced open hopper cars with the opening of the Minnesota Steel Co. steel mill, later American Steel & Wire (AS&W) at Steelton, Minnestoa, in 1915. Hopper cars were needed for hauling coal and limestone from the Duluth docks the eight miles to the mill at Steelton.

The first hopper cars ordered were 200 Class Q1 50-ton 32-foot 3-inches long 2-bay ribbed side steel open hoppers 4500–4699 with 1,676 cu. ft. capacity from Pullman Car Co. in November 1915 at a cost of $1245.22 per car. The cars were delivered between May and June 1916. Cars had seven pressed steel hat section ribs per side and rode on Andrews T-section trucks. They were immediately placed in shuttle service between the Duluth docks and the AS&W steel mill. These cars would retain their DM&N numbers after the merger.

AFE 1032 was approved on September 8, 1924, calling for substituting heavier hopper plates on 200 Q1 Class hoppers account heavy wear. Originally delivered with K brakes, replacement with AB brakes began during 1940 and was completed by 1946. When delivered these cars were painted black with the DM&N ball monogram in the far left panel with the reporting marks and number placed in the far right panel with the initials below the numbers.

On October 20, 1920, the ARA issued standards for lettering and reporting marks. This caused DM&N to move reporting marks and number to the left panel and the monogram to the right panel. Ten of the cars were sold to the DSS&A in 1953 via scrap dealer Duluth Iron & Metal per AFE 4701 March 25, 1953, and renumbered DSSA 1700–1709. The remaining cars were retired by 1955.

Unlike D&IR, DM&N would embrace open hopper car ownership with the opening of Minnesota Steel Co. at Steelton in 1915. The new mill would require coal and limestone to be hauled from docks in Duluth the eight miles to Steelton requiring hopper cars. Class Q1 4500–4699 were ordered from Pullman Car Co. in November 1915 and deliver to DM&N during May and June 1916. Delivered before the 1920 ARA lettering standards were issued, all arrived with monogram in the left panel and road number and reporting marks in the right panel. Cars would be re-lettered beginning in November 1920 with some not being completed until 1930. (Pullman Palace Photographs, Archives Center, National Museum of American History, Smithsonian Institution)

Looking for a dual purpose car, DM&N worked with Ryan Car Co. to design a hopper car that would be capable of easy unloading when handling coal, limestone and iron ore. On December 11, 1936, AFE 3251 for $196,300 was approved for the construction of 50 Class Q2 70-ton 40-foot 7-inch offset side triple hoppers 4700–4749 with 2,150 cu. ft. capacity from the Ryan Car Co. in 1937. Cars had an inside length of 34-foot 1-inch. Five cars were received in June 1937 with remainder delivered in July. They were equipped with ASF 70-ton Spring Plankless trucks. These cars were purchased on March 1, 1937, under a 15-year lease agreement between DM&N and Continental Illinois National Bank and Trust Company of Chicago at a cost of $3,930 per car.

DM&N requested Ryan Car Co. to design an open hopper car with 50-degree slope sheets that would allow a car to be used in coal, limestone and iron ore service. Ore cars had 50-degree slope sheets and during the winter coal and limestone were more easily unloaded from these cars during freezing temperatures. The resulting car, DM&N 4700, is shown in a Ryan Car Co. builder photo in 1937. Additional cars, 250 from Pressed Steel Car in 1952 and 100 from Pullman Standard in 1957 were built to this design. (University of Minnesota Duluth, Kathryn A. Martin Library Archives and Special Collections)

Class Q2 4700–4749 quad hoppers were rebuilt from offset side to ribbed side cars with Pullman-Standard supplied kits at EJ&E Joliet Shops in 1975. Note the rivet and huck bolts adjacent to the ribs on these cars, which are not present on rebuilt Class Q5 hoppers. Aurora, Illinois, May 27, 1979. (D.P. Holbrook)

These cars were unusual in they had a shortened carbody with end platforms on each end. This was caused by the slope sheets having a 50-degree incline which matched most ore car slope sheets. The steeper slope sheets had been specified as part of the design criteria because coal and limestone, when loaded in normal hopper cars, tended to freeze in the cars during the winter months. They were purchased for service from the Duluth docks to the steel mill at Steelton but with the 50-degree slope sheets they could also be used in iron ore service when needed. All Class Q2 hoppers had side sheets, slope sheets and hopper bays replaced in 1955.

Limestone and coal service was hard on the hoppers and by the early 1970s, many were out of service because of worn-out side and slope sheets. Correspondence during November 1974 indicated that increased coal and coke traffic from Duluth Works of AS&W required additional hoppers cars. During May through June 1975 the Class Q2 cars in 4700–4749 were rebuilt at EJ&E Joliet shops with Pullman-Standard-supplied kits consisting of new 11-rib riveted replacement sides for the offset sides, new end slope sheets and new hopper bottoms and doors. Retained from the original cars was the center sill, bolsters, end platforms and end support braces. All cars were completed on June 14, 1975. Rebuilding of these cars was much the same as the Q5 cars done at Proctor in 1971 to 1973 with the exception being the change from mostly welded side ribs to riveted ribs. It was noted that these cars would only have a twelve-year service life in interchange service before they would be considered overage in 1987. DM&IR considered this an investment that would allow cars to be resold on the secondhand market. At the time of the rebuilding DM&IR was leasing 59 hopper cars from B&LE and Hallett to cover on-line loadings and wished to terminate this lease. All 50 were sold to EJ&E on AFE 11296 and renumbered randomly 42200–42293 and ownership transferred between August 7, 1979 and Januaty 17, 1980.

DM&N FLAT CARS

DM&N rostered a total of 580 flat cars purchased between 1893 and 1906. These cars were purchased mainly for online lumber business. Later many were converted to pulpwood rack flats and gondolas.

DM&N acquired its first flat cars with an order placed in 1892 with Duluth Manufacturing Co. for 50 Class W 25-ton 34-foot flat cars 2000–2049 costing $420.34 per car. These were truss-rod underframe flats with 8 stake pockets per side, had cast draw heads, Westinghouse air brakes and arch-bar trucks. Desiring to keep all 200 Pullman 25-ton wood ore cars in one series caused flatcar 2000 to be renumbered 2229 in 1896.

The next group of 120 Class V 30-ton 36-

DM&N 2062, one of 120 Class V 36-foot flat cars built by Duluth Manufacturing Co. in 1893, is shown at an unknown location in April 1896. (University of Minnesota Duluth, Kathryn A. Martin Library Archives and Special Collections, S2386 Box 25a Folder 6 detail)

foot flat cars, 2050–2169, was purchased from Duluth Manufacturing Co. in 1893 at the same price as the 1892 order and identical construction details. However, these cars were unique in having side-mounted drop brake shafts and brake wheels manufactured by Universal Railway Equipment. Both groups retained their numbers after the merger and all were retired by 1960.

Again, with financial troubles plaguing the DM&N, the choice was made to purchase flat cars secondhand. The first group of 50 Class Y 30-ton 36-foot cars 2170–2219, originally built by Barney and Smith Car Co. in 1896, were purchased secondhand from the Munising Railroad (part of the Lake Superior & Ishpeming railroad in later years) in 1898 for $325 per car. These had originally been in the Munising 1–165 series. An additional group of nine Class Y 30-ton 35-foot flat cars 2220–2228 were purchased secondhand from the North West Railway in 1900 for $432.12 per car. DM&N diagrams show these cars built by Barney and Smith Car Co. in 1899. Both series were equipped with 4-foot 10-inch arch-bar trucks with 4½ x 8-inch journals and Westinghouse air brakes. DM&N equipment lists and reports list both groups of cars at being built by Barney and Smith, but builder's records do not support these statements. These cars would retain their numbers after the merger and all were retired by 1960.

The Class W 2001--2049 34-foot flat cars built in 1892 were chosen to be rebuilt to 36-feet flat cars, retaining their old numbers, starting in 1900. The cars had new 3-inch decks, draft gear and malleable stake pockets with U-bolt attachments added and placed in Class

Construction of the first wood ore dock at Duluth created this moveable wood A-frame to be built for unloading dirt to backfill around the pilings used to support the dock. Dock 1 was constructed with three tracks with the inside rail of outer tracks gauged to create the rails of the middle track. The A-frame was constructed to ride on outside rails and mount a Lidgerwood type plow below the adjustment wheel. Plow height was adjusted to deck height of flat cars and a string of flat cars loaded with dirt would be shoved through the device and dump the fill material into the water below. As needed the A-frame was moved along dock to the location fill material was needed. Flat car is DM&N 2114, one of 120 Class V 36 foot flatcars 2050–2169 built in 1893. Note the end stake pockets, link and pin couplers and air hose, indicating car was equipped with air brakes. (University of Minnesota Duluth, Kathryn A. Martin Library Archives and Special Collections, S2386 Box39 Folder 18)

Early pulpwood racks were constructed of left-over limbs from trees. The next evolution in design was railroad-assembled pulpwood racks from dimensional lumber. Class Y DM&IR 2205 is at Proctor on July 17, 1957 loaded with new railroad ties. Later pulpwood racks would be constructed from plate steel and C-channels. (Bob's Photo collection)

DM&N 2198 is one of 50 Class V 36-foot flat cars 2170–2219 acquired secondhand from Munising Railway in 1898. DM&N records show these built by Barney and Smith Car Co., but builders records conflict with this information. It's shown with Oliver Iron Mining crane at Hibbing, Minnesota, in 1922. (University of Minnesota Duluth, Kathryn A. Martin Library Archives and Special Collections, S3021 Box13 Folder 11)

V once rebuilt. All cars were rebuilt between 1900 and 1901.

Under the U.S. Steel umbrella after 1901, DM&N Proctor shops built the next 50 Class V 30-ton 36-foot flat cars 2230–2279 in 1903 at a cost of $340.70 per car. These were approved under two different appropriations; 2230–2259 were ordered in March 1903 and 2260–2279 in October 1903. They had 8 stake pockets per side with a deck 8-foot 9-inches wide and truss-rod reinforced underframe. They were equipped with 5-foot 6-inch wheelbase arch-bar trucks and Westinghouse air brakes. Continuing to build equipment, DM&N Proctor shops built identical 100 Class V 30-ton 36-foot flat cars 2280–2379 in 1905 at a cost of $329.34 per car. The entire series 2229–2379 were built using parts from dismantled wooden ore cars in the 200–1200 series. The cars would retain their numbers after the merger and all were retired by 1960.

The next group of 200 Class V1 40-ton 40-foot flat cars 2500–2699 were ordered in October 1905 from Pressed Steel Car and delivered in 1906. They had riveted fishbelly side sill and center sill, 10 stake pockets on each side and a 8-foot 10-inches wide wood deck. They were equipped with Griffin wheels, arch-bar trucks, Simplex bolsters, Westinghouse brakes, Pressed Steel Car brakebeams, Chica-

Duluth Superior Transit Co. trolley 268 is loaded on Class V1 flat car DM&N 2610 at Duluth in June 1939. Trolley is headed to Proctor shops for conversion to a jitney for hauling employees from downtown Proctor to the roundhouse and carshops. DST 268 would become DM&IR W56. (W.C. Olsen)

Class V1 flatcar DMIR 2519, part of 200 cars built by Pressed Steel Car Co. in 1906 is at Endion. Twenty six of the cars were rebuilt to gondolas in 1919 and sides were removed in 1932. It is unclear why, but 26 different cars were converted in 1932. For unexplained reasons, the 1932 cars were one number below the cars converted in 1919. (Mainline Photos, J.M. Gruber collection)

Saddled with a large roster of flat cars during the early half of the century, both D&IR and DM&N converted large numbers of flat cars to gondolas by adding sides and ends. Class V1 2588 is at Two Harbors in 1959 loaded with steam locomotive drivers. The three T-section ribs were added after World War II to strengthen the car sides. (Basgen photo, D.P. Holbrook collection)

go couplers, Railway Steel Spring Co. springs, McCord journal boxes, Westinghouse drawbar attachments and Susemihi side bearings. A number of the cars were equipped with permanent logging chains during the 1930s. Most of the cars were converted from arch-bar trucks to cast steel side frames between 1938 and 1940. Cars would have a long life on DM&N, being converted to gondolas, rack flats and eventually into MofW service flat cars. The last revenue flat cars in this series were retired in 1972.

Ninety one 2000–2379 series flat cars were sold to Duluth & Northeastern in October and November of 1929 for service on D&NE hauling pulpwood.

FLAT CAR CONVERSIONS TO GONDOLAS AND RACK FLATS

Notes in DM&N 1903 annual report indicate expenses for adding gondola sides to flat cars and this appears to be the first conversion of flat cars to gondolas. One hundred seven of the 2043–2363 flat cars were rebuilt to gondolas at the DM&N Proctor Shops in 1919 by applying gondola sides and ends to the flat cars. Many of these cars were placed into coal service. These cars were shown in the DM&N 1931 diagram book as carrying a letter suffix of "G" next to the road number. Earliest records show "G" prefix added to flat cars rebuilt with gondola sides in 1919. The prefix "G" was dropped during the 1930s as cars were rebuilt back to flat cars. Cars were again rebuilt in 1932 with most having their sides removed and new cars one number lower rebuilt with sides. A total of 89 cars were completed this way. An example was flat car 2007 had gondola sides and ends removed and the replacement car 2006 took its place. See roster notes for additional information. These gondola conversions were off the roster by 1934.

Five flat cars, 2015, 2331, 2051, 2290, 2209, were equipped with 16-inch sides and ends and signboards reading "This car to be used for coal loading for boarding cars only" in April 1923. Boarding cars were MofW bunk cars.

Beginning in 1919, the DM&N began adding gondola sideboards and ends to 26 Class V1 flat cars 2500–2699. For unknown reasons, between 1931 and 1932 the gondola

DM&N flatcar 2344 was equipped with gondola sides in 1932. G suffix was used on DM&N flatcar to gondola conversions from 1919 to 1933 after which the suffix was dropped. (MNHS OIM Collection)

DM&N flatcar 2169 was equipped with gondola sides in 1932. G-suffix was used on DM&N flatcar to gondola conversions from 1919 to 1933 after which the suffix was dropped. (MNHS OIM Collection)

sideboards and ends were removed from cars and placed on cars one number lower. Examples are gondola 2501 was replaced by 2500, See notes in the rosters for further details. The last of these conversions were retired in 1970.

Wood pulpwood racks began to be applied to flat cars as early as 1910. The first permanent steel and wood pulpwood end rack conversion of flat cars were begun around 1925. These initial end rack conversions did not have end ladders. To comply with safety appliance regulations, AFE 2389 was approved on December 16, 1930, for construction of end ladders on 55 rack flats. Work was begun on January 13, 1931, on these cars: (series 2002–2169), 2007, 2012, 2018, 2020, 2022, 2061, 2062, 2073, 2083, 2088, 2096, 2102, 2108, 2109, 2126, 2131, 2138, 2143, 2148, 2154, 2158, 2164; (series 2170–2228), 2172, 2173, 2174, 2180, 2185, 2193, 2200, 2205, 2207, 2215, 2218; (series 2230–2379), 2230, 2241, 2268, 2280, 2288, 2292, 2299, 2300, 2304, 2306, 2308, 2310, 2314, 2315, 2321, 2322, 2326, 2329, 2336, 2346, 2349, 2364. Project was completed on August 28, 1931, at a total cost of $807.54. All of these rebuilds were off the roster by 1949.

Correspondence during October 1933 indicated that wood pulpwood end racks were still in use on 33 DM&N and 17 D&IR flat cars. It was recommended that as these cars were retired that they be replaced by rebuilding 6180–6304 series gondolas by removing the gondola sides and ends and replacing them with permanent steel pulpwood racks.

Arch-bar trucks were replaced by Andrews U-section trucks between 1937 and 1939.

Beginning during the winter 1955 shopping season, DM&IR began to rebuild former DM&N 2500-series flat cars into pulpwood cars. Fifty three were completed by 1956 and an additional 23 more rebuilt during the winter of 1958-59. Faced with increasing pulpwood shipments, 39 flat cars remained in the 2500 series at the end of 1959 and the 37 best cars were rebuilt with permanent steel pulpwood end racks. The last of these was retired by 1966.

DM&N PHELAN FLATS

Forty five of the Phelan type 15-ton 30-foot flat cars were built in 1892. Diagrams from 1894 show cars were built by Duluth Manufacturing Co., however later diagrams show these cars as being built by United States

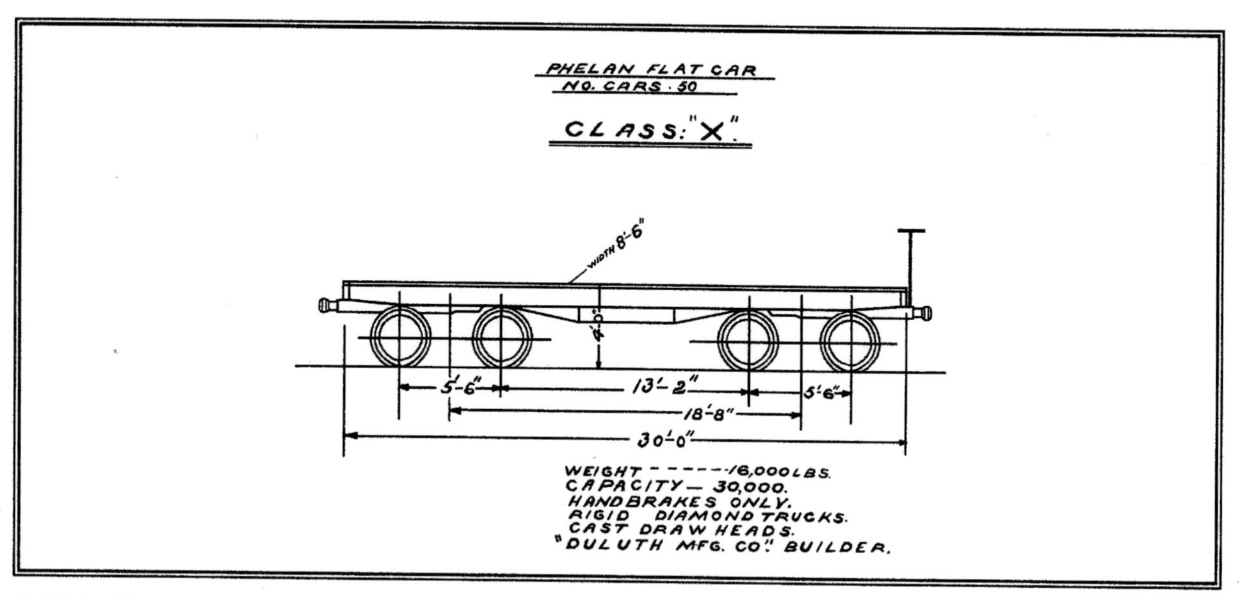

DM&N Class X Phelan flat car diagram. (D.P. Holbrook collection)

Rolling Stock Co. No number series is shown for these cars, but they were placed in Class X. Cars were used for construction of the railroad, had 3-foot 6-inches wheelbase arch-bar trucks and were delivered with link-and-pin couplers. They had 8-feet, 6-inch wide decks, a light weight of 8 tons and were equipped only with hand brakes. Diagrams from 1893 indicate 12 cars were equipped for use as boarding cars for construction crews. By 1901, only 21 cars were still on the roster with the notation that they were used for boarding cars and spreaders. By this point some had been rebuilt to cinder cars and other maintenance of way equipment. By 1905 all of the remaining cars were in MofW service.

DM&N LOG FLATS

The DM&N ordered 50 skeleton log flats from Duluth Manufacturing Co. in 1892. They carried no numbers and used trucks removed from stored wood ore cars during the winter months. These cars were unusual for the time being 17-foot 6-inches long and having 13-foot wide bunks. Russell-design log cars were 20-feet long and had bunks 10-foot wide. All 50 cars were sold during 1899.

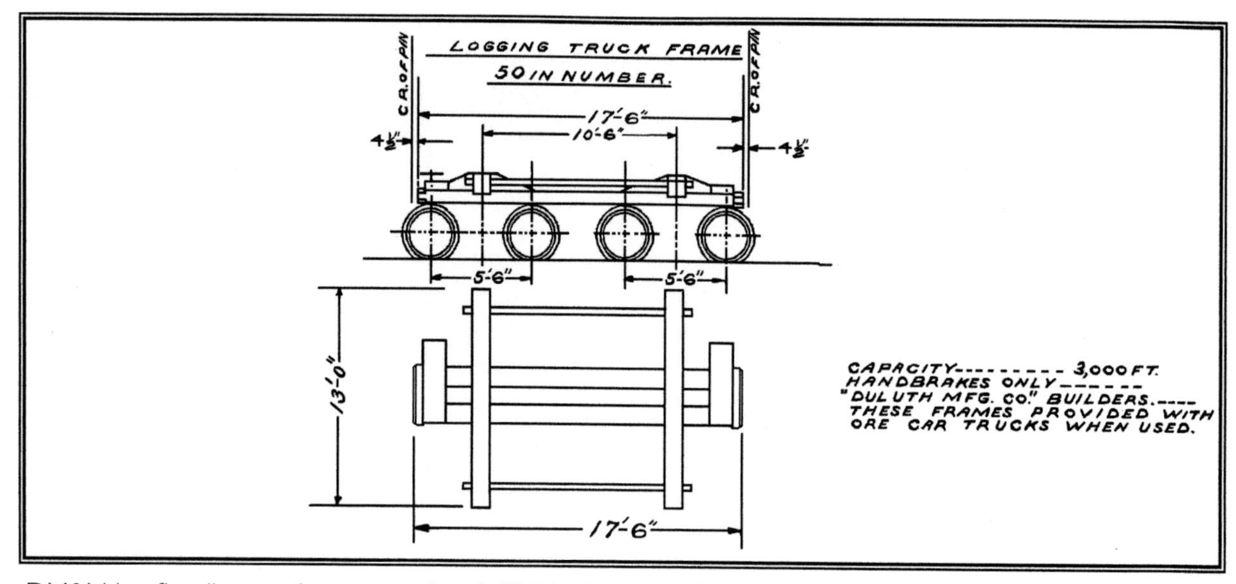

DM&N log flat diagram (no car numbers). (D.P. Holbrook collection)

DM&N CABOOSES

DM&N initially ordered six Class G 4-wheel cabooses 1–6 from Duluth Manufacturing Co. in 1892. Almost immediately orders were placed with Duluth Manufacturing in 1893 for three Class H 8-wheel cabooses 8–10 and two additional Class G 4-wheel cabooses 7 and 11. Continuing with the same builder in 1894 four additional Class G 4-wheel cabooses 12–15 were ordered. Caboose 13 was rebuilt Class I caboose (2nd)13 in spring 1895. Car had car body extended to 36-feet 1-inch equipped with two 4-wheel trucks and a side-door added in the middle of the new car body. This car was rebuilt to business car 98 during 1898. DM&N built new 4-wheel caboose (3rd)13 in 1898 and placed it in Class G. All were equipped with

Duluth, Missabe & Northern Cabooses

No	Builder	Cl	Built	Retired	O.L	Type	To DMIR	Note
1 (1st)	Duluth Mfg Co	G	1892	Retired prior to Oct. 1910	19'2"	4W		
1 (2nd)	Duluth Mfg Co	G1	Feb. 1911	12/14/1933	31'	8W		
2 (1st)	Duluth Mfg Co	G	1892	Feb. 1911	19' 2"	4W		
2 (2nd)	Duluth Mfg Co	G1	Feb. 1911	12/14/1933	31'	8W		
3 (1st)	Duluth Mfg Co	G	1892	Feb. 1911	19'2"	4W		
3 (2nd)	Duluth Mfg Co	G1	Feb. 1911	12/14/1933	31'	8W		
4 (1st)	Duluth Mfg Co	G	1892	April. 1911	19'2"	4W		
4 (2nd)	Duluth Mfg Co	G1	Feb. 1911	12/14/1933	31'	8W		
5 (1st)	Duluth Mfg Co	G	1892	Retired prior to Oct. 1910	19' 2"	4W		
5 (2nd)	Duluth Mfg Co	G1	Feb. 1911	to DM&IR	31'	8W	C-5	
6 (1st)	Duluth Mfg Co	G	1892	Retired prior to Oct. 1910	19'2"	4W		
6 (2nd)	Duluth Mfg Co	G1	Feb. 1911	Destroyed Dec. 1930	31'	8W		
7 (1st)	Duluth Mfg Co	G	1893	Destroyed 1896	19' 2"	4W		
7 (2nd)	DM&N Proctor Shops		Mar. 1900	March, 1911	31'	4W		
7 (3rd)	DM&N Proctor Shops	H	Feb. 1911	Destroyed Sept. 1929	36'	8W		
8 (1st)	Duluth Mfg Co	H	1893	Destroyed by fire 1/19/1904 Rebuilt to (2nd) 8	36'	8W		
8 (2nd)	DM&N Proctor Shops		4/5/1904	to DM&IR	36'	8W	C-8	1
9	Duluth Mfg Co	H	1893	to DM&IR	36'	8W	C-9	2
10	Duluth Mfg Co	H	1893	to DM&IR	36'	8W	C-10	3
11 (1st)	Duluth Mfg Co	G	1893	Oct. 1910	19'2"	4W		
11 (2nd)	Duluth Mfg Co	G1	Feb. 1911	Destroyed June 1919	31'	8W		
11 (3rd)	Proctor Shops	G1	1923	Jan. 1930		8W		
12 (1st)	Duluth Mfg Co	G	1894	April. 1911	19'2"	4W		
12 (2nd)	Duluth Mfg Co	G1	Feb. 1911	to DM&IR	31'	8W	C-12	
13 (1st)	Duluth Mfg Co	G	1894	Rblt to 8W caboose (2nd)13 with sidedoor 1895	19'2"	4W		
13 (2nd)	DM&N Proctor Shops	I	1895	Rebuilt to business car 98	41'11"	8W		
13 (3rd)	DM&N Proctor Shops	G	1898	Wrecked April. 1911	19'2"	4W		
13 (4th)	DM&N Proctor Shops	G1	June. 1911	Sold to D&NE. Feb. 1929	31"	8W		
14 (1st)	Duluth Mfg Co	G	1894	Oct. 1910	19'2"	4W		
14 (2nd)	Rebuilt DM&N Proctor	G1	Feb. 1911	Destroyed June. 1919	31'	8W		
14 (3rd)	DM&N Proctor Shops	G1	1923	Jan. 1930		8W		
15 (1st)	Duluth Mfg Co	G	1894	April. 1911	19'2"	4W		
15 (2nd)	Rebuilt DM&N Proctor	G1	Feb. 1911	Feb. 1929	31'	8W	C-15	4
16 (1st)	ACF Lot #513	G	1899	Oct. 1910	19'2"	4W		
16 (2nd)	Rebuilt DM&N Proctor	G1	Feb. 1911	Destroyed July. 1930	31'	8W		
17 (1st)	ACF Lot #513	G	1899	Feb. 1911	19'2"	4W		
17 (2nd)	Rebuilt DM&N Proctor	G1	Feb. 1911	Wrecked Elwood 2/14/1923	31'	8W		
17 (3rd)	DM&N Proctor Shops		1923	Retired before 1939	31'	8W		5
18 (1st)	ACFLot #513	G	1899	Oct. 1910	19'2"	4W		
18 (2nd)	Rebuilt DM&N Proctor	G1	Feb. 1911		31'	8W	C-18	
19 (1st)	ACF Lot #513	G	1899	Oct. 1910	19'2"	4W		
19 (2nd)	Rebuilt DM&N Proctor	G1	Feb. 1911	Sold to General Logging Co. Nov. 1930	31'	8W		
19 (3rd)	DM&N Proctor Shops		Unknown	Retired before 1939	31'	8W		5

No	Builder	Cl	Built	Retired	O.L	Type	To DMIR	Note
20 (1st)	ACF Lot #895	G	1900	Feb. 1911	19'2"	4W		
20 (2nd)	Rebuilt DM&N Proctor	G1	Feb. 1911	Sold to MN&S 12/22/1929	31'	8W		
21 (1st)	ACF Lot #895	G	1900	April. 1911	19'2"	4W		
21 (2nd)	Rebuilt DM&N Proctor	G1	Feb. 1911	Destroyed Oct. 1918	31'	8W		
21 (3rd)	DM&N Proctor Shops		1923	Jan. 1930	31'	8W		
22 (1st)	ACF Lot #895	G	1900	Feb. 1911	19'2"	4W		
22 (2nd)	Rebuilt DM&N Proctor	G1	Feb. 1911	Feb. 1929	31'	8W	C-22	6
23 (1st)	ACFLot #895	G	1900	Dec. 1910	19'2"	4W		
23 (2nd)	Rebuilt DM&N Proctor	G1	Feb. 1911	Sold to General Logging Co. Nov. 1930	31'	8W		
24 (1st)	ACF Lot #895	G	1900	Dec. 1910	19'2"	4W		
24 (2nd)	Rebuilt DM&N Proctor	G1	Feb. 1911	12/14/1933	31'	8W		
25 (1st)	ACF Lot #895	G	1900	Oct. 1910	19'2"	4W		
25 (2nd)	Rebuilt DM&N Proctor	G1	Feb. 1911	to DM&IR	31'	8W	C-25	
26 (1st)	ACF Lot #895	G	1900	July. 1911	19'2"	4W		
26 (2nd)	Rebuilt DM&N Proctor	G1	Feb. 1911	to DM&IR	31'	8W	C-26	
27 (1st)	ACF Lot #895	G	1900	Wrecked 1906	19'2"	4W		
27 (2nd)	D&IR Two Harbors	G	1907	Mar. 1911	19'2"	4W		
27 (3rd)	DM&N Proctor Shops	G1	Feb. 1911	Destroyed by fire at Wilpen 12/29/1922	31'	8W		
27 (4th)	DM&N Proctor Shops	G1	1923	Sold to D&NE Feb. 1929	31'	8W		
28 (1st)	ACF Lot #895	G	1900	Jul-11	19'2"	4W		
28 (2nd)	DM&N Proctor Shops	G1	Feb. 1911	12/14/1933	31'	8W		
29 (1st)	ACF Lot #895	G	1900	Off roster by 1905	19'2"	4W		
29 (2nd)	D&IR Two Harbors	G	1905	Mar. 1911	19'2"	4W		
29 (3rd)	DM&N Proctor Shops	G1	Feb. 1911	12/14/1933	31'	8W		
30 (1st)	D&IR Two Harbors	G	1905	Mar. 1911	19'2"	4W		
30 (2nd)	DM&N Proctor Shops	G1	Feb. 1911	to DM&IR	31'	8W	C-30	
31 (1st)	D&IR Two Harbors	G	1905	Feb. 1911	19'2"	4W		
31 (2nd)	DM&N Proctor Shops	G1	Feb. 1911	Sold to CGW March 1929	31'	8W		
32 (1st)	D&IR Two Harbors	G	1905	Feb. 1911	19'2"	4W		
32 (2nd)	DM&N Proctor Shops	G1	Feb. 1911	Sold to CGW March 1929	31'	8W		
33 (1st)	D&IR Two Harbors	G	1905	Destroyed July, 1909	19'2"	4W		
33 (2nd)	DM&N Proctor Shops	G1	May. 1910	12/14/1933	31'	8W		
34 (1st)	D&IR Two Harbors	G	1905	Mar. 1911	19'2"	4W		
34 (2nd)	DM&N Proctor Shops	G1	Feb. 1911	Destroyed 2/24/1925	31'	8W		
35 (1st)	D&IR Two Harbors	G	1905	Dec. 1910	19'2"	4W		
35 (2nd)	DM&N Proctor Shops	G1	Feb. 1911	Destroyed July. 1919	31'	8W		
35 (3rd)	DM&N Proctor Shops	G1	1923	Jan. 1930	31'	8W		
36 (1st)	D&IR Two Harbors	G	1905	Dec. 1910	19'2"	4W		
36 (2nd)	DM&N Proctor Shops	G1	Feb. 1911	Sold to CGW March 1929	31'	8W		
37 (1st)	D&IR Two Harbors	G	1905	Mar. 1911	19'2"	4W		
37 (2nd)	DM&N Proctor Shops	G1	Feb. 1911	Sold to D&NE Feb. 1929	31'	8W		
38 (1st)	D&IR Two Harbors	G	1905	Wrecked 1907	19'2"	4W		
38 (2nd)	DM&N Proctor Shops	G1	1907	to DM&IR	31'	8W	C-38	
39 (1st)	D&IR Two Harbors	G	1905	Dec. 1910	19'2"	4W		
39 (2nd)	DM&N Proctor Shops	G1	Feb. 1911	to DM&IR	31'	8W	C-39	
40 (1st)	D&IR Two Harbors	G	1905	Dec. 1910	19'2"	4W		
40 (2nd)	DM&N Proctor Shops	G1	Feb. 1911	to DM&IR	31'	8W	C-40	
41 (1st)	D&IR Two Harbors	G	1905	Oct. 1910	19'2"	4W		
41 (2nd)	DM&N Proctor Shops	G1	Feb. 1911	Sold to MN&S Jan. 1929	31'	8W		
42 (1st)	D&IR Two Harbors	G	1905	Feb. 1911	19'2"	4W		
42 (2nd)	DM&N Proctor Shops	G1	Feb. 1911	Destroyed by fire on EJ&E March 1913	31'	8W		
42 (3rd)	DM&N Proctor Shops	G1	Oct. 1913	12/14/1933	31'	8W		
43 (1st)	D&IR Two Harbors	G	1905	Wrecked Alborn 11/7/1906	19'2"	4W		
43 (2nd)	D&IR Two Harbors	G	1907	Feb. 1911	19'2"	4W		
43 (3rd)	DM&N Proctor Shops	G1	Feb. 1911	to DM&IR	31'	8W	C-43	
44 (1st)	D&IR Two Harbors	G	1907	Feb. 1911	19'2"	4W		
44 (2nd)	DM&N Proctor Shops	G1	Feb. 1911	to DM&IR	31'	8W	C-44	

No	Builder	Cl	Built	Retired	O.L	Type	To DMIR	Note
45 (1st)	D&IR Two Harbors	G	1907	Mar. 1911	19'2"	4W		
45 (2nd)	DM&N Proctor Shops	G1	Feb. 1911	Sold to MN&S Jan. 1929	31'	8W		
46 (1st)	D&IR Two Harbors	G	1907	Oct. 1910	19'2"	4W		
46 (2nd)	DM&N Proctor Shops	G1	Feb. 1911	12/14/1933	31'	8W		
47 (1st)	D&IR Two Harbors	G	1907	Destroyed Oct. 1909	19'2"	4W		
47 (2nd)	DM&N Proctor Shops	G1	May, 1910	Destroyed Coleraine 7/31/1935	31'	8W		
48 (1st)	D&IR Two Harbors	G	1907	Feb. 1911	19'2"	4W		
48 (2nd)	DM&N Proctor Shops	G1	Feb. 1911	Sold to CGW March 1929	31'	8W		
49 (1st)	D&IR Two Harbors	G	1907	Dec. 1910	19'2"	4W		
49 (2nd)	DM&N Proctor Shops	G1	Feb. 1911	Sold to CGW March 1929	31'	8W		
50 (1st)	D&IR Two Harbors	G	1907	Feb. 1911	19'2"	4W		
50 (2nd)	DM&N Proctor Shops	G1	Feb. 1911	to DM&IR	31'	8W	C-50	
51 (1st)	D&IR Two Harbors	G	1907	Dec. 1910	19'2"	4W		
51 (2nd)	DM&N Proctor Shops	G1	Feb. 1911	Destroyed Sept. 1918	31'	8W		
51 (3rd)	DM&N Proctor Shops	G1	1923	to DM&IR	31'	8W	C-51	
52 (1st)	D&IR Two Harbors	G	1907	Mar. 1911	19'2"	4W		
52 (2nd)	DM&N Proctor Shops	G1	Feb. 1911	to DM&IR	31'	8W	C-52	
53 (1st)	D&IR Two Harbors	G	1907	Dec. 1910	19'2"	4W		
53 (2nd)	DM&N Proctor Shops	G1	Feb. 1911	Sold to General Logging Co. Nov. 1930	31'	8W		
54 (1st)	D&IR Two Harbors	G	1907	Dec. 1910	19'2"	4W		
54 (2nd)	DM&N Proctor Shops	G1	Feb. 1911	Sold to CGW March 1929	31'	8W		
55 (1st)	D&IR Two Harbors	G	1907	Feb. 1911	19'2"	4W		
55 (2nd)	DM&N Proctor Shops	G1	Feb. 1911	Destroyed Oct. 1916	31'	8W		
55 (3rd)	DM&N Proctor Shops	G1	Aug. 1918	to DM&IR	31'	8W	C-55	7
56 (1st)	D&IR Two Harbors	G	1907	Mar. 1911	19'2"	4W		
56 (2nd)	DM&N Proctor Shops	G1	Feb. 1911	to DM&IR	31'	8W	C-56	
57 (1st)	D&IR Two Harbors	G	1907	Feb. 1911	19'2"	4W		
57 (2nd)	DM&N Proctor Shops	G1	Feb. 1911	Destroyed Oct. 1916	31'	8W		
57 (3rd)	DM&N Proctor Shops	G1	Aug. 1918	to DM&IR	31'	8W	C-57	7
58 (1st)	D&IR Two Harbors	G	1907	Feb. 1911	19'2"	4W		
58 (2nd)	DM&N Proctor Shops	G1	Feb. 1911	to DM&IR	31'	8W	C-58	
59	DM&N Proctor Shops	H1	1911	Destroyed GN wreck Bovey 6/25/1927	36'5"	8W		8
60	DM&N Proctor Shops	H1	1912	to DM&IR	36'5"	8W	C-60	8
61	DM&N Proctor Shops	H1	May. 1913	to DM&IR	36'5"	8W	C-61	8
70	DM&N Proctor Shops	G2	8/29/1924	to DM&IR	33'2"	8W	C-70	9
71	DM&N Proctor Shops	G2	Sept. 1924	to DM&IR	33'2"	8W	C-71	9
72	DM&N Proctor Shops	G2	Sept. 1924	to DM&IR	33'2"	8W	C-72	10
73	DM&N Proctor Shops	G2	Sept. 1924	to DM&IR	33'2"	8W	C-73	10
74	DM&N Proctor Shops	G2	Sept. 1924	to DM&IR	33'2"	8W	C-74	10
75	DM&N Proctor Shops	G2	Oct. 1924	to DM&IR	33'2"	8W	C-75	10
76	DM&N Proctor Shops	G2	Oct. 1924	to DM&IR	33'2"	8W	C-76	10
77	DM&N Proctor Shops	G2	5/31/1926	to DM&IR	33'2"	8W	C-77	11

TEMPORARY SIDE DOOR CABOOSES

No	Builder	Cl	Built	Retired	O.L	Type	To DMIR	Note
3010	DM&N Proctor Shops		6/9/1923	to work service 1925	40'7"	8W		12
3017	DM&N Proctor Shops		6/21/1923	to work service 1925	40'7"	8W		12
3022	DM&N Proctor Shops		6/21/1923	to work service 1925	40'7"	8W		12
3048	DM&N Proctor Shops		6/9/1923	to work service 1925	40'7"	8W		12

Notes:

1. *Painted Standard Coach color in 1900. Side door equipped.*
2. *Painted Standard Coach color in 1900 - Side door equipped, cupola in center of car.*
3. *Painted Standard Coach color in 1900 - Side door equipped, cupola removed 6/10. Cupola added ???*
4. *Sold to D&NE in Feb. 1929 per AFE 1958 4/24/1929 and numbered D&NE 5. Repurchased by DM&IR in May, 1942*
5. *SU applied 4/10/36*
6. *Sold to D&NE in Feb. 1929 per AFE 1958 4/24/1929 and numbered D&NE 3. Repurchased by DM&IR in May, 1942*
7. *Built with SU*
8. *Center cupola - Side Door equipped*
9. *SU - Equipped for Steam Heat and Train signal lines - Built under AFE 819 approved 9/6/23*
10. *SU - Built under AFE 819 approved 9/6/23*
11. *SU - AFE 1141 approved 4/6/25 to build #77 to replace destroyed caboose #34*
12. *Boxcars with same numbers with side doors applied for temporary caboose service all placed in work service in 1925*

Proctor caboose track in 1910. Caboose 7 is first in line, fourth caboose is 38, sixth caboose is 47. The seventh caboose is one of three 8-wheel cabooses 8–10 built in 1893 by Duluth Manufacturing Co. Note the narrow cupola. (MNHS)

Westinghouse air brakes. Four-wheel cabooses had 9-foot wheelbase and 8-wheel cabooses had 5-foot 6-inch arch-bar trucks. Four-wheel cabooses had center cupolas with two windows and a car body 14-foot 6-inches long by 8-foot wide with two windows per side. Class H cabooses were 36 feet long having a car body that was 30-foot 1½-inches long by 9-foot wide with an offset cupola having a single window. The cupola was unique in that it was 7-foot 3½-inches long but was only 7-foot 7-inches wide on a car body that was 9-foot wide. They also had a side door on the long end side of the car for packages to be loaded and unloaded.

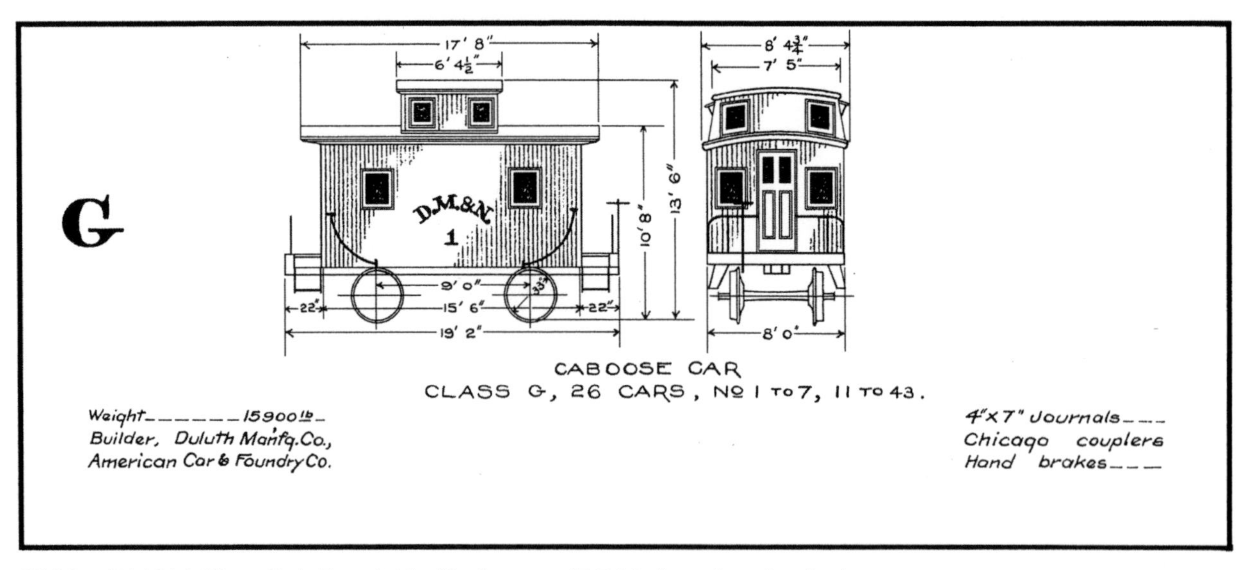

2023 – DM&N Class G 1–7 and 11–43 diagram. (D.P. Holbrook collection)

Caboose C-8 is shown at Biwabik, Minnesota, on July 28, 1969. Before LCL shipments required box cars or refrigerator cars, DM&N had acquired from Duluth Mfg. Co in 1893 three 36-foot cabooses equipped with side doors for handling LCL. Car C-8 had steel underframe applied in April, 1945 and carbody was placed on ground at T-Bird North mine in January 1966. On April 14, 1969, it was placed back in revenue service and finally retired on August 12, 1974. (Owen Leander)

Class H cabooses appear to have been purchased for mixed train service.

Growing traffic in 1899 caused DM&N to order four cabooses from American Car & Foundry under Lot number 513 numbered 16–19. Cars were of the same basic design as the previous 4-wheel cabooses. An additional ten identical cabooses were ordered from ACF in 1900 under Lot number 895 and were numbered 20–29.

In June 1900, Class H cabooses 9 and 10 were remodeled as combination cars and painted "Standard Coach Color" (probably dark green); 9 retained its cupola which was moved to the carbody center, but 10 had the cupola removed when remodeled.

Surging traffic levels required DM&N to have 14 additional 4-wheel cabooses numbered 30–43 built by D&IR Two Harbors in 1905 for $644.72 per car. Caboose 43 was destroyed in a wreck at Alborn on November 7, 1906, and was replaced by (2nd)43 built at Two Harbors in 1907. The last 4-wheel cabooses were 15 built by D&IR Two Harbors in 1907 and numbered 44–58.

It appears from photos that the Duluth Manufacturing-built cabooses had sloped sides on the cupola from the car body sides to the top of the cupola. Later 4-wheel cabooses were vertically sheathed from car body side to top of cupola. All 4-wheel cabooses were rebuilt to 8-wheel cabooses between winter 1910 and spring 1912 at a cost of $45,218.07. All were placed in Class G1 and called "ore cabooses." Markers were mounted centered on top of the cupolas of all cabooses prior to 1910.

Three new 8-wheel 36-foot 5-inch Class H1 cabooses 59–61 with centered cupolas and side doors were built at Proctor Shops, one each in 1911, 1912 and 1913. These were approved on Appropriation 302 April 18, 1911 at a cost of $3,716.10.

During the spring of 1923, a caboose shortage developed that required a quick solution for the 1923 ore shipping season. Three Class O, 3010, 3017 and 3022, and one Class M 3048 box cars were in Proctor shops being converted to tool cars. General Car Foreman W.A. Clark instructed shop forces that all four were to be fitted out as side-door cabooses

Caboose DM&IR C-5 is east of Mitchell, Minnesota, on May 11, 1960, on a mine run job. Originally built as 4-wheel caboose DM&N 5 in 1892, it was rebuilt to a 8-wheel caboose with a 29-foot 2-inch car body in February 1911. Note the steel underframe which was applied in 1939. Donated to Lake Superior Museum of Transportation at Duluth on November 8, 1977. (Bruce Meyer)

Scotchlite monogram and lettering with maroon window frames was the standard caboose paint job from 1960 on. DM&IR C-15 is at Biwabik, Minnesota, on July 28, 1969. It had a unique history, being sold to Duluth & Northeastern Railroad in 1929 and then was repurchased from D&NE on July 20, 1942, account of the caboose shortage during World War II and relettered C-15. It was sold to Hyman Michaels scrap yard on May 22, 1975. (Owen Leander, John C. La Rue, Jr. collection)

DM&IR C-30 was built by D&IR Two Harbors shops in 1905 as 4-wheel caboose and rebuilt to 29-foot 2-inch 8-wheel caboose in February 1911. It was one of 20 cabooses rebuilt with steel underframes between 1939 and 1940. Shown at Virginia, Minnesota July 27, 1969, it is still painted in the darker yellow paint with black lettering, underframe, step wells, platforms and roof, indicating it is assigned to the Iron Range Division. (Owen Leander)

DM&IR C-26 was built by American Car & Foundry in 1900 for DM&N as a 4-wheel caboose and rebuilt at Two Harbors to 29-foot 2-inch 8-wheel caboose in 1907. Rebuilt with steel underframe in 1940 it was destroyed by fire while on lease to Soo Line April 15, 1969. Proctor, April 1963, with black-white striped side grab irons evident. (Dave Repetsky)

DM&N purchased its first six 4-wheel cabooses 1–5 from Duluth Manufacturing Co. in 1892. Rebuilt to 8-wheel cabooses in February 1911, only caboose 5 would survive to the DM&IR merger. Steel underframe was applied in 1940 and Witte engine and radio were installed by1955. DM&IR C-5 shown with maroon lettering, roof and windows is seen at Biwabik July 2, 1974. It was donated to Lake Superior Railroad Museum on November 8, 1977, for use in fund raising. (R.F. Kucaba)

without end platforms for temporary service during the 1923 ore season. Cars 3048 and 3010 were completed on June 9, 1923. Cars 3017 and 3022 were completed on June 21, 1923. The cars saw service during 1923 and 1924 and once replacement cabooses were built in 1924 they were placed in work service.

Aging of the caboose fleet by 1923, the need for more cabooses with steel underframes, and replacements for the temporary boxcar cabooses saw DM&N approved AFE 819 on September 6, 1923, for the construction of seven new Class G2 cabooses. Number 70–76, all were 33-foot 2-inches with offset cupolas and steel underframes. Cabooses 70 and 71 were equipped with steam and signal lines when built, to allow them to be used in passenger service. Caboose 70 was completed on August 29, 1924, with 71–74 completed in September 1924 and 75, 76 in October 1924.

One last Class G2 caboose was constructed at Proctor before the merger. Cab 77, identical to 70–76, was built in 1926 of AFE 1141 approved on April 6, 1925, to replace 34 destroyed in November 1925.

During February 1929, DM&N cabooses 13, 15, 22 and 27 were sold to Duluth and Northeastern Railroad. Two of these, 15 and 22 would be repurchased by DM&IR in May 1942 for the surging traffic levels during World War II.

AFE 2267 was approved on March 19, 1928, to equip 23 cabooses, 56–77, and 79, with steel center sills. On June 20, 1934, the combined DM&N and D&IR had 78 cabooses in service, 44 with wooden underframes, 33 with steel center sills and one with a steel underframe. All were of 8-wheel wood-sheathed construction.

DM&N PASSENGER CARS

Access to the iron ranges of Northern Minnesota during the 1890s was via crude trails, and the opening of the railroad created a need for passenger service between Duluth and the mining communities.

DM&N initially ordered five coaches 50–54 from Ohio Falls Car Co. in 1892. Cars 50 and 51 were placed in Class B, were 58-

foot 9-inches long, and had wooden carbodies with open end platforms and 17 windows per side. Cars had a seating capacity of 54 people, a small smoking area, men's and ladies' saloons, one bathroom and a potbelly stove for heat. Class C cars 52, 53 were considered First Class and car 54, Second Class; all three were 56-foot 4-inches long, had wooden car bodies with open end platforms and 16 windows per side. Cars 50–52 were delivered in December 1892 and cars 53, 54 in October 1893 at a cost of $5,750 per car. All had a seating capacity of 58, one saloon and a potbelly stove for heat. By 1901, 54, 55 had been equipped with steam heat. All were retired between 1925 and 1933.

Ordered from Ohio Falls Car Co. in 1893 were combination cars numbered 100, 101 and baggage and mail car 102. Cars 100, 101 were placed in Class D and 102 in Class E. Combination cars were 58-foot 9-inches long, had wooden carbodies with open end platforms, a baggage door and nine windows per side. The cars had a seating capacity of 32, one saloon and a potbelly stove located in the baggage area for heat. Baggage section was 24-foot 8-inches long with passenger section being 27 feet 8 inches long. Baggage and mail car 102 was 58-foot 9-inches long, had wooden carbody with open end platforms. Baggage section was 38-foot 1- inch long and mail section was 14-foot 4-inches long. Mail section included a sorting table with one window on the sorting table side and two windows on the opposite side. Side doors were located on both sides for baggage and mail to be loaded and unloaded. All were retired in 1934.

All of the 1893-built cars were equipped with Westinghouse air brakes, Miller couplers and two standard 8-foot wheelbase 2-axle passenger car trucks.

Official car 99 was constructed by Ohio Falls Car Manufacturing Co. in 1893 and placed in Class A. Car was 58-feet 4-inches long, had a wooden carbody with open end platforms and identical observation lounges on both ends of the car. Ends had two windows adjacent to the doors with two large windows on each side of the carbody in the observation ends. Eight additional windows were spaced between the observation ends. Adjacent to the observation ends at each end were seating areas for eight people. Between these were a bathroom, potbelly stove for heat, a pantry kitchen with range and a linen closet. By 1901 one observation end had been replaced with an additional bathroom, sofa and additional seating with extension tables to serve food; this was converted back to the original observation end by 1905. One open platform end was replaced by a vestibule between 1905 and 1906. A single extension table was added to the observation end. The number 99 was dropped

DM&N private car 99 was originally built by Ohio Falls Car Co. in 1895 and renamed *Missabe* in 1905. Retired on January 17, 1931, it was sold to Foley Brothers Construction Co. and used as office at the Hoover Dam construction site. It was resold to George T. Maloy in St. Paul, in 1937 and used as a vacation home. The family donated the car to Lake Superior Museum of Transportation in 1997 and it has since been restored and preserved at the Duluth museum. Photo September 21, 2013. (D. Schauer)

Duluth, Missabe & Northern Passenger Cars

Kind	Truck	Name/No	Cl	O.L	Builder	Built	Retired
Coach (open ends)	4-axle	50	B	58' 9"	Ohio Falls Car Co	1892	Retired 10/31/28
Coach (open ends)	4-axle	51	B	58' 9"	Ohio Falls Car Co	1892	Retired 12/14/1933
Coach (open ends) (1st class)	4-axle	52	C	56'4"	Ohio Falls Car Co	1892	Sold May 1925
Coach (open ends) (1st class)	4-axle	53	C	56'4"	Ohio Falls Car Co	1893	Retired 12/14/1933
Coach (open ends) (2nd class)	4-axle	54	C	56'4"	Ohio Falls Car Co	1893	Retired 12/14/1933
Coach (open ends) (1st class)	4-axle	55	C	56'4"	Ohio Falls Car Co. Purchased secondhand from The Moles Co	1901	Rebuilt to Shop Car M55 July, 1924. Motors removed from trucks May 1940 and classified as shop train coach
Coach (open ends)	6-axle	56	B1	61'10"	Secondhand from HL&CW.	1902	Unknown builder - Sold to HL&CW Oct. 1909
Coach (open ends)	6-axle	57	C1	61'10"	Secondhand from HL&CW.	1902	Unknown builder - Sold to D&NE Feb. 1909
Coach (open ends)	6-axle	58	B1	62'8"	HL&CW	1903	Sold to Duluth Iron & Metal Sept 1912
Coach (open ends)	6-axle	59	B1	63'5"	HL&CW	1903	Sold to HL&CW Oct. 1909
Coach (open ends)	6-axle	60	B1	66'10"	HL&CW	1903	Sold to HL&CW Oct. 1909
Coach (open ends)	6-axle	61	B1	66'4"	F.M.Hicks	1906	Retired 12/14/1933
Coach (open ends)	6-axle	62	B1	66'4"	F.M.Hicks	1906	Rebuilt to YMCA car 1915. Sold for scrap to Duluth Iron & Metal 10/23/23
Coach (vestibule ends)	4-axle	63	B2	68'4"	Barney & Smith	1907	Wrecked - Retired 3/9/32
Coach (vestibule ends)	4-axle	64	B2	68'4"	Barney & Smith	1907	Sold Mar. 1929 to Minneapolis, Red Lake and Manitoba Rwy. -
Coach (vestibule ends) 1st class	4-axle	65	B3	68'3"	ACF	1908	Retired 12/14/1933
Coach (vestibule ends) 1st class	4-axle	66	B3	68'3"	ACF	1908	Retired 12/14/1933
Coach (vestibule ends) 2nd class	4-axle	67	C2	68'3"	ACF	1908	Transferred to work equipment 06/5/34 Numbered X67 then to W67
Coach (vestibule ends) 2nd class	4-axle	68	C2	68'3"	ACF	1908	Rebuilt to 114 in April 1927
Coach (vestibule ends) 2nd class	4-axle	69	C2	68'3"	ACF	1908	Retired 12/14/1933
Coach (vestibule ends) 2nd class	4-axle	70	C2	68'3"	ACF	1908	Dismantled Two Harbors 06/05/33
Coach (vestibule ends) all steel	6-axle	80	B4	70'8"	ACF	1911	to work service W80 in 4/26/65 Scrapped 1981
Coach (vestibule ends) all steel	6-axle	81	B4	70'8"	ACF	1911	Retired 7/9/62. Donated to Tower, MN in 1962
Coach (vestibule ends) all steel	6-axle	82	B4	70'8"	ACF	1911	Retired 12/14/1933
Coach (vestibule ends) all steel	6-axle	83	B4	70'8"	ACF	1911	Retired 12/14/1933
Coach (vestibule ends) all steel	6-axle	84	B4	70'8"	ACF	1911	Sold to Thunder Bay Recreation, Marquette, MI in 1962 then to Ill. Railway Museum in?
Coach (vestibule ends) all steel	6-axle	85	B4	70'8"	ACF	1911	To work service W85 in 1959 - then sold to LS&M
Cafe/Observation	6-axle	95	A2	78'6"	Barney & Smith	1907	Sold 2/29 Retired 4/24/29
Parlor/Chair/Observation	6-axle	96	A1	67'6"	HL&CW	1906	Retired 05/08/29 - Still on property Jan. 1930 - Sold to Duluth Iron & Metal 4/29
Parlor/Chair/Obs./Cafe	6-axle	97	A1	66'11"	HL&CW	1904	Retired 1929 - still on property in Jan. 1930 - Sold to Duluth Iron & Metal 4/29
Official Car	4-axle	98	A	42'4"	DM&N Proctor Shops	1898	Rebuilt from caboose (1st) 13 in Aug. 1898 - Renamed Olivette in March 1907
Official Car (open ends)	4-axle	99	A	58'4"	Ohio Falls Car Co	1895	Renamed Missabe 1905
Combination Car (open ends)	4-axle	100	D	58'9"	Ohio Falls Car Co	1893	Stored at Mitchel, MN in May 1931 Retired 12/14/1933
Combination Car (open ends)	4-axle	101	D	58'9"	Ohio Falls Car Co	1893	Retired 12/14/1933
Baggage/RPO/Express (open ends)	4-axle	102	E	58'9"	Ohio Falls Car Co	1893	Retired 04/18/34 Sold for scrap to Duluth Iron & Metal
Combination Car (open ends)	4-axle	103	D1	56'4"	2nd hand from HL&CW - Unknown builder	1902	Retired 12/31/23 Sold for scrap to Duluth Iron & Metal 10/23/23
Baggage/RPO/Express (no ends)	6-axle	104	E1	63'5"	Secondhand from F.M. Hicks	1903	Sold to HL&CW Oct. 1910

Kind	Truck	Name/No	Cl	O.L	Builder	Built	Retired
Baggage/RPO (no ends)	6-axle	105	E1	62'4"	HiL&CW	1905	Sold to Duluth Iron & Metal Dec. 1912
Baggage/RPO (no ends)	6-axle	106	E1	62'4"	F.M. Hicks	1906	Sold to South Dakota Central Rwy May 1916
Combination Car (open ends)	6-axle	107	D1	66'4"	HL&CW	1906	Retired 12/31/23 Sold for scrap to Duluth Iron & Metal 10/23/23
Combination Car (vestibule one end)	4-axle	108 (1st)	D2	66'	ACF	1908	Rebuilt to Motor Car M108 at Proctor in 6/26
Combination Car (vestibule one end)	4-axle	108 (2nd)	D2	66'	Rebuilt Proctor		Converted from Motor Car M-108 to Combination Car (2nd) 108 in 04/04/33. To work equip W108 1958
Combination Car (vestibule one end)	4-axle	109	D2	66'	ACF	1908	To work equip W109 1958
Baggage (no ends) wood	4-axle	110	F1	63'4"	ACF	1908	To work equip. sleeping car W110 7/20/37 06/17/37
Baggage (no ends) wood	4-axle	111	F1	63'4"	ACF	1908	To work equp. Bunk car W111 2/06/40
Baggage/RPO (no ends) all-steel	6-axle	112 (1st)	E2	70'8"	ACF	1911	Conv. To (2nd) 112 05/03/35
Baggage (no ends) all-steel	6-axle	112 (2nd)	F2	70'8"	Rebuilt Proctor	R1935	to W112 5/17/65 Then sold to WGN Spooner, WI in 1997
Baggage/RPO (no ends) all-steel	6-axle	113 (1st)	E2	70'8"	ACF	1911	To (2nd)113 in 1935
Baggage(no ends) all-steel	6-axle	113 (2nd)	F2	70'8"	Rebuilt Proctor	R1935	Rebuilt to (3rd)113 11/6/46
Baggage/RPO (no ends) all-steel	6-axle	113 (3rd)		70'8"	Rebuilt Proctor	R1946	Converted to W113 10/18/60 then renumbered W806
Combination Car (vestibule ends)	4-axle	114	D3	68'3"	Rebuilt DM&N Proctor from 68 in 4/27		To work service bunk/Diner W114 06/7/48. Rebuilt back to coach in ??. Donated to LSMT in 1982.
Private Car (Open Platforms)	4-axle	***Missabe***	A	58'4"	Ohio Falls Car Co	1893	Renumbered from 99 in 1905 - Retired 01/17/31 Sold to Foley Brothers Construction - Resold to George T Malloy St. Paul in 1937. Donated to LSMT in 1997 and restored
Private Car (both ends open)	4-axle	***Olivette***	A	42'4"	DM&N Proctor Shops	1898	Renumbered from 98 in March 1907. Retired 1919
Official Car (vestibule/Open platform)	6-axle	***Frontenac***	A	69'9"	Pullman	circa 1898	Purchased from Minn. Iron Co. 1920 and stenciled DM&N 4/20 - To Bunk car W803 8/8/41
Official Car (vestibule/Open platform)	6-axle	***Soudan***	A	61'	Ohio Falls Car Co	1905	Sold to D&IR March, 1917 -To MofW Engineer's car 184 by 1931
Official Car (vestibule/open platform) all steel	6-axle	***Northland***	A	71'2"	Pullman	1917	Air conditioned in 1950 - Sold LSMT on 08/03/03
Coach	4-axle	***Minnesota II***		85'	Pullman	1946	Purchased from BN 06/06/75 former NP 517. Donated to LSMT 7/31/98

***Abbreviations**:*

CL. *Class*
O.L. *Outside Length over strikers*
HL&CW *Hicks Locomotive & Car Works*
D&NE *Duluth & Northeastern*
LSMT *Lake Superior Museum of Transportation*
WGN *Wisconsin Great Northern, Spooner, WI*
BN *Burlington Northern*
NP *Northern Pacific*

in 1905 and car named *Missabe*. The car was sold in 1931 to Foley Brothers Construction and used as an office at the Boulder Dam construction site. In 1937 George T. Maloy, an executive with Foley Brothers, bought the car and moved it to a location along the St. Croix River in Hudson, Wisconsin for use as a summer home. The car was donated by the family to Lake Superior Museum of Transportation in 1997. It has been restored and is in the museum collection.

An additional first-class coach numbered 55 was purchased secondhand in 1901 from The Males Co., a dealer from Ohio, for $2750. The car originally was built by Ohio Falls Car Co. and was identical to cars 52–54. It would later be rebuilt to motor car M55 in July 1924. Motors were removed from trucks in May 1940, renumbered to 55, and used as a shop train coach at Proctor.

Class A official car 98 was rebuilt from 8-wheel Class I caboose 13 by DM&N Proctor Shops and released in August 1898. It had a wooden carbody with open end platforms on both ends with a total length of 42-foot 4-inches. One end had an observation platform with two windows to the rear and one large and two small windows per side. Two berths were located in the observation area. The remainder of car had five windows per side, a kitchen, two closets, a bathroom and a porter's area on the opposite end as the observation section with a potbelly stove for heat. The car had 6-foot 4-inch wheelbase passenger car trucks and was equipped with Westinghouse air brakes. Ornamental railings and a sofa bed were installed in early 1904. Number 98 was dropped in March 1907 and it was named *Olivette.* The car was retired in 1919.

Expanding passenger business caused DM&N to purchase two additional coaches, Class B1 56, Class C1 57 and one Class D1 combination car 103 in 1902. All were delivered in November 1902. All were secondhand cars purchased from Hicks Locomotive & Car Works (HL&CW) with wooden carbodies and open platforms on both ends at a cost of $9,164.80. Coaches were 61-foot 10-inches long and the combination car 56-foot 4-inches long. Coach 56 had nine paired windows per side, a seating capacity of 61 with two bathrooms located at opposite ends of the car, a small five-person section at one end, steam heat, signal lines and 12-foot 6-inch wheel base 3-axle passenger trucks.

Car 57 was identical to 56 but did not have a smoking section and had seating for 64. Combination car 103 had an 18-foot 2-inch baggage section and 30-foot 11-inch coach section with seating capacity for 36 people, a single bathroom and 6-foot 6-inches wheelbase 2-axle trucks. Car 56 was sold back to HL&CW in October 1909, 57 was sold to Duluth & Northeastern in February 1908 and 103 was retired December 31, 1923.

DM&N returned to HL&CW in 1903 and purchased three secondhand Class B1 coaches 58–60 and Class D1 baggage/mail car 104 for $126,000. All had wooden carbodies and open platforms on both ends on coaches and no platforms or vestibules on car 104. The three coaches were all ordered at the same time, but all three were different. Car 58 had a seating capacity of 61, with nine paired windows per side. Car 59 had a seating capacity of 65 with eight paired windows and one three section window. Car 60 had a seating capacity 69, with seven paired windows and two three section windows. All three floor plans showed bathrooms at both ends of car and a five-person section at one end. All had Westinghouse air brakes and steam lines. Cars 59, 60 were sold back to HL&CW in October 1909 and car 58 was off the roster in September 1912. Baggage and mail car 104 had a 56-foot 4-inch long wooden carbody, end platforms on both ends and equipped with two 6-foot 6-inch wheelbase 3-axle trucks. The baggage section was 18-foot 2-inches long with a side door on each side. The mail section was 16-foot long with two windows on either side of side door, a sorting table and a toilet. The car was sold back to HL&CW in March 1910.

It is unknown why DM&N sold cars 56, 59, 60, 104 back to HL&CW. All future cars came equipped with steam and signal lines.

Prosperity of the iron ranges brought increased demands for luxury. DM&N ordered a single, more luxurious 60-foot long Class A1 observation/cafe car 97 from F.M. Hicks, the predecessor of Hicks Locomotive & Car Works, which was delivered in August 1904 for $6,066. The car had a wooden carbody, open platforms, two 10-foot wheelbase 3-axle trucks, and a seating capacity of 28. Six of these seats were in a section on one end and the remainder, with aisle facing seats in the rear parlor area which had two rearward facing windows. The car had a men's bathroom on one end and a large women's bathroom in the center of the car. It was rebuilt with a small kitchen in 1906 located next the women's bathroom. One side had nine paired windows and the opposite side having an additional single window. All windows had arched frames. Car was retired in 1929 and off roster by early 1930.

Two baggage Class E1 mail cars were ordered from F. M. Hicks, 105 in 1905, and 106 in 1906. The cars were identical, having 60-foot long wooden carbodies, two 10-foot 6-inch wheelbase 3-axle trucks and no end platforms. Baggage section was 39-foot 4-inches long with a single window and side door on each side. Mail section was 20-foot long with two windows on either side of the side door. Car 105 was off roster in 1912 and 106 in 1916.

Official car *Soudan* was received from Ohio Falls Car Co. in 1905. It was 61-foot long, had a wooden carbody, truss rod underframe with open platform on rear and an enclosed vestibule. A 22-foot 9-inch observation section seating 12 people occupied the rear of car. Two bathrooms, a kitchen and a drawing room occupied the remainder of the car. It was equipped with steam and signal lines. It was sold to Duluth & Iron Range in March 1920, placed in MofW service as an engineer's car and renumbered 184 by September 1931.

Apparently satisfied with Hicks, DM&N purchased two additional Class B1 coaches 61-62 an additional Class A1 observation/cafe car 96 and Class D1 combination car 107 in 1906 all with wooden carbodies and open platforms. Coaches were 60-foot, had 10-foot wheelbase 6-wheel trucks, a seating capacity of 70 with two bathrooms at opposite ends of the car. Cars had nine paired and one single window per side with arched frames. Car 96 was 60-foot long, had two 10-foot wheelbase 3-axle trucks, and a seating capacity of 60 people. One end had a section for six people, with a small four-person dining room with table located adjacent to a very small kitchen. The rear observation area had full-height windows facing the observation platform with aisle facing seats. Car 107 was 60-foot long, had a seating capacity of 44, a baggage section 22-foot 9-inches long with a single side door and a coach section 36-foot 7-inches long having five paired windows and one single window all with arched frames. Car 62 was rebuilt to a YMCA car before 1917 by removing all seating to create a dance floor in the car; it was retired in 1923. Car 61 was retired in 1933, 96 in 1929 and 107 in 1923.

Expansion of the iron ranges caused an ever increasing flow of people from Duluth and the need to expand passenger operations. Two Class B2 coaches 63 and 64 were delivered from Barney & Smith in 1907. These had 68-foot 1-inch long wooden carbodies, truss rod underframes and rode on 8-foot wheelbase trucks. Each sat 76 people, had nine paired windows with one single window, all with arched frames. These were he first cars delivered with closed vestibules on both ends.

Ordered from B&S at the same time was Class A2 Café/Observation 95. It had a 78-foot 6-inch wooden carbody, truss rod underframe, rode on two 10-foot 6-inch wheelbase Commonwealth cast steel 3-axle trucks, a closed vestibule, and an observation platform. The car was pure luxury, having a 25-foot rear seating area, five large arched windows on each side with full height arched windows on either side of the observation platform door allowing an unobstructed view. The center of the car had a 6-person seating section and two bathrooms. The rear of the car had four dining tables that sat 12 people and a kitchen area. Car 64 was retired in 1929, 63 in 1932, and 95 in 1919.

A large order was placed with ACF in 1908 for six Class B3 coaches 65–70, two Class D2 combination cars, 108, 109, and two Class F1 baggage cars, 110, 111, at a cost of $71,194.66. The coaches were purchased to replace cars 56–60. Class B3 coaches 65, 66 were considered first-class cars and Class C2 67–70 second-class cars. All were identical 68-foot 3-inches long wooden carbody truss-rod underframe cars

Six all-steel coaches, DM&N 80–85, were purchased from ACF in 1911. Car 85 is pictured at an unknown location in the mid-1940s before being renumbered to W59 in 1959 and placed in work service. Car was sold to Lake Superior & Mississippi tourist railroad at West Duluth where it is preserved (W.C. Olsen)

riding on two 8-foot wheelbase 2-axle trucks. They had seating capacity of 74, enclosed vestibule ends and were almost identical to B&S cars built in 1907. Coach 67 was converted to Proctor Shops jitney car on AFE 2915 June 5, 1934, and restencilled X67. Car 68 was rebuilt to combination car 114 in April 1927, to work service W114 in June 1948 and finally donated to Lake Superior Museum of Transportation in 1982. Remaining cars were retired between 1927 and 1933.

Class D2 combination cars 108 and 109 were 66-foot long wooden carbody truss-rod cars riding on two 8-foot wheelbase Commonwealth cast steel 2-axle trucks. They had a 28-foot long seating area for 34 with vestibule end. At the opposite end of car was a 32-foot baggage section with single side door on each side and no vestibule. Car 108 would be converted to gasoline powered motor car M108 at Proctor in 1926, then back to an unpowered combination car in 1933. Both cars would be placed in MofW service in 1958.

Class F1 baggage cars 110 and 111 were 63-foot 4-inch long wooden carbody truss-rod cars riding on two 8-foot wheelbase Commonwealth cast steel 2-axle trucks. Cars had one large and one small baggage door per side and no vestibules. Both cars were retired by 1940.

The year 1911 found DM&N with an aging passenger car roster of 17 coaches, three observation cars, six combination cars, three baggage/RPO cars and two baggage cars. Five coaches and two combination cars dated from 1893. A strike by RPO clerks in early 1911 got the attention of Congress and on March 4, 1911, they mandated that all future RPO cars must be of all-steel construction. It further required that all wooden RPO cars be retired by July 1, 1916.

To fulfill the RPO mandate and further upgrade the passenger car roster, DM&N ordered six coaches and two baggage/RPO cars from AC& F in 1911. These were the first all-steel constructed passenger cars for DM&N.

Class B4 coaches 80–85 were 70-foot 8-inches long all-steel cars riding on two 10-foot 6-inch wheelbase Commonwealth cast steel 3-axle trucks. Cars 80–83 were delivered in December 1911, 84 in January 1912 and 85 in February 1912. Cars had enclosed vestibules on each end; two bathrooms located one on each end of the car and a seating capacity of 84. Ten paired windows per side were flanked by one single window on each end. Cars 82 and 83 had electrical equipment removed per AFE 2746 March 13, 1933, and installed in D&IR coaches 33 and 34. Cars were all off roster or in MofW service by 1962.

Class E2 baggage/RPO cars 112, 113 were 70-foot 8-inches long all-steel cars riding on two 10-foot 6-inch wheelbase Commonwealth cast steel 3-axle trucks with both being delivered during January 1912. Cars had no vestibules, a 45-foot 3-inch baggage section

Coach DM&N 81 was one of six, built by ACF in 1911. It was donated to Tower, Minnesota, and is preserved there. (American Car & Foundry collection, John W. Barriger III National Railroad Library at UMSL)

DM&N 112 Baggage-Express-RPO, one of two, 112, 113, built by ACF in 1911. Number 112 would be rebuilt into a plain baggage car at Proctor Shops in 1935. (American Car & Foundry collection, John W. Barriger III National Railroad Library at UMSL)

and a 25-foot 5-inch RPO section, each with a single door per side and RPO section having three windows per side. Electrical equipment was removed from both cars per AFE 2750 May 25, 1933, and installed in D&IR mail cars 8 and 9. Declining mail business and consolidation of operations with D&IR allowed both cars to be rebuilt to baggage cars at Proctor in 1935 by removing the RPO section. Windows were blanked and RPO side door was retained and car changed from Class E2 to Class F2. Car 113 would be rebuilt again in 1947 back to a baggage/RPO on AFE 5906 dated November 6, 1946. A 15-foot RPO section with one window and door per side was added. Both cars were placed in MofW service by 1962.

Updating the aging three wooden official cars, two dating from 1890s, caused DM&N to order all-steel 71-foot 2-inch long business car *Northland* from Pullman in 1916 with car being delivered in January 1917 at a cost of $37,067.34. The car was pure luxury for its time, having a 23-foot 9-inch observation section that seated 22. A dining area 15-foot 6-inches seated 11 people at the table, two bathrooms, a kitchen, a full 9-foot bedroom section and two smaller compartments completed the car. It rode on two 11-foot wheelbase 3-axle trucks and had a light weight of 180,000 pounds. The car body was painted Pullman green and roof was painted with one coat of Illinois Steel Co. #2720 Canadian Oxide followed by two coats

Private car *Northland* was purchased from Pullman Car Co. in 1917 becoming the fourth and only steel private car on the DM&N after the *Missabe* built in 1893, *Olivette* built in 1898 and *Frontenac* purchased from Minnesota Iron Co. in 1920. *Northland* was purchased by Lake Superior Museum of Transportation in 2003 and is preserved at the museum in Duluth. Endion, Minnesota, early 1960s. (Ted Schnepf collection)

of Illinois Steel Co. #320 Black Graphite. The car was purchased by the Lake Superior Museum of Transportation on August 3, 2003, and is still in serviceable condition at the museum.

An additional business car *Frontenac* was purchased from Minnesota Iron Co. in March, 1917 for $15,000. DM&N records are sketchy but show the car originally built by Pullman in about 1898. Constructed with an all-wooden carbody, it was 69-foot 7-inches long and rode on two 9-foot 6-inch wheelbase 3-axle trucks. It had a vestibule end with open platform observation end, a 23-foot 11-inch observation section, two bathrooms and two sleeping sections. The car was converted to bunk car W803 per AFE 4664 dated July 29, 1941.

During 1927, Proctor shops rebuilt coach 68 to combination car 114. This was caused by the need for an additional combination car account combination car 101 being retired. The car had a 26-foot baggage section added and retained a seating capacity of 40 people. It was placed in work service in 1948 and numbered W804. Parlor car service was discontinued in 1927.

At the time of merger DM&N rostered 12 passenger cars, including the business cars.

EQUIPMENT LEASED BY DM&N

During October, 1893, DM&N leased Duluth & Winnipeg caboose 03 and coach 100 for one month. Both cars were used on special passenger train from Cloquet, Minnesota via Stony Brook Jct. to Mt. Iron, Minnesota, and return on October 18 and 19, 1893.

Shortages of wood ore cars and increasing ore shipments caused DM&N to lease from Wisconsin Central 400 wooden ore cars during 1897, 200 during 1899 and 200 during 1902 ore shipping seasons for about three months. Fifty flat cars were leased during 1900 from St. Paul and Duluth RR. Fifty logging bunks were leased from T.S.R.L. Co during 1900. During 1902 passenger cars GN 56 and CNW 118 were leased for expanding passenger service.

The DM&N 1903 annual report indicate a severe shortage of equipment with cabooses leased from Dakota & Iowa, refrigerator cars from Armour Refrigerator Lines and flat cars leased from both Chicago Great Western and Minneapolis & St.Louis. ■

Proctor Yard September 1974. A virtual sea of ore cars. Unusual is caboose C-135 normally assigned to T-Bird service at Virginia, and a string of leased former B&LE ore cars re-stencilled DMIR tied on to MofW service passenger cars. (D.P. Holbrook Collection)

CHAPTER FOUR
DULUTH, MISSABE & IRON RANGE
ROLLING STOCK

The Duluth, Missabe & Northern and the Spirit Lake Transfer Railroad were merged in July 1, 1937, to become the Duluth, Missabe & Iron Range and on March 21, 1938, the Duluth & Iron Range and Duluth, Missabe & Iron Range were consolidated into the Duluth, Missabe & Iron Range Railway. These consolidations included the rosters of equipment to which the combined railway continued to maintain and make additions and dispositions.

DM&IR OPEN HOPPERS

The 50-ton DM&N 4500–4699 open hoppers were wearing out in the early 1950s and the DM&IR went shopping for replacement hopper cars to cover movement of coal, limestone and aggregates. An order was placed with Pressed Steel Car for 250 Class Q5 70-ton 40-foot 7-inch hopper cars 2700–2949 on AFE 7079 for $1,950,000 approved on September 27, 1951, and built in 1952. These were identical cars to the former DM&N Class Q2 4700–4749 cars built by Ryan Car Co in 1937. The war years had been hard on the Class Q2 hoppers and AFE 7826 was approved May 27, 1955, for replacement of sides, ends and hopper sheets on these cars.

Unable to secure car builder's shop time, the DM&IR was short of open hoppers by 1953. AFE 7510 for $119,560 was approved on October 15, 1953, for the purchase of 40 second-hand Class Q6 70-ton 40-foot 5-inch rib side open hoppers 4800–4839 from Ortner Car Co. These were originally built by Standard Steel Car Co. in 1924 for NYC in the 900000 series. This purchase allowed the retirement of all Class Q1 hoppers in the 4500–4699 series. A November 9, 1954, inspection of these cars showed 32 cars were serviceable for two additional years and the other eight cars were considered unserviceable. Once the Class Q7 cars arrived on the roster these cars were retired in 1959.

Shop time was finally secured in 1957 and an additional 100 Class Q7 70-ton 40-foot 7-inch hopper cars 2950–3049 from Pullman Standard Car was placed in 1957. Cars were identical to the Class Q5 cars purchased from Ryan Car Co. in 1937. Cars were delivered with journal box lids painted yellow in accordance with then current AAR requirements to indicate that the journal boxes were equipped with lubricating pads. This was important for carmen that were used to using cotton waste for journal box packing. DM&IR was an early advocate of journal box lubricating pads and tested more than a dozen different manufac-

-continued on page 180

DM&IR Freight Car Roster

DM&IR Numbers	Cl	Former Numbers	Old Cl	Description	Total	O.L	C.F	Lbs	Builder	Blt	Trucks	Note
100–849	U4	DMN 100–849	U4	Steel Ore Car	750	22'1"	689	100,000	Pressed Steel Car Co	1906	Arch Bar & A. T-Section	
1000–1249	Q5	DMIR 2700–2949	Q5	Quad hopper	250	40'7"	2,609	154,000	Pressed Steel Car Co	1952	ASF A3 RC	1
2001–2169	V	DMN 2001–2169	V	Flat Car	170	36'11"		60,000	DMC	1892/93	Arch Bar	
2170–2229	Y	DMN 2170–2219	Y	Flat Car	50	36'11"		60,000	Barney and Smith Car Co	1898	Arch Bar	
2220–2228	Y	DMN 2220–2228	Y	Flat Car	9	36'11"		60,000	Barney and Smith Car Co	1899	Arch Bar	
2229–2279	V	DMN 2229–2279	V	Flat Car	51	36'11"		60,000	DM&N Proctor Shops	1903	Arch Bar	
2280–2379	V	DMN 2280–2379	V	Flat Car	100	36'11"		60,000	DM&N Proctor Shops	1905	Arch Bar	
2500–2699	V1	DMN 2500–2699	V1	Flat Car	200	41'		80,000	Pressed Steel Car Co	1906	Arch Bar	
2500–2699	V1	DMN 2500–2699	V1	Gondola	26	41'	978	80,000	Pressed Steel Car Co	1906	Arch Bar	2, 31
2700–2949	Q5			Quad hopper	250	40'7"	2,150	140,000	Pressed Steel Car Co	1952	ASF A3 RC	
2950–3049	Q7			Quad hopper	100	40'7"	2,150	140,000	PS	1957	ASF A3 RC	3
3100–3149	P	DMN 3100–3149	P	Box Car	50	41'4"	2,921	80,000	WSCF	1906	Arch Bar	
3150–3249	P1	DMN 3150–3249	P1	Box Car	100	41'9"	2,804	80,000	Standard Steel Car	1907	Arch Bar	
3150–3154	P1	DMN 3150–3154	P1	Automobile Car	5	41'9"	2,804	80,000	Standard Steel Car	1907	Arch Bar	4
3300–3399	P2	DMN 3300–3399	P2	Box Car	100	42'1"	3,153	80,000	ACF	1923	A. U-Section	
3379	P2		P2	Box Car	1	42'1"	3,153	80,000	ACF	1923	A. U-Section	5
3500–3699	V6	EJE		Gondola	200	50'6"	1,360	110,000	Mt. Vernon Car Mfg. Co	1937	Barber S1L	6
3700–3824	V7	EJE 34200–34899		Gondola	125	51'	1,360	110,000	ACF & Ralston Steel Car Co	1947/48	Barber S1L & S2A	7
3900–3980	V8	PLE 9000–10499		Gondola	81	55'1"	1,912	154,000	NYC Despatch Shops	1946/53	Barber S2A	8
4002–4003	P	DMN 4002–4003	P	Stock Car	2	41'4"	2,921	80,000	WSCF	1906	Arch Bar	9.
4026–4084	Q	DMN 4026–4084	Q	Gondola	59	42'5"	1,075	100,000	Rogers Ballast Car Co	1907	Arch Bar	10
4100–4199	V2			Gondola	100	41'5"	1,959	100,000	ACF Lot #2199	1941	Barber S2	11
4201–4220	H	DIR 201–220	H	Drop Bottom Gon	20	41'1"	1,546	100,000	Pressed Steel Car Co	1908	Arch Bar	
4221–4235	H	DIR 221–235	H	Drop Bottom Gon	15	41'1"	1,546	100,000	Pressed Steel Car Co	1907	Arch Bar	
4251–4260	H	DIR 251–260	H	Drop Bottom Gon	10	41'8"	1,546	100,000	Pressed Steel Car Co	1908	Arch Bar	
4261–4310	H1	DIR 261–310	H	Drop Bottom Gon	50	41'8"	1,546	100,000	WSC&F	1918	Andrews U-Section	
4200–4239(2nd)	V9	WM 55336–55848		Gondola	40	54'6"	1,745	154,000	Bethlehem Steel	1951/54	ASF A3 & Barber S2A	12
4300–4339(2nd)	V9	WM 55336–55648		Gondola	10	54'6"	1,745	154,000	Bethlehem Steel	1951/54	ASF A3 & Barber S2A	13
4350–4359	V3	EJE 90000–90099		Mill Gondola	50	67'	1,761	150,000	Mt. Vernon Car Mfg. Co	1937	Barber	14
		EJE 4350–4374		Gondola	25	53'11"	1,995	200,000	GSCC	A1974	Barber S2A	15
4375–4424	V4	URR 8000–8965		Gondola	50	44'9"	1,529	154,000	GSCC	1953		16
		EJE 4410–4497		Gondola	25	53'11"	1,995	200,000	GSCC	A1974	Barber S2A	17
4500–4699	Q1	DMN 4500–4699	Q1	Twin hopper	200	32'3"	1,565	100,000	Pullman Car Co. Lot #5269	1916	A. T-Section	18
4500–4649(2nd)	V5	EJE 33750–34189		Gondola	150	51'1"	1,360	110,000	Ralston Steel Car Co	1943/4	Barber S1L	19
4700–4749	Q2	DMN 4700–4749	Q2	Quad hopper	50	40'7"	2,150	140,000	The Ryan Car Co	1937	ASF SP	
4700–4749	Q2			Quad hopper	50	40'7"	2,627	154,000	The Ryan Car Co	1937	ASF SP	20
4800–4839	Q6	NYC 900000 series		Triple hopper	40	40'5"	2,700	140,000	Standard Steel Car	1924		21
4950–4959	Q3			Covered Hopper	10	35'2"	1,958	140,000	ACF Lot #2120	1941	National B1	22
4960–4969	Q4			Covered Hopper	10	35'3"	1,958	140,000	ACF Lot #3674	1952	ASF A3 RC	23
4970–4989	Q8			Covered Hopper	20	35'3"	2,003	140,000	PS Lot #8331C	1956	ASF A3 RC	24
5000–5048e	F	DIR 5000–5048e	F	Box Car	24	37'7"	2,194	60,000	DMC	1895	Arch Bar	
5033–5061o	K	DIR 5033–5061o	I	Flat Car	15	37'7"		48,000	DIR Two Harbors	1887	Arch Bar	
5050–5120e	F	DIR 5050–5120e	F	Box Car	11	37'7"	2,194	60,000	Wells & French Co	1886	Arch Bar	
5063–5145o	K	DIR 5063–5145o	I	Flat Car	45	37'7"		60,000	D&IR Two Harbors	1887	Arch Bar	31
5122–5132e	F	DIR 5122–5132e	F	Box Car	6	37'7"	2,342	60,000	D&IR Two Harbors	1883	Arch Bar	
5134–5136e	M	DIR 5134–5136e	M	Stock Car	2	37'4"	2,284	60,000	D&IR Two Harbors	1884	Arch Bar	25
5140–5142e	F	DIR 5140–5142e	F	Box Car	2	37'7"	2,342	60,000	D&IR Two Harbors	1904	Arch Bar	
5146	F	DIR 5146	F	Box Car	1	35'5"	2,194	60,000	D&IR Two Harbors	1904	Arch Bar	
5147–5541o	K	DIR 5147–5541o	I	Flat Car	200	37'7"		60,000	Lafayette	1888	Arch Bar	31
5148	F	DIR 5148	F	Box Car	1	37'7"	2,342	60,000	D&IR Two Harbors	1904	Arch Bar	
5152	F	DIR 5152	F	Box Car	1	35'5"	2,194	60,000	D&IR Two Harbors	1904	Arch Bar	
5160	F	DIR 5160	F	Box Car	1	37'7"	2,342	60,000	D&IR Two Harbors	1904	Arch Bar	
5174–5216e	F	DIR 5174–5216e	F	Box Car	25	38'3"	2,389	60,000	WSCF	1903	Arch Bar	
5176	M	DIR 5176	M	Stock Car	1	37'4"	2,448	50,000	D&IR Two Harbors	1903	Arch Bar	
5200–5202e	S	DIR 5200–5202e	S	Automobile Car	2	38'3"	2,414	60,000	WSCF	1903	Arch Bar	26

DM&IR Numbers	Cl	Former Numbers	Old Cl	Description	Total	O.L	C.F	Lbs	Builder	Blt	Trucks	Note
5224–5260e	F1	DIR 5224–5260e	F1	Box Car	20	38'3"	2,508	60,000	WSCF	1908	Arch Bar	
5264–5340e	F	DIR 5264–5340e	F	Box Car	37	37'5"	2,318	60,000	D&IR Two Harbors	1908	Arch Bar	
5342–5390	F2	DIR 5342–5390	F2	Box Car	25	37'6"	2,448	88,000	Standard Steel Car Co	1918	A. U-Section	
5543–5641o	K	DIR 5543–5561o	I	Flat Car	50	37'7"		60,000	Illinois Car Co	1899	Arch Bar	31
5643–5741o	K	DIR 5643–5699o	I	Flat Car	50	37'7"		60,000	WSCF	1903	Arch Bar	31
5743–5825o	K	DIR 5743–5825o	I	Flat Car	42	37'7"		60,000	D&IR Two Harbors	1905	Arch Bar	31
5827–5841o	K	DIR 5827–5841o	I	Flat Car	8	37'7"		60,000	D&IR Two Harbors	1906	Arch Bar	
5000–5008(2nd)	Q9			Covered Hopper	9	43'	3,560	200,000	ACF	1968	ASF RC	
5009–5012(2nd)	Q10			Covered Hopper	4	43'	3,560	200,000	ACF	1968	ASF RC	
5013–5017(2nd)	Q11			Covered Hopper	5	43'	3,560	200,000	ACF	1969	ASF RC	
5018(2nd)	Q12			Covered Hopper	1	43'	3,560	200,000	ACF	1970	ASF RC	
5019–5026(2nd)	Q13			Covered Hopper	8	43'	3,560	200,000	ACF	1970	ASF RC	
5027–5039(2nd)	Q14	HTCX 9851–9863		Covered Hopper	13	43'	3,560	200,000	ACF	1972	ASF RC	27
5900–5919	P3	ACL 14100–14199		Box Car	20	51'8"	4,680	110,000	ACF	1941	National B1	28
6000–6049	K1	DIR 6000–6049	I	Flat Car	50	37'7"		60,000	D&IR Two Harbors	1900	Arch Bar	31
6050–6099	K1	DIR 6050–6099	I	Flat Car	50	37'7"		60,000	D&IR Two Harbors	1910	Arch Bar	
6100–6179	G1	DIR 6100–6179	G	Gondola	80	37'11″	953	60,000	WSCF	1907	Arch Bar	29
6135/6144	G1	DIR 6135/6144	G	Gondola	2	37'11″	1,050	60,000	WSCF	1907	Arch Bar	30
6180–6279	K2	DIR 6180–6279	I	Flat Car	100	37'11"		60,000	ACF Lot #7934	1916	Arch Bar	31
6180–6279	G2	DIR 6180–6279	G2	Gondola	100	37'11"	956	60,000	ACF Lot #7934	1916	Arch Bar	31, 32
6280–6304	K2	DIR 6280–6304	I	Flat Car	25	37'11"		60,000	ACF Lot #8406	1917	A. U-Section	31
6305, 6306	K3	DIR 6305, 6306		Flat Car	2	37'6"		80,000	Standard Steel Car Co	1918	A. U-Section	33
6310	K4	DMIR 5248		Flat Car	1	38'3"		60,000	WSCF	1907	Arch Bar	34
6315	K5	EJE 6803		Flat Car	1	54'2"		152,000	Ralston Steel Car Co	1943	Barber	35
6316	K5	EJE 6818		Flat Car	1	54'2"		154,000	Ralston Steel Car Co	1943	Barber	36
6316 (2nd)	K5	EJE 6800 series		Flat Car	1	54'2"		154,000	Ralston Steel Car Co	1943	Barber	37
6325–6339	K6	EJE 6300-6500 series		Flat Car	15	54'2"		110,000	Ralston Steel Car Co	1943	Barber	38
6325/6339	K7	EJE 6300-6500 series		Flat car	5	54'2"		110,000	Ralston Steel Car Co	1943	Barber	39
6340–6354	K7	EJE 6300-6500 series		Flat Car	15	54'2"		110,000	Ralston Steel Car Co	1943	Barber	40
7022–7033	J1	DMN 5022–5033	J1	Refrigerator Car	12	37'11"	1,904	50,000	ACF	1906	Arch Bar	41
7034–7058	J2	DMN 5034–5058	J2	Refrigerator Car	25	37'11"	1,904	50,000	Standard Steel Car Co	1908	Arch Bar	42
7059–7083	J3	DMN 5059–5083	J3	Refrigerator Car	25	37'11"	1,904	50,000	Peteler Car Co	1910	Arch Bar	43
7122–7130	R	DIR 8022–8030	R	Refrigerator Car	11	38'2"	1,904	80,000	Peteler Car Co	1912	Arch Bar	44
7131	R	DIR 8031	R	Refrigerator Car	1	38'2"	1,583	80,000	Peteler Car Co	1912	Arch Bar	45
7132–7137	R1	DIR 8032–8037	R1	Refrigerator Car	6	38'2"	1,583	80,000	D&IR Two Harbors	1918	A. T-Section	46
7138, 7139	R1	DIR 8038, 8039	R1	Refrigerator Car	2	38'2"	1,583	80,000	D&IR Two Harbors	1918	A. T-Section	47
7140	R1	DIR 8040	R1	Refrigerator Car	1	38'2"	1,583	80,000	D&IR Two Harbors	1918	A. T-Section	
8005–8354	U1	DMN 8001–8354	U1	Steel Ore Car	350	22'1"	680	100,000	Pressed Steel Car Co	1901	Arch Bar	
8405–9704	U3	DMN 8405–9704	U3	Steel Ore Car	1300	22'1"	689	100,000	Standard Steel Car Co	1905	Arch Bar	
9705–9954	U3	DMN 9705–9954	U3	Steel Ore Car	250	22'1"	689	100,000	Standard Steel Car Co	1906	Arch Bar	
9955–10554	U3	DMN 9955–10554	U3	Steel Ore Car	600	22'1"	689	100,000	Standard Steel Car Co	1907	Arch Bar	
10555–11704	U4	DMN 10555–11704	U4	Steel Ore Car	1150	22'1"	689	100,000	Pressed Steel Car Co	1907	Arch Bar & A.L-Section	
11004/11247 (2nd)	Q5	DMIR 2700/2949	Q5	Quad hopper	101	40'7"	2,609	154,000	Pressed Steel Car Co	1952	ASF A3 RC	48
12000–12149	Q15	NYC & BA see text	Q15	Triple hopper	150	41'8"	2,662	154,000	Various - See Text	1956-1961	ASF A3 RC	49
14000–14149	U6	DMN 14000–14149	U6	Steel Ore Car	150	22'1"	669	100,000	WSCF	1910	Arch Bar	
14500–15499	U9	DMN 14500–15499	U9	Steel Ore Car	1000	22'1"	669	100,000	WSCF	1913	A. T-Section	
15026/15098		BLE 15000–15099		Gondola	10	54'2"	1,976	185,000	GSCC Order #962	1968	Buckeye	50
16000–16749	U7	DMN 16000–16749	U7	Steel Ore Car	750	22'	667	100,000	Standard Steel Car Co	1910	Arch Bar	
19000–19999	U10	DMN 19000–19999	U10	Steel Ore Car	1000	22'1"	704	100,000	WSCF	1914	A. T-Section	
20000		DMN 20000		Steel Ore Car	1	22'	654	100,000	Ralston Steel Car Co	1916	A. T-Section	51
20001–21000	U11	DMN 20001–21000	U11	Steel Ore Car	1000	22'	704	100,000	WSCF	1916	A. T-Section	
20001–20700		BLE 20000–20700		Steel Ore Car	61	20'7″	1,085	154,000	GSCC	1952	ASF RC	52
21001–21025	U12	DMN 21001–21025	U12	Steel Ore Car	25	21'5"	935	140,000	ACF Lot #9907	1925	A. U-section	53
21026	U12	DMN 21026	U12	Steel Ore Car	1	21'5"	935	140,000	ACF Lot #9532	1929	A. U-section	54

DM&IR Numbers	Cl	Former Numbers	Old Cl	Description	Total	O.L	C.F	Lbs	Builder	Blt	Trucks	Note
21101–21125	U13	DMN 21101–21125	U13	Steel Ore Car	25	21'6"	1,046	140,000	PS Lot #5384	1925	Pflager	55
21126	U13	DMN 21126	U13	Steel Ore Car	1	21'6"	965	140,000	GATC	1931	Pflager	56
22001–22125	U14	DMN 22001–22125	U14	Steel Ore Car	125	21'5"	935	140,000	GATC	1928	Dahlmen 2-Level	57
23001–23125	U15	DMN 23001–23125	U15	Steel Ore Car	125	21'5"	943	140,000	Standard Steel Car Co	1927	Dahlmen 2-Level	
23171–23175	U16	DMN 23171–23175	U16	Steel Ore Car	5	21'6"	1,080	140,000	Standard Steel Car Co	1925	Pflager	58
23176, 23177	U16	DMN 23176, 23177	U16	Steel Ore Car	2	21'6"	1,055	140,000	Standard Steel Car Co	1925	Verona	59
23975–23999	U17		U17	Steel Ore Car	25	21'6"	1,000	140,000	PS Lot #5583	1938	ASF	60
24000–24799	U17	DMN 24000–24799	U17	Steel Ore Car	800	21'6"	1,000	140,000	PS Lot #5554	1937	ASF	
24800–24999	U17	DMN 24800–24999	U17	Steel Ore Car	200	21'6"	1,000	140,000	PS Lot #5554	1938	National B1	61
25000–25499	U18	DMN 25000–25499	U18	Steel Ore Car	500	21'6"	1,000	140,000	GACC	1937	ASF	
25500–25999	U19			Steel Ore Car	500	21'6"	1,000	140,000	PS Lot #5723	1942	Buckeye	62
26000–26499	U20			Steel Ore Car	500	21'6"	1,000	140,000	GACC	1942	ASF/National B1	63
26500–26999	U21			Steel Ore Car	500	21'6"	1,000	140,000	ACF Lot #2513	1942	ASF SP	64
27000–27499	U22			Steel Ore Car	500	21'6"	1,000	140,000	Pressed Steel Car Co	1943	ASF SP	
27500–27999	U23			Steel Ore Car	500	21'6"	1,000	140,000	PS Lot #5903	1948	ASF A3 RC	65
28000–28499	U24			Steel Ore Car	500	21'6"	1,000	140,000	ACF Lot #3285	1948	ASF A3 RC	66
28500–28999	U25			Steel Ore Car	500	21'6"	1,000	140,000	Pressed Steel Car Co	1948	ASF A3 RC	
29000–29499	U26			Steel Ore Car	500	21'6"	1,000	140,000	GACC	1948	ASF A3 RC	
29500–30499	U27			Steel Ore Car	1000	21'6"	1,000	140,000	PS Lot #5921	1949	ASF A3 RC	67
30500–30999	U28			Steel Ore Car	500	21'6"	1,000	140,000	GACC	1949	ASF A3 RC	
30025/30189 (2nd)		BLE 30025/30189		Gondola	10	53'11"	1,985	198,000	ACF Lot #11-09601	1980		68
31000–32498	U29			Steel Ore Car	1499	21'6"	1,000	140,000	PS. Lot No #8041	1952	ASF A3 RC	69
32499	U29			Steel Ore Car	1	21'6"	1,000	140,000	PS Lot No #8041	1952	ASF A3 RC	70
32500–32999	U30			Steel Ore Car	500	21'6"	1,000	140,000	PS Lot No #8125	1953	ASF A3 RC	71
33000–33499	U31			Steel Ore Car	500	21'6"	1,000	140,000	ACF Lot 01-4974	1957	ASF A3 RC	72
33900–34399	D3	DIR 3900–4399	D3	Steel Ore Car	500	22'1"	689	100,000	Standard Steel Car Co	1905	Arch Bar	
34400–34899	D4	DIR 4400–4899	D4	Steel Ore Car	500	22'1"	689	100,000	Pressed Steel Car Co	1906	Arch Bar	
39000–39699	D4	DIR 9000–9699	D4	Steel Ore Car	700	22'1"	689	100,000	Pressed Steel Car Co	1907	Arch Bar	
40500–41249	E7	DIR 10500–11249	E7	Steel Ore Car	750	22'	667	100,000	Standard Steel Car Co	1910	Arch Bar	
41250–41399	E6	DIR 11250–11399	E6	Steel Ore Car	150	22'1"	669	100,000	Pressed Steel Car Co	1910	Arch Bar	
41500–42299	E8	DIR 11500–12299	E8	Steel Ore Car	800	22'	695	100,000	Standard Steel Car Co	1913	Arch Bar	
42300–42499	E9	DIR 12300–12499	E9	Steel Ore Car	200	21'11"	685	100,000	ACF	1913	Arch Bar	
42500–42749	E10	DIR 12500–12749	E10	Steel Ore Car	250	21'11"	685	100,000	ACF	1916	A. T-Section	
42750–43249	E11	DIR 12750–13249	E11	Steel Ore Car	500	21'11"	695	100,000	Standard Steel Car Co	1916	A. T-Section	
40000–40167	VAR	DMIR 29500/33499	VAR	HiSide Crude Tac	168	21'6"	1,314	154,000	See individual classes	R1976	VAR	73
40168–40271	VAR	DMIR 29500/33499	VAR	HiSide Crude Tac	104	21'6"	1,314	154,000	See individual classes	R1976	VAR	74
40272–40279	VAR	DMIR 29500/33499	VAR	HiSide Crude Tac	8	21'6"	1,314	154,000	See individual classes	R1978	VAR	74
40280–40460	VAR	DMIR 29500/33499	VAR	HiSide Crude Tac	181	21'6	1,314	154,000	See individual classes	R1982	VAR	74
40461–40510(2nd)	VAR	DMIR 29500/33499	VAR	HiSide Crude Tac	50	21'6"	1,314	154,000	See individual classes	R1988	VAR	74
40511–40522(2nd)	VAR	DMIR 29500/33499	VAR	HiSide Crude Tac	12	21'6"	1,314	154,000	See individual classes	R1989	VAR	74
40523–40533(2nd)	VAR	DMIR 29500/33499	VAR	HiSide Crude Tac	11	21'6"	1,314	154,000	See individual classes	R1994	VAR	74
40534–40562(2nd)	VAR	DMIR 29500/33499	VAR	HiSide Crude Tac	29	21'6"	1,314	154,000	See individual classes	R1994	VAR	74
40563–40587(2nd)	VAR	DMIR 29500/33499	VAR	HiSide Crude Tac	25	21'6"	1,314	154,000	See individual classes	R1994	VAR	74
40588–40612(2nd)	VAR	DMIR 29500/33499	VAR	HiSide Crude Tac	25	21'6"	1,314	154,000	See individual classes	R1995	VAR	74
40613–40712(2nd)	VAR	DMIR 29500/33499	VAR	HiSide Crude Tac	100	21'6"	1,314	154,000	See individual classes	R1997	VAR	74
40713–40714(2nd)	VAR	DMIR 29500/33499	VAR	HiSide Crude Tac	2	21'6"	1,314	154,000	See individual classes	R1997	VAR	74
40715–40718(2nd)	VAR	DMIR 29500/33499	VAR	HiSide Crude Tac	4	21'6"	1,314	154,000	See individual classes	R1998	VAR	74

DM&IR Numbers	Cl	Former Numbers	Old Cl	Description	Total	O.L	C.F	Lbs	Builder	Blt	Trucks	Note
40719–40727(2nd)	VAR	DMIR 29500/33499	VAR	HiSide Crude Tac	9	21'6"	1,314	154,000	See individual classes	R1998	VAR	74
40728–40797(2nd)	VAR	DMIR 29500/33499	VAR	HiSide Crude Tac	70	21'6"	1,314	154,000	See individual classes	R1999	VAR	74
40798–40807(2nd)	VAR	DMIR 29500/33499	VAR	HiSide Crude Tac	10	21'6"	1,314	154,000	See individual classes	R2002	VAR	74
40808–40831(2nd)	VAR	DMIR 29500/33499	VAR	HiSide Crude Tac	24	21'6"	1,314	154,000	See individual classes	R2004	VAR	74
44100–44124		PLE 44100–44124		Gondola	25	57'4"	1,912	200,000	GSCC 0.0. 868	1965		75
45001–45050	VAR	DMIR 29500/33499	VAR	HiSide Crude Tac	48	21'6"	1,314	154,000	See individual classes	R1986/87	VAR	76
45051–45052	VAR	DMIR 29500/33499	VAR	HiSide Crude Tac	2	21'6"	1,314	154,000	See individual classes	R1989	VAR	76
45053	VAR	DMIR 29500/33499	VAR	HiSide Crude Tac	1	21'6"	1,314	154,000	See individual classes	R1994	VAR	76
49001–49008	VAR	DMIR 29500/33499	VAR	Crude Tac Idler car	8	21'6"	1,314	154,000	See individual classes	R1989	VAR	77
(2nd)49001/49008	VAR	DMIR 29500/33499	VAR	Crude Tac Idler car	5	21'6"	1,314	154,000	See individual classes	R1998	VAR	78
(2nd)49004/49006	VAR	DMIR 29500/33499	VAR	Crude Tac Idler car	2	21'6"	1,314	154,000	See individual classes	R1999	VAR	79
50031/59991	VAR	DMIR 29500/33499	VAR	HiSide Tac Car	388	21'6"	1,314	154,000	See individual classes	R1964	VAR	80
51001/53498	VAR	DMIR 31001/33498	VAR	Mini-HiSide	356	21'6"	1,157	154,000	See individual classes	R1971–72	VAR	81
51001/53498	VAR	DMIR 29500/33499	VAR	Mini-quad sets	1560	21'6"	1,157	154,000	See individual classes	R1971–75	VAR	82
60000–61142	VAR	DMIR 29500–33500	VAR	Steel Ore Car	465	21'6"	1,000	154,000	Various – See Text	VAR	VAR	83
		EJE 73101–73200		Coke Hopper	100	60'10"	4,415	146,000	GSCC	1973	Barber S2A	84
	Q15	EJE 74000–74349		Triple Hopper	350	41'8"	2,662	154,000	Various – See Text	1956–1961	ASF A3 RC	85
77054/78500		BLE 92076–95944		Triple Hopper	10	44'2"	2,775	154,000	ACF	1936		86
83053/83629		EJE 83053–83629		Gondola	25	57'1	1,912	154,000				87
92076/95944		BLE 92000–96999		Triple Hopper	34	44'2"	2,775	200,000	PS & Pressed Steel Car	VAR		88
529615/533499	VAR	DMIR 29500/33499	VAR	HiSide Crude Tac	69	21'6"	1,314	154,000	See individual classes	R1964	VAR	89
529615/533499	VAR	DMIR 29500/33499	VAR	HiSide Tac Car	224	21'6"	1,314	140,000	See individual classes	VAR	VAR	90
529615/533499	VAR	DMIR 29500/33499	VAR	HiSide Tac Car	5	21'6"	1,444	140,000	See individual classes	VAR	VAR	91
87347/88013		EJE 87347/88013		Gondola	15	56'7"	1,995	194,000	ACF	1980		92
		EJE 88000–88199		Gondola	200	52'6"	1,995	200,000	GSCC	A1974	Barber S2A	93
24100	U17	DMIR 24100	U17	Ballast car	1	21'6"	1,000	154,000	PS Lot #5554	1937	ASF A3 RC	94
24350	U17	DMIR 24350	U17	Ballast car	1	21'6"	1,000	154,000	PS Lot #5554	1937	ASF A3 RC	95
32675	U30	DMIR 32675	U30	Ballast car	1	21'6"	1,157	154,000	PS Lot No #8125	1952	ASF A3 RC	96
32880	U30	DMIR 32880	U30	Ballast car	1	21'6"	1,157	154,000	PS Lot No #8125	1952	ASF A3 RC	97
24158	U17	DMIR 24158	U17	Covered hopper	1	21'6"	1,386	154,000	PS Lot #5554	1937	ASF A3 RC	98
24187	U17	DMIR 24187	U17	Covered hopper	1	21'6"	1,386	154,000	PS Lot #5554	1937	ASF A3 RC	99
24456	U17	DMIR 24456	U17	Covered hopper	1	21'6"	1,386	154,000	PS Lot #5554	1937	ASF A3 RC	100
24899	U17	DMIR 24899	U17	Covered hopper	1	21'6"	1,386	154,000	PS Lot #5554	1938	ASF A3 RC	101
28214	U24	DMIR 28214	U24	Covered hopper	1	21'6"	1,386	154,000	ACF. Lot #3285	1948	ASF A3 RC	102
1400–1501	VAR		VAR	Ballast car	102	21'6"	1,157	154,000	Various	VAR	ASF A3 RC	103

Abbreviations

/	Not continuous numbers
A	Acquired
A	Andrews
ACF	American Car & Foundry
DMC	Duluth Manufacturing Co
E	Even numbers only
GACC	General American Car Co
GATC	General American Tank Car Co
GSCC	Greenville Steel Car Co
O	Odd numbers only
PS	Pullman Standard
R	Rebuilt
RC	Ride Control
SP	Spring Plankless
WSCF	Western Steel Car & Foundry

DM&IR FREIGHT CAR ROSTER NOTES

Note 1. Rebuilt by DMIR Proctor shops with Pullman Standard parts betwwen 1971-73 from offset side to rib side

Note 2. DM&N flat cars converted to gondolas in 1919. All on roster from 1919 until 1932 when converted back to flat cars. 2501, 2505, 2515, 2516, 2529, 2535, 2546, 2558, 2564, 2565, 2576, 2580, 2581, 2587, 2589, 2592, 2596, 2624, 2625, 2629, 2635, 2650, 2567, 2571, 2685, 2691. DM&N flat cars converted to gondolas in 1932: 2500, 2504, 2514, 2515, 2528, 2534, 2545, 2557, 2563, 2564, 2575, 2579, 2580, 2586, 2588, 2591, 2595, 2623, 2624, 2628, 2634, 2649, 2656, 2670, 2684, 2690

Note 3. All sold to EJE in Feb. 1975

Note 4. Rebuilt with 12' Door opening in 1910 for Automobile service

Note 5. Rebuilt by DMIR Proctor Shops with Steel sheathing in March 1941

Note 6. Received from EJE in 1969

Note 7. Received from EJE in 1972

Note 8. Former PLE 9000-10499 purchased by EJE from PLE. Per AFE 11037 09-28-73 Received from LS&I via EJ&E in 1973. (EJE Joliet shops reconditioned cars before delivery to LS&I for exchange to DM&IR for ore cars) - Per AFE 11037 06-28-76 40 cars sold to EJE and Renumbered EJE 84354-84393

Note 9. Rebuilt from box cars 3148, 3149 in 1906 3149 back to box car ?? To stock 4003 5-35, to box car 3149 12-13-37.

Note 10. Rebuilt to solid bottom and permanent sides in 1922, then in 1935 some rebuilt to drop-side cars

Note 11. Ordered 12-19-1940 ACF Chicago plant. Cor-Ten steel

Note 12. Bought second hand from WM in 1980 - 7 cars rebuilt to rack gons 1981, 33 cars reblt to rack gons 1982

Note 13. Bought second hand from WM in 1980 - 7 cars rebuilt to rack gons 1981, 33 cars reblt to rack gons 1982 -Ten cars to 4300 series pulpwood end extensions removed 3-20-92

Note 14. Received from EJE 1965 - Former EJE 90002, 90066, 90080, 90093, 90095, 90025,90034, 90075, 90040, 90006 renumbered in that order

Note 15. Owned by DM&IR - Leased to EJE - Equipped with troughs for handling steel coils(Leased to EJE 1974)

Note 16. Received from URR 1966

Note 17. Owned by DM&IR - Leased to EJE - Equipped with troughs for handling steel coils(Leased to EJE 1974)

Note 18. Ordered 11-15. Built 5-6 1916

Note 19. Received from EJE 1966(End racks applied by EJE then delivered and repainted at Proctor)

Note 20. Rebuilt by EJE Joliet shops in 1975 from DMIR 4700-4749(rebuilt from offset side to rib side cars)

Note 21. Received second hand from Ortner Car Co 1954

Note 22. Ordered 9-18-40 ACF Madison, IL Plant

Note 23.Ordered 12-20-50 Built at ACF Berwick, PA plant

Note 24. Ordered 12-55 Built at PS Butler, PA plant 12-56

Note 25. Rebuilt to stock cars 1899 - See D&IR roster for details

Note 26. Rebuilt with 10' door openings (5200 in 1910, 5202 in 1912) 5200 Sold to D&NE 8/10/1972

Note 27. Acquired secondhand from Hallett Minerals April 26, 1972

Note 28. Purchased second hand from Chicago Freight Car Co in 1966 (Reconditioned by CFC before deliver/converted from double door)

Note 29. Converted from flatcars to wood sided gondolas at Two Harbors (25 cars in 1914 and 51 cars in 1929)

Note 30. Converted from flatcars to steel sided gondolas at Two Harbors (6144 in 1922 and 6135 in 1930)

Note 31. Cars converted to rack flats: Cars in series 6001-6047 temporarily equipped with end racks for the handling of pulpwood and differ in inside dimensions from other cars in the same series. Inside length 33' 4". Extreme height 10' 8"

Car No.	Date Equip.	Retired	Car No.	Date Equip.	Retired
6001	1936	1965	6034	1936	1964
6002	1936	1964	6038	1936	1965
6021	1936	1964	6039	1936	1959
6026	1936	1959	6042	1936	1962

Cars in series 5089 to 5835 temporarily equipped with end racks for the handling of pulpwood and differ in inside dimensions from other cars in the same series. Inside length 33' 4". Extreme height 10' 8"

Car No.	Date Equip.	Retired	Car No.	Date Equip.	Retired
5137	1936	1959	5633	1936	1962
5183	1936	1959	5647	1936	1959
5317	1936	1962	5717	1936	1964
5459	1936	1959	5725	1936	1964
5463	1936	1960	5737	1936	1945
5537	1936	1964	5789	1936	1962
5835	1936	1964			

Cars in series 6001-6304 temporarily equipped with end racks for the handling of pulpwood and differ in inside dimensions from other cars in the same series. Inside Length 34' 5". Extreme height 10' 5"

Car No.	Date Equip	Retired	Car No.	Date Equip	Retired
6188	1966	1970	6249	1935	1966
6212	1935	1937	6250	1935	1970
6214	1935	1937	6252	1936	1970
6216	1935	1965	6254	1935	1970
6217	1935	1970	6255	1966	1970
6219	1935	1971	6256	1935	1968
6220	1935	1970	6258	1935	1970
6222	1935	1970	6260	1935	1970
6223	1935	1970	6261	1935	1970
6224	1935	1970	6263	1935	1964
6225	1935	1971	6264	1935	1964
6226	1935	1968	6265	1966	1970
6228	1935	1964	6266	1935	1971
6229	1935	1970	6267	1935	1970
6230	1935	1968	6268	1935	1965
6231	1935	1970	6269	1934	1970
6232	1935	1972	6270	1934	1971
6233	1935	1970	6271	1934	1970
6234	1935	1970	6272	1934	1965
6236	1936	1970	6273	1934	1968
6237	1935	1970	6274	1934	1970
6239	1935	1971	6275	1934	1968
6241	1935	1968	6276	1934	1970
6242	1935	1964	6277	1934	1971
6245	1935	1970	6278	1934	1964
6246	1935	1970	6283	1966	1970
6247	1935	1966	6284	1966	1970
6248	1935	1970	6297	1966	1970

Cars in series 6001 to 6304 temporarily equipped with end racks for handling pulpwood and differ in inside dimensions from other cars in the same series. Inside Length 33' 4". Extreme height 10' 5".

Car No.	Date Equip.	Retired	Car No.	Date Equip.	Retired
6207	1964/5	1970	6288	1964/5	1970
6215	1964/5	1966	6289	1964/5	1970
6218	1964/5	1970	6290	1964/5	1970
6221	1964/5	1970	6291	1964/5	1970
6227	1964/5	1970	6292	1964/5	1970
6240	1964/5	1970	6295	1964/5	1970
6251	1964/5	1970	6300	1964/5	1970
6281	1964/5	1971	6282	1964/5	1970
6286	1964/5	1970	6287	1964/5	1970

Cars in series 2500 to 2699 temporarily equipped with end racks for handling pulpwood and differ in inside dimensions from other cars in the same series. Inside Length 38' 9". Extreme height 10' 8"

Car No.	Date Equip.	Retired	Car No.	Date Equip.	Retired
2503	1956	1962	2600	1956	1970
2507	1956	1968	2602	1956	1962
2508	1956	1969	2603	1956	1968
2509	1956	1968	2604	1960	1968
2517	1960	1970	2606	1960	1964
2521	1956	1965	2608	1956	1971
2525	1956	1968	2613	1956	1966

2530	1956	1962	2614	1956	1965
2532	1956	1968	2615	1956	1969
2535	1956	1962	2616	1960	1969
2537	1956	1970	2619	1956	1965
2541	1956	1970	2620	1960	1962
2543	1960	1964	2621	1956	1968
2548	1956	1968	2626	1960	1970
2549	1960	1965	2631	1956	1968
2550	1956	1968	2632	1956	1964
2559	1956	1968	2635	1956	1970
2560	1960	1965	2637	1960	1964
2562	1960	1966	2638	1956	1968
2565	1956	1968	2640	1960	1968
2567	1960	1964	2642	1956	1964
2568	1956	1970	2643	1956	1965
2569	1956	1968	2645	1960	1969
2570	1960	1964	2646	1956	1971
2576	1956	1962	2651	1956	1968
2577	1956	1965	2653	1960	1968
2582	1956	1966	2660	1956	1968
2583	1956	1965	2662	1960	1966
2584	1956	1970	2663	1956	1959
2585	1956	1968	2665	1956	1964
2594	1960	1966	2666	1956	1966
2597	1956	1966	2667	1956	1960
2598	1960	1962	2668	1956	1965
2673	1960	1968	2674	1960	1965
2676	1960	1966	2680	1956	1968
2681	1960	1965	2682	1956	1964
2685	1956	1970	2689	1956	1964
2691	1956	1965	2692	1956	
1965					

Note 32. 77 converted to gondolas when delivered

Note 33. Rebuilt from boxcars 5348 and 5384 per AFE IR3506 April 11, 1935

Note 34. Rebuilt from boxcar 5248 in 1955 after fire destroyed boxcar body

Note 35. Received from EJE 1966, Rebuilt with double floor deck and 12' wide floor in 1967 by DMIR

Note 36. Received EJE 1966, car did not have 10'6" floor and was restenciled at Proctor and returned to EJE

Note 37. Received from EJE June 2, 1966 with 10'6" floor. Rebuilt with double deck floor and 12' wide floor in 1967 by DMIR

Note 38. Received from EJE July and August1969

Note 39. Five cars, 6328, 6330, 6333, 6334, 6335 rebuilt Two Harbors with racks March 1971

Note 40. Received from EJE Sept. thru Oct. 1969. Received from EJE with racks already installed

Note 41. Rebuilt 1938

Note 42. Rebuilt 1936, 1938, 1941, 1942

Note 43. Rebuilt 1937, 1938, 1941

Note 44. Rebuilt Two Harbors 1939

Note 45. Rebuilt 1951

Note 46. Rebuilt 1951(except 7136)

Note 47. Rebuilt Two Harbors 1949

Note 48. Rebuilt by DMIR Proctor shops with Pullman Standard kits in 1971-73 from offset side to rib side as 1000-1249. 101 cars sold to Illinois Hopper Car Co and then renumbered and leased back to DM&IR(Re#ed to 11004-11247 by adding "1" to original numbers) Car numbers: 11004, 11007-11008, 11011-11012, 11014, 11017, 11020, 11023, 11024, 11025, 11027-11029, 11031-11034, 11037, 11039, 11045-11046, 11049, 11050-11053, 11056-11058, 11062, 11064-11066, 11068, 11070-11071, 11074, 11076, 11078, 11080, 11083, 11093-11095, 11102, 11103-11109, 11111, 11116, 11120-11124, 11127, 11132, 11134, 11138-11140, 11142, 11145, 11147, 11151, 11153, 11158-11160, 11162-11163, 11166, 11170, 11171, 11173-11174, 11176, 11179, 11181, 111083, 11187, 11190, 11194, 11195, 11198, 11200, 11203, 11206, 11210, 11217, 11221, 11222, 11224, 11228, 11229, 11233, 11235, 11239, 11241, 11242, 11247,

Note 49. Former B&A/NYC cars rebuilt at PC Samuel Rea Shops with Bethlehem Steel kits in June thru August 1975 before being acquired by DM&IR

Note 50. Purchased from B&LE 5-2003. Car numbers: 15026, 15027, 15033, 15036, 15049, 15060, 15068, 15078, 15098, 15099. Rejected and returned 9-2003

Note 51. Former Ralston Steel demo car

Note 52. Leased from BLE in 1974 - Returned to BLE 1976

Note 53. Built with fabricated underframes

Note 54. Former demo car ACFX 9001. Originally DM&N 22000, then DM&N 21026

Note 55. 21101 built 11-24. 21102 and 21108 Special Pflager type trucks - Built with GSC cast underframes GSC drawing 12987

Note 56. Former GATX 1000 demonstator car acquired in 1931. Originally DM&N 21999, then DM&N 21126

Note 57. Built with GSC cast underframes GSC drawing 17646

Note 58. Purchased second hand 1931. Former Standard Steel Car demo cars STDX 171-175

Note 59. Former Standard Steel Car demonstrator cars STDX 176-177-Purchased 1931

Note 60. Built at PS Michigan City plant. 23975-23999 ordered by DM&N - Delivered as DM&IR

Note 61. Cor-Ten steel

Note 62. Built at PS Michigan City plant. Built 9-10/1942

Note 63. 26000-26249 ASF trucks - 26250-26499 National B1 trucks

Note 64. Ordered 4-6-1942 ACF Huntington Plant Lot 2513

Note 65. Built at PS Butler, PA plant. Built 2-5/48

Note 66. Ordered 6-3-1947 ACF Berwick Plant

Note 67. Built 10-48 to 2-49

Note 68. Acquired from BLE 2003. Numbers 30025, 30085, 30091, 30123, 30127, 30139, 30154, 30158, 30163, 30189

Note 69. Built at PS Michigan City plant. All built between 1-4/52(All riveted)

Note 70. Built at PS Michigan City plant. One car, built 5-52(Only car in order built as welded car)

Note 71. Built at PS Michigan City plant

Note 72. ACF Lot #01-4974 Ordered 10-23-56. Built at St. Louis

Note 73. Ore cars equipped with 19 -1/2 inch extensions for crude taconite loading between 1964-1967 - T-Bird cars renumbered 1976 from 50031-59991 series - All have shaker pockets

Note 74. Ore cars equipped with 19-1/2 inch extensions for crude taconite loading - T-Bird cars - All have shaker pockets

Note 75. Acquired from P&LE AFE 11009 1/9/76. Sold to EJE August 1978 to EJE 4006-4109 random numbers. Wood floors

Note 76. Ore cars equipped with 19-1/2 inch extensions for crude taconite loading - T-Bird cars with straight air - All have shaker pockets

Note 77. T-Bird idler cars - Rebuilt WITHOUT HiSides - Cars had permanent ETD brakets installed for hanging ETD

Note 78. Cars 49001, 49003, 49005, 49007, 49008 rebuilt to replace original cars - Rebuilt WITH HiSides

Note 79. Cars 49004, 49006 rebuilt to replace original cars - Rebuilt WITH HiSides

Note 80. Ore cars equipped with 19-1/2 inch extensions for Taconite loading with NO shaker pockets - Converted Proctor 1964

Note 81. Hi-Side taconite cars converted to 9-3/4 inch Mini Hi-Side extensions for taconite loading. Beginning in summer 1972 cars began to be permanently coupled into 4 cars mini-quad sets

Note 82. Ore cars rebuilt with 9-3/4 inch Mini Hi-Side extensions for Taconite loading from former Hi-Side taconite cars - Converted Proctor 1971 through 1975 and placed in four-car drawbar coupled sets

Note 83. Ore cars rebuilt with special angle iron reinforcements on doors to keep doors tight for loading taconite pellets

Note 84. Never carried DM&IR reporting marks. Purchased by DM&IR but leased to EJE - All sold to EJ&E on 12/31/81 - Cor-Ten steel

Note 85. Former B&A/NYC hoppers Rebuilt in 1975 by Samuel Rea shops with Bethlehem Steel kits prior to purchase. Owned by DM&IR leased to EJ&E

Note 86. Former BLE 92076-95944 series that were rebuilt during 1960's by Greenville Steel Car

Note 87. Former EJE 83053, 83120, 83216, 83219, 83248, 83274, 83301, 83338, 83343, 83371, 83428, 83433, 83455, 83461, 83483, 83522, 83530, 83539, 83540, 83576, 83580, 83588, 83608, 83619, 83629. Leased from EJE 10/4/72. Returned to EJE

10/18/73

Note 88. Former BLE 92000-96999 series that were rebuilt during 1960's by Greenville Steel Car

Note 89. Rebuilt for T-Bird crude ore service: 14 Class U27 29615, 29648, 529780, 529991, 30031, 30058, 30143, 30195, 30230, 30245, 30346, 30356, 30481, 30491, 2 Class U28 30675, 30690, 19 Class U30 32531, 32547, 32552, 32591, 32611, 32699, 32740, 32771, 32778, 32798, 32791, 32811, 32814, 32824, 32833, 32840, 32841, 32874, 32896, 32951, 34 Class U31, 33000, 33021, 33037, 33043, 33044, 33048, 33092, 33097, 33113, 33120, 33157, 33174, 33186, 33193, 33205, 33216, 33220, 33236, 33247, 33254, 33345, 33354, 33357, 33373, 33375, 33380, 33400, 33407, 33423, 33425, 33432, 33441, 33487, 33497 with 5 added ahead of old number. All equipped with shaker pockets. All renumbered into 40000 series in 1976

Note 90. Rebuilt Hi-Side Tac cars renumbered by adding "5" to original number of car. Carried on roster from 1964 to 1967 - then re#ed to 50031-59991 series. None of these cars equipped with shaker pockets

Note 91. Rebuilt at Proctor in 1961 with 33 inch extensions for concentrate loading at Pilotac. Numbers, 533002, 533073, 533098, 533106, 533143 - Rebuilt back to 19 1/2 extensions 8/9/66

Note 92. Leased EJE gons: 87347, 87422, 87426, 87533, 87643,87647,87663,87742,87797,87809,87922,87933,87945,87990,88013. Returned to EJE 06-2003

Note 93. Owned by DM&IR leased to EJE. EJE 88175-88199 renumbered to EJE 4350-4374 after delivery account being equipped with coil troughs. Cars sold to EJE 10/2/81 on AFE 11508 and AFE 11507 12/31/81

Note 94. Converted to Ballast cars in 1972 at Proctor Shops With modified MK Ballast doors - Renumbered to 1400 series in Nov. 1977 (9-3/4 inch extensions added 1978)

Note 95. Converted to Ballast cars in 1972 at Proctor Shops With modified MK Ballast doors - Renumbered to 1400 series in Nov. 1977 (9-3/4 inch extensions added 1978)

Note 96. Converted to Ballast cars in 1972 at Proctor Shops With 14" side extensions and modified MK ballast doors - Renumbered to 1400 series Nov. 1977

Note 97. Converted to Ballast cars in 1972 at Proctor Shops With 14" side extensions and modified MK ballast doors - Renumbered to 1400 series Nov. 1977

Note 98. Converted to covered hopper(sand car)with 24-1/4 inch extensions in 1965

Note 99. Converted to covered hopper(sand car)with 24-1/4 inch extensions in Sept. 29, 1967

Note 100. Converted to covered hopper(sand car)with 24-1/4 inch extensions in Oct. 31, 1967

Note 101. Converted to covered hopper(sand car)with 24-1/4 inch extensions in May 24, 1965

Note 102. Converted to covered hopper(sand car)with 24-1/4 inch extensions in Sept.1974

Note 103. 4 cars converted in 1972(see above 4 cars) Remaining cars converted with 9-3/4 inch extensions, modified MK ballast doors and roller bearings, 66 in 1978 and 34 in 1985

-continued from page 173
turers pads. None of these were rebuilt and all were sold to EJ&E on April 21, 1975.

During 1970 DM&IR began to handle coal for Minnesota Power & Light that was transloaded from BN unit trains at their Clay Boswell power plant at Cohassett, Minnesota and shipped to the MP&L Colby, Minnesota power plant located on the DM&IR. BN, because of the joint line haul of this traffic was not happy about supplying all the cars for this movement and was requesting DM&IR provide cars on a 50/50 basis. The shortage of equipment to handle coal reached critical levels in 1969 and the majority of coal from Duluth Docks to MP&L at Colby was handled in ore cars. Ore cars were tried in October 1970 for the movement of coal from MP&L at Cohassett to Colby but this was not acceptable.

The situation during late 1970 at MP&L Colby became so bad that the plant only had a two-day supply of coal which was considered an emergency. BN placed ballast cars in this service and DM&IR leased 16 coal hoppers. Four hundred open hoppers were owned in on August 31, 1969, with 84 held out of service. Total out of service hoppers climbed to 191 by June 15, 1970. Seventy one percent of DM&IR hopper cars were out of service in March 1971 for repairs. The hopper car shortage had become critical. Because of this shortage, DM&IR had lost the coal business to Virginia Water and Light to DW&P and Hibbing Public Utilities to BN. A March 11, 1971 study by Financial Department showed coal being handled to MP&L at Colby, Northwest Paper at Cloquet via D&NE at Saginaw, Minnesota, Two Harbors, city power plant, dolomite for AS&W and limestone for UAC both at Steelton, Minnesota. DM&IR was also handling about 80,000 tons of crude clay to Burnett, Minnesota, in per diem cars.

From October 1970 until March 1971 DM&IR short-term leased 45 C&NW and 24 Hallett Minerals open hoppers and used per diem cars to cover the shortage of equipment. Car shortages nationwide caused ICC to arrange with DM&IR to send 96 Q-class hoppers to RI for sugar beet service during early 1970s. These cars could be used for the larger sugar beets but were not capable of hauling coal or limestone. During 1967 testing of B&LE hoppers in winter weather showed that the 30-degree slope sheets, compared to the 50-degree slope sheets on DM&IR cars, made it difficult for cars to be unloaded and custom-

DM&IR, in 1953, was short of open hoppers for handling coal shipments and was unable to secure car builder shop time until 1957 for Class Q7 hoppers 2950–3049. The short-term solution was to purchase 40 secondhand NYC 900000 series hoppers cars built by Standard Steel Car Co. in 1924 from Ortner Car Co. These Class Q6 70-ton hoppers arrived on DM&IR in 1953 numbered 4800–4839. All were retired by 1959. Proctor, 1959. (Russ Porter, D.P. Holbrook collection)

After the merger DM&IR would acquire another 350 end-platform quad hopper cars, 250 Class Q5 4700–2949 from Pressed Steel Car in 1952 and 100 Class Q7 2950–3049 from Pullman Standard in 1957. Beat up from constant heavy loading of limestone and coal, most would be rebuilt with rib-side kits from Pullman Standard during the 1970s. Proctor, July 17, 1957. (Bob's Photo collection)

DM&IR 2991 on EJ&E at Waukegan, Illinois, March 4, 1975. Not useable on DM&IR for coal service account side and slope sheet condition, a national hopper car shortage caused ICC to arrange to send 96 Class Q hopper cars to Rock Island in early 1970s for fall sugar-beet loading. Note the stencil on the side of car "SUGAR BEET LOADING ONLY CRIP RY." (D.P. Holbrook)

Many Class Q2, Q5 and Q7 hopper cars were out of service by 1970. The first two of 50 rebuilt Class Q5, cars 2805 and 2808, were completed in March, 1971. Eight cars received their old numbers before a decision was made to place cars in the 1000 series, making it easier to identify rebuilt cars. 2805 would become 1105 and 2808 the 1108. (University of Minnesota Duluth, Kathryn A. Martin Library Archives and Special Collections)

Rebuilt Class Q5 1043 former 2743 is at Biwabik, MN in the early 1970s loaded with coal for Minnesota Power & Light Colby power plant. (Ed Stoll)

ers advised DM&IR that they would not accept the B&LE cars. Testing with B&LE cars was again tried in late 1969 with the same results. Funds for rebuilding 50 of the DM&IR Class Q2 and Q5 battleship hoppers were approved in 1970 but it was determined that at least 200 cars needed to be rebuilt to fulfill customer requirements.

A decision was made to buy Pullman Standard parts and rebuild the Class Q5 cars 2700–2949. The first 50 rebuilds were approved on AFE 10519 on May 8, 1970. Parts were ordered from Pullman Standard's Bessemer, Alabama, plant in September 1970 and shipped by rail to Proctor. Rebuilding was started in late February 1971 at Proctor and Two Harbors with the project completed on Aug 13, 1971. Cars were stripped of their offset-side car bodies and the frame, trucks and door frames were retained. Attached to these were new rib-side car sides and the cars were rerated to 77-ton capacity. Cubic capacity was increased from 2,150 to 2,609 during the rebuilding. On March 19, 1971, a letter from Supt. Motive Power and Cars B.E. Lewis, to R.L. Vanneste stated that; "The first car outshopped, 2808, was to be painted maroon and gold for publicity purposes but the balance of the cars would be painted ore car brown with white stencils. Six barrels of ore-car brown are in stock at Proctor Car Shop Stores and this material should be used first." On April 6, 1971, a follow-up letter stated only first car would be painted "maroon and gold" and other cars were to be left unpainted (an aluminum primer color), stenciled with reporting marks and numbers and placed in service because of the severe hopper car shortage. As time permitted during the summer of 1971 they would be painted maroon and gold. The first car completed was 2805 on March 4, 1971; however car 2808 was chosen to be repainted in maroon and gold and was released from Proctor Shops on March 19, 1971. A decision was made after the first eight cars had been re-stenciled to renumber all the rebuilt cars in the 1000–1249 series. These eight were renumbered, 2805 to 1105, 2808 to 1108, 2835 to 1135, 2841 to 1141, 2854 to 1154, 2919 to 1219, 2943 to 1243 and 2949 to 1249 between March 4 and April 11, 1971.

A second group of 200 Class Q5 70-ton 40-foot 7-inch hopper cars 2700–2949 was approved for rebuilding on AFE 10572 in 1971. The DM&IR started the rebuilding program at Proctor and Two Harbors in 1972 lasting thru 1973. Cars were stripped of their offset-side car bodies and the frame, trucks and door frames were retained. These were placed in the 1000–1249 series retaining Class Q5. Attached to these were new rib-side car sides and the cars were rerated to 77-ton capacity. Capacity of the cars was increased from 2,150 cu.ft. to 2,609 cu.ft. during the rebuilding. Painting of these cars in maroon paint with yellow lettering was accomplished outdoors at Proctor carshops.

On January 12, 1973, the DM&IR Board of Directors approved AFE 10737 for the sale of 101 Class Q5 2700–2949 70-ton open-top hopper cars. These cars were sold to Illinois Hopper Car Co., rebuilt by DM&IR to 1000–1249 series and immediately leased by DM&IR for a 15-year period at $71,000 per year. AAR Circular OT-37 provided an incentive through increased per diem rates for the rebuild by an owner of a secondhand car versus the original owner. If DM&IR retained ownership the average annual per diem over a 15-year period was $400, but sale and lease back would increase this to $655. From June through October 1973 cars were shopped at Proctor and renumbered by adding the numeral "1" in

End-platform Class Q5 open hoppers built in 1952 by Pressed Steel Car Co. were rebuilt with Pullman-Standard rib side kits in 1971. Cars were renumbered from 2700–2949 series to 1000–1249 series when rebuilt. 101 of these would be sold to Illinois Hopper Car Co.in 1973 and leased back to DM&IR and be renumbered 11004–11247. Mt. Iron, August 8, 1985. (Doug Buell)

front of the DM&IR road number. This made cars owned by Illinois Hopper Car Co. to be easily identified. Cars were random numbers within the 250-car series, with the remaining Class Q5 hoppers retaining their four-digit road numbers. All rebuilding and renumbering of the Class Q5 hoppers was completed on October 19, 1973. The 15-year lease expired in August 1988. DM&IR, because of decreased need for open hoppers, terminated the lease agreement on October 30, 1987 and returned 50 cars to Illinois Hopper Car Co., and kept 51 cars under lease.

Of the returned cars, 25 were resold to Railmark for service at Tilcon Tomasso Quarries in North Branford, Connecticut, with reporting marks changed to TLTX and renumbered into the 87000 series. In the remaining cars, 65 were sold in 1986 to Soneco Service Inc. of Groton, CT, for aggregate service in New England and re-stencilled SONX 1001–1065 at Proctor car shops. The remaining Class Q5 hoppers owned, not leased, by DM&IR were sold to EJ&E: 100 in 1977 to EJ&E 44000–44099, and 43 in 1979/80, EJ&E 42200–42293.

During January 1973 DM&IR leased 44 B&LE 77-ton 2,775 cu. ft capacity offset-side hopper cars built by ACF, Pullman Standard and Pressed Steel Car during the 1930s. Cars had random numbers between 77054–78500 (10 cars) and 92076–95944 (34 cars). A shortage of hopper cars had developed because most Class Q7 hoppers were out of service for repairs to side and end slope sheets. Cars kept their B&LE numbers but reporting marks were changed to DM&IR. No car class was assigned. The reporting mark changes were made to make sure cars, once off-line, got returned to DM&IR and not the B&LE. Once older Class Q2, Q5 and Q7 hopper cars were rebuilt they were placed in this service. The B&LE cars were returned in late 1975.

Increased loading of coke at the U.S. Steel Duluth Works for Chicago area U.S. Steel facilities in early 1970 caused a shortage of hopper cars. AFE 10954 was approved April 29, 1975, to cover the purchase of 500 77-ton hopper cars. Classed Q15 by DMIR, the first 150 cars, DM&IR 12000–12149, were received during June through August 1975. The remaining 350 cars, EJ&E 74000–74349, were received during August through October 1975. Both groups were placed in coke service. These had originally been NYC 900000–922068 and B&A 910000 series built by ACF, Greenville, PS, GATC and NYC DSI shops between 1956 and 1960. Cars were rebuilt at the Penn Central Samuel Rea shops with Bethlehem Steel

EJ&E 73094, one of 201 coke cars purchased by EJ&E and DM&IR in 1973. The first 101, EJ&E 73000–73100 were owned by EJ&E, 73101–73200 were owned by DM&IR. DM&IR sold 73101–73200 to EJ&E on December 31, 1981. (Bob's Photo collection)

kits between June and October 1975. Once coke shipments were discontinued from U.S. Steel Duluth Works, the 146 Class Q15 hopper cars, DM&IR 12000–12149 (four cars had been destroyed on C&NW), were leased to EJ&E between January 6, 1976, and December 30, 1977, and renumbered to EJE 74350–74495. Re-stenciling of these cars took place at Proctor. All were sold to EJE on December 31, 1981 on AFE 11507.

DM&IR purchased 100 coke cars in January 1973 from Greenville Steel Car Co. numbered EJE 73101–73200 and they were immediately leased to EJE for an 18-year period. All were delivered by May 1973. Initially these were ordered to protect coke movements from AS&W Steelton. Before delivery, the coke plant at AS&W was being phased out by U.S. Steel and the decision to lease these cars to EJE was made. These cars did see service with EJE reporting marks, moving coke from Steelton to other U.S. Steel mills. Cars were sold to EJE on AFE 11507 on December 31, 1981, under Clayton Act Contract EJ&E 4-1981.

Between August 6 and December 11 of 1979, 43 of the 1000–1249 series cars were sold to EJE on AFE 11295 and renumbered randomly 42200–42293. Twenty cars were sold to Garrett Railroad Car on January 21, 1983, who repaired cars and resold them to Roberval and Saguenay Railway Co. (RS). Remaining cars were "stored for disposal" per Car Dept. Bulletin 86-3 on Februry 3, 1986. Fourteen of these cars were sold to Railmark on October 6,1986, and restenciled RS. Of the remaining cars, 21 were sold to Citirail, Inc. Pittsburgh, and restenciled "RT" per AFE 11506 on December 28, 1988. The original rebuilding of these cars did not comply with AAR Rule 88 (rebuilt equipment) and cars would no longer be interchange-able after 1993. Many of the cars were sold off to secondhand owners during the late 1980s.

Two hundred 100-ton open quad-hoppers were acquired under AFE 11852 in 1982 from the B&LE. These were former B&LE 65000–65205, built by Pullman Standard at their Bessemer, Alabama, plant under Lot No. 9904 in 1975 and given B&LE Class HT48. Cars were received during October 1982, leased directly to B&LE, and kept B&LE reporting marks and numbers.

DM&IR GONDOLAS

The newly formed DM&IR received 164 gondolas from D&IR and 81 from DM&N.

The first gondolas acquired by the newly formed company were 100 Class V2 50-ton 41-foot 6-inch gondolas 4100–4199 approved on AFE 4399 July 27, 1940, from American Car and Foundry under Lot No. 2199 in September 1941. Carbody was built up from Cor-Ten steel supported by nine rolled T-section ribs and 5-rib dreadnaught ends with the top rib being a half dreadnaught. Cars were purchased for the all-rail movement of iron ore.

Ten Class V3 70-ton, 65-foot gondolas 4350–4359 were purchased from EJE in 1965 per DM&IR AFE 10070 for $20,730, approved December 9. Cars were originally EJE 90000–90099 built in 1937 by Mt. Vernon Car Co. They were purchased to protect shipping of reinforcing rods manufactured at American Steel & Wire (AS&W) at Steelton. At the time of acquisition 100 percent of the cars used for these shipments were foreign-line cars which were always in short supply. Projected off-line movements in 1965 were 75,000 tons with on-line shipments to Range destinations expected to reach 13,000 tons. Before acquisition EJE removed the drop ends from these cars. AFE 10146 was approved on June 3, 1966, for the purchase and installation of fixed ends to these cars. This project was completed on November 23, 1966. During their service lives, the cars did travel off-line as indicated by AFE 10401 March 24, 1968, for the retirement of one Class V3 gondolas destroyed in derailment on C&NW at McIntire, Iowa on January 22, 1969.

Another 50 Class V4 70-ton 42-foot gon-

First steel gondolas purchased after D&IR and DM&N merger were 100 Class V2 41-foot cars from ACF. Initially used for movement of iron ore during World War II, they later were used for pulpwood and scrap iron service. (American Car & Foundry collection, John W. Barriger III National Railroad Library at UMSL)

Ten 65-foot Class V3 4350–4359 gondolas were purchased secondhand from EJ&E in 1965 for movement of reinforcing rods produced at American Steel & Wire, Steelton. (Dick Kulebs, Ed Stoll collection)

dolas 4375–4424 with 1,529 cu. ft. capacity were purchased for $95,000.00 under AFE 10089 of December 30, 1965, to protect the interline forwarded movement of billets from AS&W Steelton to U.S. Steel in the Chicago area. When acquired, DM&IR supplied none of the cars for this movement for which the DM&IR received from 15 to 20 percent of the revenue depending on exact destination. The cars resulted in an annual decrease of 2,426 foreign car-days on line and an annual increase of 13,343 per diem income days. Once in service, this provided about 30 percent of the hauling capacity required for the normal volume of billets. Because the billet loads originated on DM&IR the cars could be assigned to Steelton for billet loading. These were former Union Railroad 8000–8965 series 77-ton 44-foot 9-inch gondolas built by Greenville Steel Car in 1953. AFE 10090 for $90,000 was approved on December 30, 1965, for reconditioning of these gondolas to make them suitable for billet loading. Cars were rebuilt by Garrett Railroad Car & Equipment at the Union Railroad Shops. Rebuilding of cars consisted of straightening sides and ends, new floors and new sides sills as necessary. Cars were on roster from 1965 until 1977.

Continued shortages of pulpwood gondolas caused DM&IR to purchase 150 Class V5 50-ton 49-foot 6-inch gondolas 4500–4649 from EJE. Originally built by Ralston Steel Car Co. in 1943 as EJE 33750–34189, DM&IR purchased 100 gondolas per AFE 10088 for $250,000, and 50 gondolas per AFE 10091 for $40,000.00 on December 10, 1965. All were

Saving money during the economic downturn in the mid-1960s, DM&IR turned to purchasing secondhand equipment. Replacing overage rack flats, 150 Class V5 4500–4649 former EJ&E 33750–34189 were purchased in 1965. The first 100 were equipped by EJ&E Joliet shops with pulpwood racks, with the remaining 50 cars having racks installed at Proctor. (University of Minnesota Duluth, Kathryn A. Martin Library Archives and Special Collections)

originally built with composite wood and steel construction, having a Pratt steel truss side-rib arrangement. Cars were rebuilt by EJE Joliet shops during 1948 and 1949 with steel sides and fixed ends. One hundred of these cars were equipped by EJE Joliet shops with pulpwood racks. All 100 were received by January 6, 1965, with EJE reporting marks and re-stenciled with DMIR reporting marks at Proctor. The remaining 50 cars were delivered without pulpwood end racks. AFE 10092 for $102,650 was approved on December 30, 1965, for Proctor shops to install pulpwood racks and repaint these 50 cars in the maroon and gold scheme. During a board meeting on September 22, 1965, painting and renumbering of the first 100 rack gondolas purchased from the EJ&E was discussed. At that time it was decided that numbering and stenciling would be changed and paint would be touched up only. Repainting in DM&IR colors would be on an as-needed basis. This purchase allowed an annual decrease of 10,101 foreign car-days on line and annual increase of 17,749 per diem income days for DM&IR.

Securing pulpwood cars from connecting carriers had become increasing difficult and C&NW, SOO and MILW were reluctant to allocate cars to DM&IR because many times they were also short of cars for their own on-line customers. Cars were purchased for pulpwood service on DM&IR for shipments to Minnesota and Wisconsin mills and could haul approximately 26 cords of pulpwood. Historically, loading patterns showed that during April, May and June each year, pulpwood loading was not sufficient to fully utilize these cars and they

Retirements of large groups of 40-foot pulpwood rack flats caused DM&IR, to purchase a large number of gondolas from EJ&E starting in 1965. DM&IR purchased 200 Class V6 3500–3699 gondolas in 1969, equipping them with pulpwood end extensions at Two Harbors. DM&IR 3550 was at Schiller Park, Illinois, August 14, 1979 (R.F. Kucaba)

should be considered for billet loading from U.S. Steel, Duluth Works during those months to avoid using foreign gondolas in that service.

Retirement of early 40-foot rack flats caused the DM&IR again to look for additional pulpwood gondolas in 1969. Turning again to sister railroad EJ&E, AFE 10417 was approved on April 18, 1969, for $630,940. to acquire 200 Class V6 55-ton 50-foot 6-inch gondolas 3500–3699 with 1,344 cu. ft. Originally these had been built by Mt. Vernon Car Co. in July 1937, as drop-end gondolas and numbered in EJE 31000–31549 series. Proctor already had a full heavy-repair schedule when the cars arrived and a temporary production line rebuilding area was established at Two Harbors. Cars had reinforced draft sills and body bolsters applied, drop ends welded in upright position and new pulpwood end extensions applied to make the car suitable for pulpwood loading. Rebuilding of cars was started in October 1969 and lasted until December. 1970. Once completed the cars had their former EJE lettering painted out with black paint and new DM&IR reporting marks applied. All were off the roster by 1983.

During 1972 an additional 125 Class V7 55-ton 51-foot gondolas 3700-3824 were acquired secondhand. Cars were former EJE 34200-34899 built by ACF and Ralston Steel Car Co. in 1947 and 1948. EJE Joliet Shops changed the reporting marks and renumbered cars before shipment to DM&IR. After arrival on DM&IR, pulpwood end extension kits were made by Paper Calmenson in Duluth beginning in August 1972. Parts were shipped to Proctor car shops for application to cars.

Unable to keep up with demand for gondolas for pulpwood service, DM&IR leased 25 70-ton 57-foot gondolas in EJE 83000–83662 series on October 4, 1972. These were originally P&LE cars built in late 1940s or early 1950s before being purchased by EJ&E. Cars were received between November 6, 1972, and March 6, 1973, and reporting marks changed to DM&IR. The purchase of the 81 Class V8 gondolas in 1973 caused the lease to be terminated on October 18, 1973.

Continuing to replace rack flats with gondolas, 81 Class V8 77-ton 52-foot 6-inch gondolas 3900–3980 were acquired from the LS&I in 1973. These had originally been built by New York Central Despatch Shops as P&LE 9000–10499 series during 1946–1953. Cars were purchased by EJ&E from P&LE for rebuilding and then sold to LS&I in 1973. They were reconditioned by EJ&E Joliet Shops and did not have the pulpwood racks or end extensions applied. Once cars had been reconditioned, DM&IR traded 170 Class U17 ore cars to LS&I for the 81 gondolas in November 1973. Under AFE 11037, 40 cars were

Multiple groups of secondhand gondolas were purchased during the 1960s and early 1970s. 125 Class V7 gondolas 3700–3824 were acquired from EJ&E during 1972. Former EJ&E 34200–34899 had pulpwood end extensions applied at Proctor carshops. Allen Jct. August 10, 1985. (Doug Buell)

Eighty-one additional secondhand gondolas Class V8 3900–3980 were acquired from Lake Superior & Ishpeming RR in 1973. Originally built by NYC Despatch Shops for P&LE, cars were purchased and rebuilt by EJ&E and then sold to LS&I. LS&I traded these gondolas to DM&IR for ore cars. Ironically, 43 of these cars would be leased by DM&IR to EJ&E in 1976. Allen Jct. MN September, 1974. (Doug Buell)

leased by DM&IR to EJ&E between July and November 1976 and renumbered EJE 84354–84393 at Proctor Shops. Repainting and renumbering only involved a patch paint job with initial and number changed but cars retained DM&IR maroon paint. The remaining cars were off roster by 1993.

Finding other sources of revenue, DM&IR turned to acquiring and leasing out of equipment. An order for 200 100-ton 52-foot 6-inch gondolas with 1,995 cu. ft. capacity was placed with Greenville Steel Car and delivered in September and October 1974 on AFE 10858 at a cost of $4 million. These cars were numbered EJ&E 88000–88199. EJE 88175–88199 were renumbered after delivery as 4350–4374 account being equipped with coil troughs. Cars were sold to EJ&E on AFE 11507 on December 31, 1981.

DMIR acquired 16 gondolas through "Involuntary Conversion Funds and Exchange" during January 1976. Salvage funds of $294,844.91 from the sale of 82 ore cars, two gondolas and 39 covered hoppers, all retired and sold for salvage, plus payment of $60,700. This money allowed the DM&IR to acquire 16 reconditioned secondhand 100-ton 60-foot P&LE gondolas 44100–44124 built by Greenville Steel Car (GSC) August 1965 under O.O. 868 these were renumbered DM&IR 44100–44124 series. All 16 cars were leased to EJE under Clayton Act contract EJE1-1981 in August 1978 and renumbered in random numbers between EJE 4006–4109. Cars were sold to EJ&E on AFE 11508 on October 2, 1981.

Finding money to be made, DM&IR returned to acquisition and leasing of equipment in 1979. Cars were acquired under Clayton Act Contracts which allowed money from the sale of retired equipment to be used for purchase of secondhand and new equipment with no tax implications. The first purchase was 127 77-ton gondolas purchased from GTW under AFE 11271 and Clayton Act Contract EJ&E 5-1981 on May 15, 1979. These were former GTW 145700–146099 built by GATC November 1953 through January 1954 under Builder Order (B.O.) 8036 and GTW 146100–146299 built by GATC Mar. 1954 under B.O. 8036A. Cars were renumbered EJ&E 86000–86126 and leased to EJ&E on January 17, 1980. All were sold to EJ&E under AFE 11507 on December 31, 1981.

LS&I acquired from EJE 196 70-ton 69-foot gondolas from series 91000–91199 built by Bethlehem Steel November 1957 under OO 878-G. These cars had originally been P&LE 16000–16499 that EJ&E acquired in 1978. LS&I then traded these cars to DM&IR for ore cars under AFE 11319 of October 12, 1979. Gondolas were then leased by DM&IR

The majority of gondolas owned by DM&IR were becoming overage by 1980, causing a gondola shortage for pulpwood service. DM&IR acquired 152 gondolas from Western Maryland in 1980, all originally built by Bethlehem Steel between 1951 and 1954. The best 40 of these were rebuilt with pulpwood extensions, repainted, placed in Class V7 and numbered 4200–4239. A sudden decline in pulpwood shipments in 1982 caused the remaining WM gons to be sold for scrap. DM&IR 4218, former WM 55535, at Chisholm, September 22, 1986. (D.P. Holbrook)

DM&IR 4201, former WM 55620, at Missabe Jct. Yard, Duluth, September 30, 1990. (S.D. Lorenz)

to EJ&E on November, 10, 1979, with same numbers under Clayton Act Contract EJ&E 7-1979. These retained EJ&E reporting marks all through the exchange process. Cars were never stenciled DM&IR and were sold to EJE per AFE 11511 on December 31, 1981.

After acquiring the Class V7 and V8 gondolas in 1972, DM&IR owned 680 gondolas. Retirements through 1979 had reduced the fleet size to 363 cars. On January 1, 1980, this fleet declined further with the retirement of 127 Class V6 gondolas account being overage. The remaining fleet was insufficient to protect pulpwood loadings and DM&IR purchased 152 second-hand 77-ton 57-foot gondolas from the Western Maryland (WM) in 1980 under AFEs 11102, 11241 and 11071. Cars were, 33 Class G-36 WM 55336–55478, 85 Class G-42 WM 55486–55701 and 34 Class G-43 WM 55707–55848 originally built by Bethlehem Steel in 1951 to 1954 as 1,745 cu. ft. gondolas. Eighty one of the cars had 3-foot 6-inch interior heights with remainder being 3-foot 3-inch. The best 40 of these cars were rebuilt as DM&IR Class V9 4200–4239 with pulpwood end extensions in 1980 and 1981.

DM&IR 87647, Duluth, April 22, 1996, one of 15 EJE gons leased by DM&IR from EJ&E in 1993. (D.P. Holbrook)

A sudden decline in pulpwood shipments in 1982 caused the rebuild program to be suspended. The remaining 112 cars were sold for scrap to Azcon at Duluth in 1984. During March 1992, 10 of these cars had the pulpwood end extensions removed and were renumbered into the 4300 series: 4306, 4307, 4310, 4320, 4323, 4326, 4328, 4331, 4335, and 4339. Five of these cars, 4205, 4214, 4218, 4222, 4255, were converted to scale monitor cars in 1987 and equipped with rounded covers and kept at Minntac to monitor the scales. On January 1, 1992, all 40 cars were overage and prohibited in interchange after which they were used in MofW service.

Returning to purchasing and leasing equipment in 1982, under AFE 11582, DM&IR acquired 200 B&LE Class GB34 99-ton 53-foot 11-inch gondolas B&LE 30000–30199 built by ACF 1980 under Lot 11-09601 with 1,995 cu. ft. capacity. Cars were received in October 1982 and leased back to B&LE keeping the same numbers.

The remaining secondhand gondolas were overage by 1992. Needing gondolas for scrap, pulpwood and landscaping boulder service DM&IR leased 15 100-ton 56-foot 7-inch gondolas built by ACF in 1980 as EJE 87000–88149. Individual car numbers were: 87347, 87422, 87426, 87533, 87643, 87647, 87663, 87742, 87797, 87509, 87922, 87933, 87945, 87990, and 88013 from 1993 until 14 cars returned in May 2003 and last car returned on June 11, 2003.

The return of these 15 leased cars left DM&IR with no interchange allowable gondolas for the movement of taconite concentrate and other commodities. AFE 5236 requested on May 14, 2003, for the purchase of ten gondolas from B&LE. Initially, B&LE supplied ten 92-ton 53-foot 6-inch gondolas built by Greenville Steel Car on Order No. 962 in 1968 as B&LE 15000–15099. Individual car numbers were: 15026, 15027, 15033, 15036, 15049, 15060, 15068, 15078, 15098, 15099. Cars were re-stenciled by B&LE to DM&IR reporting marks before shipment. On arrival at Proctor all ten cars were found to require excessive amounts of repair to make them acceptable and they were rejected and returned to B&LE.

On return to B&LE, by July 2003 all had been re-stenciled back to B&LE reporting marks. B&LE in June 2003 supplied ten different cars from series 30000–30199. These had been built by ACF on Lot No. 11-09601 in 1980 as 99-ton 53-foot 11-inch cars with 1,985 cu. ft. capacity. Individual car numbers were: 30025, 30085, 30091, 30123, 30127, 30139, 30154, 30158, 30163, and 30189. All had originally been purchased by DM&IR and leased back to B&LE in 1982 and then purchased by B&LE. All ten would be on the roster when CN purchased the DM&IR.

One order of covered hoppers was acquired before World War II, Class Q3 4950–4959 from ACF in 1941. War Production Board limits on steel use during the war and steel shortages after the war caused DM&IR to have to wait until 1952 to acquire more covered hoppers. (ACF Industries, Hawkins/Wide/Long collection)

DM&IR COVERED HOPPERS

By 1940, road construction, industrial development and home building were expanding in Minnesota. To fulfill the needs of bulk shipments of cement to ready-mix plants, DM&IR approved AFE 4399 on July 27, 1940, and ordered ten Class Q3 70-ton 35-foot 2-inch covered hoppers 4950–4959 from American Car & Foundry Lot 2120 on September 18, 1940. All were built during April 1941 and received from NP at Pokegama, Wisconsin, on April 21, 1941. Cars had 1,958 cu. ft. capacity, Miner power hand brakes, rode on National B trucks and were equipped with Apex Tri-lok running board and brake steps. Painted Scully Steel Products Battleship Gray, they were stenciled with black Gothic data lettering with reporting marks in Railroad Roman and a DM&IR safety-first monogram. The cars were a success, but limitations by WPB on steel and car construction during World War II would prevent further cars from being ordered until 1950. The 4950–4959, excluding 4952 destroyed on Soo Line in May 1972, were sold to Joliet Equipment Co., McCook, Illinois on June 21, 1976.

Development of the covered hopper saw DM&IR order 40 70 ton covered hoppers, 10 Class Q3 4950–4959 from ACF in 1941, 10 Class Q4 4960–4969 from ACF in 1952 and 20 Class Q8 4970–4989 from PS in 1956. Dark gray paint with white lettering was used only for ACF builder photos. Cars were delivered in light gray paint with black lettering.(ACF Industries, Hawkins/Wide/Long collection)

The next group of ten Class Q4 70-ton 35-foot 2-covered hoppers 4960–4969 were ordered on December 20, 1950 from ACF Berwick, PA plant under Lot 3674. The AFE 6930 for purchase was approved on January 13, 1951, and all cars were constructed during December 1951 and delivered in January 1952. Cars were identical to the Class Q3 cars

Increasing cement car loadings caused DM&IR to order 20 Class Q8 70-ton covered hoppers 4970–4989 from Pullman-Standard in 1956. Note the yellow journal box lids per AAR requirements indicating these cars were equipped with lubricating pads and not with normal cotton waste. (Pullman-Standard photo, Jim Kinkaid collection)

except they rode on ASF A3 Ride Control trucks and had Universal power hand brakes. All were painted DuPont Acid and Fume Resisting Gray on sides and ends. The roofs, underbody, hopper bottoms, trucks and lettering were black. They used Railroad Roman data lettering and reporting marks with a DM&IR safety-first monogram. All ten cars were sold to American Allied Railway Equipment of Washington, Illinois, on June 21, 1976.

Continued expansion of road construction and ready-mix plants and the ease of loading and unloading the covered hoppers, caused the DM&IR to order 20 additional covered hoppers in December 1955. Class Q8 70-ton 35-foot 3-inch PS-2 covered hoppers 4970–4989 were built by Pullman Standard under Lot 8331C in December. 1956. Cars had 2,003 cu. ft. capacity and rode on ASF A3 Ride Control trucks. Per letter to Pullman-Standard on April 19, 1956, these cars were to have the journal box lids painted yellow in accordance with AAR requirements indicating cars were equipped with journal box lubricating pads instead of the more common cotton waste packing. Eleven of these cars, 4970–4980, were retired and sold to American Allied Railway Equipment Co. on June 21, 1976. Four of the cars were converted to company sand service, 4988 and 4989 in December 1976 on AFE 11006 and 4986 and 4987 in November 1981.

Initially all of the covered hoppers were used by Universal Atlas Cement (UAC), but by 1960, some of these cars were in pool service, being used to cover shipments of cement from Huron Cement at Superior, Wisconsin. During 1964 five cars were placed in assigned hydrated lime service between Duluth and the U.S. Steel Extaca Plant at Virginia. Soon after this assignment, Extaca indicated that they were having difficulty unloading the cars and felt four pipe nipple connections, one on each side of each bottom bay, would solve the problem. The pumping connections were applied to 4951, 4954, 4956 and 4957 during January 1965 and cars were stenciled "For Extaca Lime Service Only."

A 1968 study of monthly cement loads showed during May through September, UAC shipped 3 to 4 carloads to Range destinations, outbound interchange was 35 to 53 cars and 2 to 3 carloads were received from connecting carriers. As cement traffic declined, some of these cars were used in bentonite service between Burnett, Minnesota, and taconite plants. UAC closed on January 1, 1976. After 1984, four cars were carried on the roster for company sand service until these were retired late in 1994.

During the steam era, locomotive sand had been hauled in ore cars and box cars, requiring the need for drying plants to eliminate moisture. A decision was made in 1965 to equip two ore cars, 24158 and 24899, with 24¼-inch side extensions, a roof and three roof hatches per side to be assigned for locomotive sand service. Two additional cars, 24187 and 24456 were rebuilt in 1967 and one, 28214, in 1974. All retained their revenue numbers because cars were being loaded off-line. All had a "W" prefix added to the number in early 1980s, indicating MofW service. By this time sand was being trucked into most facilities and the sand cars were being used for storage.

The need for additional covered hoppers was spurred by the development of the taconite industry. One of the ingredients used in the production of taconite pellets was bentonite clay as a binding agent. Processed bentonite required covered hoppers for shipment. Initially, some DM&IR Q3, Q4 and Q5 covered hoppers, DM&IR 4950–4989 series, removed

Five ore cars were converted to covered ore cars for company sand service between 1965 and 1974 and retained their revenue numbers until a "W" for work service was added in early 1980s. DM&IR W24187 with the "W" recently added at Keenan, May 27, 1983. (D.P. Holbrook)

from the UAC service, were used to ship the processed bentonite.

DM&IR Profit Improvement Project, MS-1077 was submitted by Frank King on Feb 28, 1967, concerning the need for covered hoppers for the shipment of bentonite. The traffic department purposed purchasing 15 100-ton capacity covered hoppers to handle the initial volumes from Hallett Minerals' new plant at Burnett. It was noted that the cost of purchasing this equipment was a significant portion of the overall expense in connection with bentonite movement. As such, Frank King proposed in Study 493 that excess standard ore cars be modified for bentonite service with the addition of 48- to 54-inch sideboards, enclosed top hatches and bottom discharge gates. This would provide a substantial reduction in the initial expense to provide equipment for the Burnett service.

Additionally, King proposed converting Q3, Q4 and Q5 open hopper cars, with 2-foot sideboard extensions, adding 550 cubic feet to the car, and equipping it with round hatches and two bottom discharge gates. Further correspondence on March 8, 1967, from J.P. Keeney, Jr, manager of sales, to J. Halley, chief industrial engineer, noted that: "The DM&IR Traffic Department request for evaluation of acquisition of fifteen 100-ton covered hoppers cars is extremely urgent. The basic specification for the cars results from careful analysis of the requirements of both the shipper and receivers of bentonite. The large capacity of the suggested cars will also permit competitive pricing should the motor carriers become active in their bid for this short-haul traffic. For these reasons, we request that there be no delay in the evaluation of the recommended equipment pending the results of your study on the conversion of ore cars to covered hopper equipment. I am not suggesting that you table Study No. 493 for we are still far short of our ownership requirements of covered hopper equipment for cement handling, in fact the feasibility study on the converted ore cars should include cement handling as well as bentonite."

Producing taconite pellets required bentonite clay as a binder. To haul that clay DM&IR purchased 40 ACF 3,560 cu. ft. covered hoppers between 1968 and 1972 and assigned them to the Hallett Minerals bentonite processing plant at Burnett. Class Q13 DM&IR 5020 is at Proctor on July 3, 1987. (Bob's Photo collection)

Both studies were tabled and discussions took place in September 1967 concerning the purchase of 15 100-ton 4,600 cu. ft. capacity covered hoppers. This specification was revised on September 25, 1967, to purchase ACF 100-ton 3,560 cu. ft., 3-bay covered hoppers equipped with 5 round hatches and roller bearing trucks. By 1968, tax laws allowed DM&IR to sell destroyed or retired equipment and convert the proceeds into similar property within two years of destruction and avoid tax on the financial gain. This was a win-win for ACF and DM&IR.

Two orders were placed with ACF for 100-ton 43-foot 3,560 cu. ft. capacity covered hoppers equipped with gravity outlet gates and 30-inch diameter hatches to be delivered in 1968. The first 9 cars were ordered March 26, 1968, under AFE 10324, Class Q9 100-ton 43-foot covered hoppers 5000–5008 were delivered during April 1968. These cars were purchased by trading in 44 Class U14 and 55 Class U15 70 ton ore cars and a payment of $5,070. Cars were delivered without the new "Arrowhead" logo. Painting of the new logo took place at Proctor between June 3, 1968, and October, 1968 with 5008 being the first car completed. As Hallett Minerals shipments increased from Burnett each year, additional equipment was acquired as needed. All Q9 through Q12 class cars were delivered with minimal lettering and DM&IR initials and numbers. Once on hand, Proctor mechanical department forces applied the distinctive "Arrowhead" logo and "Specialized Service for Industry" stenciling. This lettering was a simplified logo from the winning entry in an employee contest to design a more appropriate identifying monogram.

The second order was for four Class Q10 100-ton 43-foot covered hoppers 5009–5012. These cars were purchased by trading in 25 70-ton ore cars under AFE 10328 approved on June 14, 1968. All 25 ore cars were located on B&LE at time of trade in and ACF resold 18 of these cars to Bethlehem Steel at Bethlehem, Pennsylvania. All four covered hoppers were received from Soo Line at Missabe Jct. during September 1968.

DM&IR continued the trade-in and payment philosophy and purchased three more identical orders in 1969 and 1970. Five cars Class Q11 5013–5017, identical to Class Q10 cars, were delivered in 1969. AFE 10431 was approved August 19, 1969, for exchanging 18 70-ton ore cars and a payment of $8,350 for the five Class Q11 cars.

AFE 10480 was approved on March 11, 1970, for selling 13 Class U17 ore cars to

Hallett Minerals opened their bentonite processing plant at Burnett in 1967. A total of 40 Class Q9 through Q14, 100-ton 3,560 cu. ft, capacity covered hoppers, 5000–5039, would be acquired from ACF between 1968 and 1972. DM&IR 5028 is at Congress Park, Illinois, on December 16, 1977. This car had been wrecked on C&NW in South Dakota and repaired at C&NW Clinton, Iowa, shops, causing it to have a CLN light weight location. (R.F. Kucaba)

Becker Sand and Gravel and using proceeds for payment on one additional car Class Q12 5018 built in May 1970 and placed in service on May 14, 1970.

AFE 10487 was approved on April 7, 1970, for exchanging ore cars and payment of $124,800 for eight additional cars. Built by ACF, Class Q13 5019-5026, built during June 1970, were identical to Class Q12 car.

Working with the DM&IR, Hallett Minerals acquired 13 identical cars from ACF, lettered HCTX 9851–9863 built in January, 1972. These were acquired under AFE 10623 from Hallett on April 26, 1972 and renumbered DM&IR 5027–5039 and placed in Class Q14. Financing was the same as previous Class Q hopper purchases using the proceeds from selling 100 U17 and U18 ore cars to LS&I to finance this purchase. Some of these cars were repainted with the "Arrowhead" logo. Major difference on these cars was DM&IR Class Q9 through Q13 cars had been delivered from ACF with Keystone discharge gates. The Hallett cars were equipped with ACF gates. Taconite plants were having difficulty using the as-built ACF gates and DM&IR replaced these with Keystone Portoloc 13 x 42-inch center discharge gates in 1973.

After the Hallett Minerals (Federal Bentonite) plant at Burnett closed in September 1986, most of the Q9 through Q13 class covered hoppers sat unused on yard tracks at Steelton. Some were assigned and used at various bentonite plants on the CNW and BN between 1986 and 1988. In April 1988 cars were leased to Helm Financial and then sold to Helm on December 31, 1990. Helm in 1991 sold cars to the Detroit and Mackinac Railroad (D&M). All were simply relettered by painting out the "IR" in the "DMIR" reporting marks. Cars were used by D&M in cement service from 1988 through 1992. During 1992 cars were resold to David J. Joseph Company and re-intitialed with DJJX reporting marks and repainted. These cars remain in service with DJJX marks.

DM&IR FLAT CARS

The newly formed DM&IR found itself with 98 D&IR flat cars, 71 D&IR rack flats and 303 DM&N flat cars. These were predominantly being used for pulpwood, steel and similar commodities. From the merger until 1955 these cars fulfilled the needs for flat cars.

In 1955, one Class P2 box car, 5248, was rebuilt into rack flat 6310. When rebuilt, this car was equipped with pulpwood end racks and placed in Class K4. This would be the only box car rebuilt to flat car and the sole car in Class K4.

Increased pulpwood loading and decreased car ownership for allocation to pulpwood service caused DM&IR to consider the purchase of second-hand rack flats. Arrangements were made to inspect 100 rack flats being offered for sale by Kimberly-Clark Co. of Neenah, Wisconsin during May 1963. These cars were former C&NW 40001–41419 series cars built by Western Steel Car & Foundry in 1922. Kimberly-Clark purchased cars in 1951 and then further overhauled them during 1961. Lettered WBLX 100–200 (123 retired), they had an outside length of 42-feet 1-inch and 8-foot high end racks and appeared to be a perfect fit in the DM&IR roster. Nine of the cars were shipped to DM&IR in June 1963 with four cars being set out, bad order, at Two Harbors and the remaining five sent up the Wales Branch for loading. Further inspection indicated that the cars had a possible service life of 10 years but would require extensive rebuilding to keep them in service. Cars were returned the following month and the possible purchase was cancelled.

Each winter, as pulpwood shipments increased, DM&IR requested connecting carriers to supply foreign ownership 40-foot and 50-foot gondolas to load pulpwood. This eliminated the need for having a large roster of equipment for handling pulpwood during the peak winter months, but also caused excessive per-diem and car hire payments. A financial evaluation was completed in November 1964 addressing the use of converted rack flats for pulpwood movement between DW&P interchange at Rainy Jct. (Virginia) and Cloquet, Minnesota via Saginaw, Minnesota interchange to D&NE to replace D&NE flat cars being used at that time. Evaluation was also made at this time using these cars for pulpwood movement between Wales Branch and Cloquet via the D&NE interchange at Saginaw. Evaluation showed many of the Class V1 and Class K2 flat cars badly deteriorated, with center plates rusted through, intermediate body supports failing, center sills and body bolsters rusted and deteriorating. Retirement of the remaining 57 year old Class V1 flatcars was recommended. Consideration was given to rebuilding Class G1 and G2 gondolas, but these cars were 56 and 47 years old respectively and underframes were in poor condition.

AFE 9867 for $6,100 was approved on December 18, 1964, to rebuild 20 of the best condition Class K2 flat cars 6207, 6215, 6218, 6221, 6227, 6240, 6251, 6257, 6281, 6282, 6286, 6287, 6288, 6289, 6290, 6291, 6292, 6295, 6300, and 6301 with wood and steel pulpwood racks. The converted cars had a capacity of 18.5 cords of pulpwood vs. gondolas having a 21-cord capacity. Rebuilding was completed on March 22, 1965. Eight of the remaining Class K2 rack flats, 6283, 6287 through 6292 and 6295, were sold to Reserve Mining Co. on November 18, 1970.

Mining company equipment became larger and more transient during the 1960s and from time to time was moved from one mine site to another. Most of this equipment did not fit on standard 10-foot flatcar decks and needed decks 10-foot 6-inches or wider for movement. Local movements of mining equipment amounted to 54 carloads per year and involved borrowing a flat car from a connecting carrier to protect this load, decreasing the profit margin on the movement and higher per-diem costs. It was expected during 1966 that 24 additional loads between T-Bird Mine and Eveleth Taconite at Fairlane, Minnesota and 30 loads between Erie Mining Co. and South Hibbing, Minnesota, totaling 108 carloads would be handled in 1966. A proposal was made to acquire two 53-foot 6-inch 70-ton flat cars with

Mining equipment was always increasing in size and by the late 1960s DM&IR needed flat cars with 10-foot 6-inch wide decks for moving this equipment. Two former EJ&E flat cars, Class K5 6315 and 6316, were acquired in 1966. Restencilled at EJ&E Joliet Shops it retains its EJ&E green and orange paint job. DM&IR 6316 at Rainy Jct. Yard, Virginia, July 27, 1969. (Owen Leander)

a deck height not higher than 3-foot 9-inches and a nailable steel floor. It was also specified that preference would be to acquire cars with 10-foot 8-inch wide deck and a deck height of 3-foot 4-inches above rail.

This need was fulfilled with approval of AFE 10077 for $6,800 on December 17, 1965, for the purchase of two 70-ton 53-foot 6-inch long flat cars from EJE. Approved on the same date was AFE 10078 for reconditioning of both cars. Initially EJE offered 10-foot wide deck flat cars from EJE 6000–6074 series. DM&IR rejected the 10-foot wide deck cars account the need for a 10-foot 6-inch deck. Two cars originally built by Ralston Steel Car in March 1943 were renumbered at EJE Joliet shops, EJE 6803 to DMIR 6315 and EJE 6818 to DMIR 6316. Both were placed in Class K5 and shipped from Joliet on March 24, 1966. The 6316 was immediately shipped to Rainy Junction yard at Virginia for loading. On arrival at Virginia it was determined that car had a 10-foot wide deck that would not accommodate 54-B mining shovels or 65-ton mining trucks. Car was returned to Proctor for disposition. Proctor shops forces re-stenciled this car EJE 6818 and shipped it back to EJE on May 27, 1966. The second DM&IR 6316, former EJE 6800 series, arrived Proctor on June 2, 1966, with the correct width deck and was placed in service. Car had been painted EJ&E green and had EJ&E orange DMIR stenciling.

As mining equipment increased in size even the 10-foot 6-inch decks were not wide enough for newer mining equipment. During 1967 car 6315 was rebuilt with two floors each 3-inches thick, the deck extended to 12-foot wide and steel supports applied in between the stake pockets to support the new floor. The 12-foot deck was outside of AAR allowed interchange rules and this car was restricted to DM&IR on-line service only. Car 6315 was renumbered W6315 and was still on roster as of June 2004 and 6316 was sold to Azcon, Duluth on September 7, 1989.

Retirement of older 40-foot flat cars and rack flats caused DM&IR to purchase 30 secondhand EJ&E 53-foot 6-inch flat cars numbered DM&IR 6325–6354 in April 1969 to fulfill traffic requirements. Cars 6340–6354 were rebuilt by EJ&E Joliet Shops, painted maroon, stenciled and equipped with pulpwood racks before delivery to DM&IR. An additional five cars, 6328, 6330, 6333-6334, 6335 were converted by Two Harbors shops during March 1971. DM&IR 6344, September 23, 1989. (D.P. Holbrook)

Continued retirement of the aging 40-foot rack flats and regular flats caused a shortage of cars for loading. To supply cars for this loading, DM&IR again went to sister EJ&E, and under AFE 10416 approved April 18, 1969, purchased 15 55-ton 53-foot 6-inch flat cars, former EJE 6556, 6574, 6570, 6565, 6527, 6517, 6445, 6444, 6394, 6496, 6466, 6461, 6409, 6421 and 6493 renumbered in this order to 6325–6339 and placed in Class K6. Cars were built by Ralston Steel Car Co. in April and May 1943. Five of these, 6328, 6330, 6333, 6334, 6335, were converted to rack flats at Two Harbors under AFE 10565 approved on November 3, 1970, completed on March 15, 1971, and placed in Class K7. The conversion allowed these to be used for company tie service five months of the year and pulpwood service the remaining seven months. Conversion of these cars took place during March 1971. All were painted standard DM&IR maroon color with yellow Railroad Roman stenciling.

Fifteen additional flat cars were purchased second hand under AFE 10416 approved on April 18, 1969. Former EJE 6571, 6529, 6401, 6451, 6478, 6386, 6490, 6443, 6398, 6418, 6435, 6442, 6494, 6564 and 6378 renumbered in this order to 6340–6354 and placed in Class K7. All were 55-ton 53-foot 6-inch flat cars built by Ralston Steel Car Co. in April and May 1943, and were converted to rack flats at EJE Joliet shops for pulpwood loading before delivery to DM&IR in September and October 1969. Cars were painted DMIR maroon with yellow stenciling. Joliet shops lacked the proper Railroad Roman stencils and cars received Gothic lettering. Pulpwood traffic declined by early 1990s and some of the cars had their racks removed making them more suitable for MofW service.

Effective January 1, 1984, all Class K6 and K7 flat cars were overage and prohibited in interchange. Some went to MofW service.

Over the years, many of the remaining 40-foot flat cars were converted to MofW service. The most interesting of these conversions was AFE 9095 on April 4, 1965, for the conversion of Class V1 flat car 2607 into a mobile ramp car for use in piggyback service. Car was completed on August 15, 1961, renumbered W2607, and when needed, moved to various team track locations that had the ability to end-load or unload trailer on flat car (TOFC) loads. This was in response to requests for TOFC shipments to various locations on DM&IR. At the time, neighboring Great Northern was already handling TOFC to Virginia, Hibbing, and other iron-range communities.

DM&IR BOX CARS

At the time of the merger, D&IR provided 101 box cars and DM&N 242 box cars to the combined roster. Most of these cars were wood cars built between 1884 and 1923. Some had already been rebuilt with steel centersills, AB brakes and more modern safety appliances.

During 1941 P2 box car 3399 was rebuilt to a steel box car at Proctor shops. Rebuilding involved replacing the single-sheathed sides with new 5-panel riveted sides. Car retained its 7/8 Murphy ends and 6-foot Youngstown door. Additional rebuilds were planned, but the outbreak of World War II and rationing of steel by the WPB curtailed any further rebuilds. Car was retired during February 1970.

On July 22, 1941, AFE M4526 was approved for the installation of steel running boards on 25 box cars to further modernize them. The P2 box cars were delivered with K-brakes in 1923. AFE M4868 was approved May 25, 1942, for the purchase and installation of AB brakes on 40 Class P2 box cars. Modernization continued with AFE 5455 February 6, 1945, for the installation of steel running boards on 10 Class P1 box cars. AFE 9904 was approved on March 11, 1965, for the application of AAR-mandated safety hangers

During March 1941, Class P2 box car 3379 was rebuilt with steel sheathing as a trial car for an eventual rebuilding program. World War II and its associated steel shortages caused the program never to proceed beyond this single car. The only known photo of this car is presented showing it at Steelton, sandwiched between two un-rebuilt Class P2 box cars outside the AS&W steel mill in early 1960. (Glen Blomeke)

During 1966, on-line shipper J.C. Campbell approached DM&IR about the movement of wood chips from their facility at Waldo, to Kaukauna, Wisconsin. Not rostering suitable equipment for this movement, DM&IR purchased secondhand from Chicago Freight Car Co. 20 former ACL Class O-22 automobile cars, had them rebuilt from 14-ft. double doors to 10-ft. single doors and placed them in Class P3 5900–5919. (D.P. Holbrook collection)

to side doors on 120 interchange boxcars.

Declining merchandise shipments caused DM&IR to not purchase any additional box cars until 1966.

During 1966 the J.C. Campbell Co. lumber mill began to ship wood chips from their plant at Waldo, Minnesota to Kaukauna, Wisconsin. To prevent loss of lading, the wood chips were being loaded in box cars. The Waldo facility did not have any storage capacity and DM&IR was forced to request cars from connecting carriers. When cars were in short supply this caused the plant to either shut down or divert wood chips to a waste burner. This meant the loss of product for shipment. DM&IR did not have suitable box cars for this service and AFE 10139 was approved on March 3, 1966, to purchased 20 Class P3 55-ton 50-foot secondhand box cars 5900–5919 from Chicago Freight Car Co. These were former ACL Class O-22 Furniture-Automobile boxcars 14100–14199 built in 1941 by Mt. Vernon Car Co. with 5/4 square corner-post dreadnaught ends, panel roof, 4,680 cu. ft. capacity with National B-Trucks. AFE 10140 was approved on March 3, 1966, for modification by Chicago Freight car of the 14-foot double door opening with a new 10-foot door and additional side sheathing modifications to the carbody to center the door on the rebuilt car. Also included was the elimination of running boards, revision of ladders and installation of stabilized plain journal bearings. The financial evaluation also identified that these cars would be used for interline forwarded movement of: sulphate of ammonium, bagged cement, and wire and nails from U.S. Steel Duluth Works. Cars were purchased and rebuilt for $121,800. Received in late 1966, the DM&IR rostered these cars until 1980.

DM&IR ORE CARS

At the time of the merger the D&IR had 3470 50-ton ore cars and no 70-ton ore cars. DM&N had 7145 50-ton ore cars and 1809 80-ton to 95-ton ore cars for a combined roster of 10625 ore cars. Standardization of the ore car fleet had begun with the DM&N order for 1000 Class U17 cars. Not seeing an operating advantages to the 70-ton to 95-ton cars, DM&IR standardized on a 70-ton capacity ore car for all future orders. The appearance of classes U17 through U31 would remain basically the same with very small variations. The side sheet bottom was raised 3-inches on the Class U19 through Class U22 built in 1942 and 1943 to cut the amount of steel required for the side sheet during World War II. This design change gave the side sheet a small 6-foot long fishbelly appearance in the middle of the car. Similar GN cars built during the war had this design modification. Once the war was over, the side sheet was extended back to its original design.

The first cars purchased by the merged

company were 25 Class U17 70-ton 22-foot ore cars 23975–23999 built by Pullman Standard on Lot #5583 in 1938. These were an add-on order to the 1000 DM&N U17 cars 24000–24999 built in 1937. It is unclear why only 25 cars were ordered.

The outbreak of World War II caused a surge in iron ore demand. The aging fleet of ore cars was insufficient to cover this demand

An outside car shop repaired and repainted DM&IR 24054 causing this distinctive paint job. Proctor, September 1974. (Doug Buell)

and the War Production Board determined that the steel required for building ore cars was necessary to provide the transportation of ore for the war effort. The car builders were not able to meet the demand on an individual basis and the 2,000 ore cars were split equally between Pullman Standard, General American Car, Pressed Steel Car and American Car and Foundry. These orders were all placed on April 6, 1942. Initially this order was for 1,500 cars approved on AFE M4878 on June 4, 1942, for the Pullman, GATC and ACF orders. The Pressed Steel car order was approved on AFE M4946 on November 6, 1942.

The 500 Class U19 70-ton 21-foot 6-inch ore cars 25500–25999 were built by Pullman Standard on Lot #5723 during September and October 1942. Cars had 1,000 cu. ft. capacity and were equipped with Buckeye trucks.

The 500 Class U20 70-ton 21-foot 6-inch ore cars 26000–26499 were built by General American Car during 1942. Cars had 1,000 cu.ft. capacity. The first 250 cars 26000–26249 were equipped with ASF trucks and the last 250 cars, 26250–26499, were equipped with National trucks. The first Class U20 cars were

Pullman-Standard builder photo DMIR 25561, (Class U19 25500–25999) and B-end, right. (Pullman-Standard photos, Jim Kinkaid collection)

Class U21 26532 is at Mitchell, Minnesota, on May 11, 1960, showing off the distinctive World War II-era side sheet construction, causing the car to have a fishbelly type drop in the side sheet in the middle of the car. (Bruce Meyer)

received from Soo Line at Ambridge, Wisconsin, on September 29, 1942, and the final 15 cars received from CMO at South Itasca on November 17, 1942.

The 500 Class U21 70-ton 21-foot 6-inch ore cars 26500–26999 were built by American Car & Foundry on Lot 2513 in 1942. Cars had 1,000 cu. ft. capacity and were equipped with ASF trucks.

The 500 Class U22 70-ton 21-foot 6-inch ore cars 27000–27499 were built by Pressed Steel Car in early in 1943. Cars had 1,000

Class U21 DM&IR 26599 showing Railroad Roman type lettering applied on the right side of car. (American Car & Foundry Collection, John W. Barriger III National Railroad Library at UMSL)

cu. ft. capacity and were equipped with ASF trucks.

This influx of 2,000 70-ton cars was the key element in DM&IR's ability to handle war tonnages during World War II. At the start of 1946 DM&IR rostered 10,153 50-ton and 3914 larger-capacity ore cars. The 50-ton cars had soldiered on during ther war but all were at the end of their useful lives in 1946

Class U22 27196 and Class U28 30593 loaded with iron ore at Proctor in 1957. Compare the World War II car body on 27196 with the post-WWII car body on 30593. This reduction in side sheet height saved steel during the war years. (D.P. Holbrook collection)

The lack of a "stencil group" on the side sill indicates this photo was made before 1952. Class U-23 27772 and Class U-24 28004 are at Proctor. The letter "C" next to the class number indicates these cars have been inspected and are "clean and tight" allowing them to be used for washed iron ore. Note the "mine card" bracket on both cars had the cardboard "mine way-bill tag" inserted. (D.P. Holbrook collection)

Class U-23 27760 was part of 500 cars 27500–27999 built by Pullman Standard Car Co. in 1948. Note attached to the left truck directly below the sill step is the empty-load air brake sensor. Mitchell, Minnesota, May 11, 1960. (Bruce Meyer)

Class U24 28190 is at Duluth, July 3, 1955 with stencil group B applied during winter 1953-1954 shopping season is barely visible showing how quickly ore dust covered the lettering on ore cars. (Bruce Meyer)

with most needing to have K-brakes replaced with AB brakes. AFE 5769 was approved on April 3, 1946, for the purchase of 3,500 70-ton steel ore cars. Car builders lacked the steel and the capacity to build cars after the war and this AFE was cancelled. The aging 50-ton cars were in dire need of replacement after the war and the effort to replace them began in 1947.

During June 1947 DM&IR approved AFE 6069 on June 7, 1947, to order 3,000 70-ton ore cars split between Pullman Standard Car Co., 1500 cars, American Car and Foundry Co., 500 cars, Pressed Steel Car Co., 500 cars and General American Car Co., 500 cars.

The first group of 500 Class U23 70-ton 21-foot 6-inch ore cars 27500–27999 were built by Pullman Standard Car Co. on Lot #5903 between February and May, 1948. Cars had 1,000 cu. ft. capacity and were equipped with ASF A3 Ride Control trucks.

The next group of 500 Class U24 70-ton 21-foot 6-inch ore cars 28000–28499 were built by American Car & Foundry on Lot #3285 at the Berwick, Pennsylvania, plant during 1948. Cars had 1,000 cu. ft capacity and were equipped with ASF A3 Ride Control trucks.

Pressed Steel Car delivered the next group of 500 Class U25 70-ton 21-foot 6-inch ore cars 28500–28999 during 1948. Cars had 1,000 cu. ft. capacity and were equipped with ASF A3 Ride Control trucks. Correspondence with Pressed Steel Car on February 20, 1948,

General American Car Corp. built 50 Class U26 70-ton ore cars 29000–29499 in 1948. (GATC Photo, D.P. Holbrook collection)

advised them to deliver cars in 25-car lots to B&O via P&LE to Demmler Transfer, Pennsylvania for loading. Once empty, cars were to be routed B&O-EJE-Barrington, IL-CNW-CMO-Duluth-DM&IR. It was also noted that when necessary, cars were to be delivered to PRR via P&LE to Scully, Pennsylvania, for loading. Once empty, cars were to be routed PRR-Clark, IN-EJE-Barrington, IL-CNW-CMO-Duluth-DM&IR. Because cars were not needed during the winter months on DM&IR this allowed DM&IR to earn a little per-diem before they were delivered.

Class U26 29278 at Conneaut, Ohio, on lease to B&LE April 7, 1968. (Bob's Photo collection)

The heaped iron ore created the appearance of an overloaded car. Class U28 30547 is at Proctor, June 16, 1959. Note: car had previously had a different stencil group C which was painted out and new D stencil applied when air brakes were tested outside of the normal shopping plan. (Bruce Meyer)

Class U27 29555 has had letters X placed on either side of the monogram for testing paint used on the "X" stenciling when exposed to heat in the infrared ore thawing sheds. Car originally had a stencil group of D but was placed in stencil group A when it was shopped on Jan 15, 1956. (Basgen photo. D.P. Holbrook collection)

The last group of 500 Class U26 70-ton 21-foot 6-inch ore cars 29000–29499 were built by General American Car during 1948 on Order #3003. Cars had 1,000 cu. ft. capacity and were equipped with ASF A3 Ride Control trucks.

Even before these four groups of cars were delivered AFE 6184 was approved on December 12, 1947, for an additional 1500 70-ton ore cars for delivery during 1949. Pullman Standard was chosen to build 1,000 cars and General American Car to build the remaining 500 cars. Delivered during 1949 the PS built Class U27 70-ton 21-foot 6-inch ore cars 29500–30499 were built under Lot #5921. The GATC built Class U28 70-ton 21-foot 6-inch ore cars 30500–30999. All had 1,000 cu. ft. capacity and were equipped with ASF A3 Ride Control trucks.

By 1950 the DMIR rostered 5621 50-ton and 7372 70-ton ore cars. Many of the remaining 50-ton cars were equipped with K-brakes which were in the process of being outlawed. Iron ore traffic remained heavy and car shortages during the summer shipping season caused the DMIR to start equipping many of the remaining serviceable 50-ton ore cars with AB brakes. Additional 70-ton ore cars were ordered in January 1951 to assist with tonnage and allow continued retirement of some of the remaining 50-ton ore cars.

The 1500 Class U29 70-ton 21-foot 6-inch

Photographed from the highway bridge at the north end of Proctor on June 16, 1959, we see Class U30 32903 and Class U20 26110. Note the side sheet differences. Cars built during World War II were modified to allow the bottom of the side sheet to be a little higher, reducing the necessary steel for construction. These gave the WWII-built cars a distinctive drop in the middle of the side sheets. (Bruce Meyer)

Pullman Standard builder photo of Class U29 31016. (Pullman-Standard photo, Jim Kinkaid collection)

ore cars 31000–32499 were ordered from Pullman Standard under Lot No. 8041 and built at their Michigan City, Indiana, plant. AFE 6944 was approved for this purchased on January 29, 1951. All were delivered in 1952. When ordered all cars were to be built of riveted construction, have 1000 cu. ft. capacity and ride on ASF A3 Ride Control trucks. Pullman asked the DM&IR for permission to build the last car 32499 as an all-welded car for testing future car designs which was approved. Pullman, yearly sent an inspector from their Michigan City plant to perform an inspection of the

Class U29 31854 is at Mitchell, Minnesota, May 11, 1960. (Bruce Meyer)

Pullman-Standard delivered 1500 Class U29 70-ton ore cars 31000–32499 in 1952. DM&IR 32082 at Proctor, April 1963. (D. Repetsky)

Elimination of K-brakes and 50-ton ore cars caused DM&IR to place multiple orders for 70-ton ore cars during the 1950s. DM&IR 32887, one of 500 32500–32999 Class U30 70-ton ore cars built by Pullman-Standard in 1953, is shown in a PS builder photo. (Jim Kinkaid collection)

car. When delivered in 1952, 32499 was stenciled "For Use between Mitchell and Steelton Only." Correspondence indicated this was for the movement of ore from the Godfrey Mine to AS&W at Steelton. Every attempt was made to keep this car in service year around, even when the shipping season was over. The Superintendent directed yardmasters to make sure this car was expedited whenever possible. During the December 15, 1959, inspection showed longitudinal welds of the striker wing assembly were fractured at four locations. PS recommended repairs when practical and for car to remain in service. These inspections continued up until the mid-1960s. It was rebuilt to scale test car W32499 on May 10, 1985, and then quickly renumbered 132499 a few months later and is still in service.

During September 1952, the DM&IR placed an additional order for 500 identical Class U30 70-ton 21-foot 6-inch ore cars 32500–32999 of riveted construction, from Pullman Standard under Lot No. 8125 and built at their Michigan City plant. All had 1,000 cu. ft. capacity and rode on ASF A3 Ride Control trucks.

Four years would pass before one last order of 70-ton 21-foot 6-inch ore cars would be

The final order of 70-ton ore cars, 500 Class U31 33300–33499, would be constructed by ACF in 1957. The majority of these cars would be rebuilt to taconite service during the late 1960s and early 70s. DM&IR 33199 would be rebuilt to DM&IR 53199 in November, 1967 with the addition of 19-inch high-side extensions. The hi-side extension would be cut down to 9½-inch mini hi-side extensions on May 19, 1972. ACF Builders photo. (Jim Kinkaid collection)

placed. The final order was for 500 Class U31 70-ton 21-foot 6-inch ore cars 33000–33499, ordered on October 23, 1956 on AFE 8133 for $47,625. All were of riveted construction, 1,000 cu. ft. capacity and rode on ASF A3 Ride Control trucks.

At the time the order was placed, further development in AB brake systems was under way. The updated AB control valve was introduced in 1955, and classified as an "AC" brake system. The AC system incorporated diaphragm-operated piston, rubber "O" rings and spool valves that replaced slide valves wherever possible. All 500 cars were delivered with AC brakes. (See section C, air brakes, appendix, for more about AC brakes.) Prior to this order all ore cars had been painted ore car brown, but AC brake systems had not been approved for interchange, so all Class U31 cars, when delivered, were painted black to allow easy identification of their non-interchange status.

The U31 order would close out the 70-ton ore car roster on the DM&IR. A total of 9515 ore cars between Class U17 through U31 created a standardized roster. All of the cars, with very minor detail variations, were identical.

A shortage of serviceable open hopper cars caused DM&IR to lease a number of open hoppers and ore cars for coal service during 1974. DMIR 20001–21000 were 77-ton 27-foot 2-inch ore cars B&LE 20000–20700 Class HMA1 built by Greenville on Order No. O.O.603 in 1952 with 1,085 cu. ft. capacity. A total of 61 cars with random numbers were leased. Cars retained their former B&LE numbers but were re-stencilled with DM&IR reporting marks to assure if they got off-line that they would be returned to DM&IR and not B&LE. The 27-foot 2-inch outside length of these cars was longer than the 22-foot standard DM&IR ore car length making them unsuitable for ore service. Cars were on roster during 1974 and 1975.

U.S. Steel built their pre-taconite mine, called Pilotac at Mt. Iron and the processing plant called Extaca at Virginia during 1958.

Detail photo of 70-ton ore car underframe. Proctor, April 22, 1996 (D.P. Holbrook)

Class U31 33363 has had test paint and lettering applied in this 1959 photo. The X on either side of the monogram was painted with two different manufacturers paint to test longevity when exposed to heat in the infrared thawing sheds. Note the "Test car do not touch up or paint this car until Jan. 1960" in the upper left corner of side sheet. (Basgen Photo)

DM&IR was requested to supply ore cars assigned to this service. Some 75 70-ton ore cars were initially assigned in 1958, with cars being stenciled "For Pilotac Service Only." Retirements caused DM&IR in February 1961 to assign 125 Class U11 50-ton ore cars to this service, which allowed the 70-ton cars to be placed in regular ore service. The same stenciling was applied to these cars and removed from the 70-ton cars at this time. Once they were placed in this service it was determined

A shortage of coal hoppers caused DM&IR to lease open hoppers and ore cars during 1974 and 1975. A total of 61 B&LE 27-foot ore cars in 20000–20700 series were leased and reporting marks changed to DM&IR. Cars could not be mixed with DM&IR ore cars because B&LE cars were longer cars which would have caused spacing issues on the ore docks when unloading. Biwabik, November, 1974 with load of coal for Minnesota Power & Lite plant at Colby. (Doug Buell)

that the wooden poking platforms on each end of the 50-ton cars created a slipping hazard from spilled concentrate. By September 1961 all 125 cars had had the wooden end platforms replaced with Apex steel-grid platforms.

Coal hoppers were always in short supply during the late 1950s and many of the 50- ton ore cars were downgrade to coal, ballast and aggregate service. One hundred and fifty Class U10 ore cars were placed in coal service during April and May 1960. These cars had eight-inch "COAL" lettering applied to indicate their service. By 1951, 199 Class E7 and U7 50-ton ore cars, of the poorest condition, were assigned to Arcturus mine service. Early in 1952 Oliver Iron Mining (OIM) requested that these cars be replaced. Initially 200 Class E8 and E11 50-ton ore cars were selected as replacements. Additional correspondence indicated that 200 Class E7 and U7 cars would be selected and sent to Coleraine.

Ore shipments continued to decline in the 1960s. On September 21, 1962, a decision was made to store 800 70-ton ore cars for the 1963 season. Two hundred cars from each stencil group were chosen. This was further increased to 1,000 70-ton ore cars on December 13, 1962, split evenly within stencil groups. Per D.J. Smith, president of DM&IR, none were to be returned to service without his approval.

Declining ore shipments in 1963 caused 1,100 70-ton ore cars to be held in dead storage. Four hundred 70-ton ore cars were assigned to hauling coal between Duluth and Minnesota Power and Light at Colby, and 300 ore cars were assigned to shuttle service hauling crude ore between Rust Crusher and Fraser yards and the Sherman Mine concentrator.

Effective with the April 30, 1975, DM&IR Equipment list, 70-ton ore cars began to be listed as: Regulars, Mini Hi-Sides, Shaker Pocket Hi-Sides and ballast. This was at the recommendation of R.A. Huttely, manager industrial engineering, to R.L. Wagner, superintendent, in a written memo during March 1975. Prior to this time ore cars were tracked as 50-ton and 70-ton ore cars. After this date non-taconite service ore cars were referred to as "regulars" in correspondence and radio conversations.

AFE 11071 was made out on September 15, 1976, for the retirement of 133 50-ton 4-door ore cars that had been placed in ballast service. It also authorized the conversion of 66 regular ore cars with modified Morrison Knudson (MK) ballast doors and 14-inch side extensions. Numbered 1404–1469, all were painted yellow with maroon lettering. Rebuilding started in November 1977 and was completed on July 21, 1978. Prior to this, regular ore cars 24100, 24350, 32675 and 32880 had been equipped with MK ballast doors in 1972, the first two with no side extensions and the last two with 9¾-inch side extensions. Side extensions were added to 24100 and 24350 in 1975. These four cars were renumbered 1400–1403 in 1976. An additional 34 cars identical ballast cars, 1470–1503 were rebuilt and released for service on November 14, 1979. None of this group was repainted.

Expanding taconite tonnage and lack of

DM&IR 1483 was one of 34 ore cars converted to ballast cars 1470–1503 in 1985. Unlike earlier conversions none of these cars was repainted yellow with maroon lettering. Biwabik, September 22, 1986. (D.P. Holbrook)

Declining natural ore shipments and increasing taconite pellet production caused DM&IR to rebuild and renumber 224 regular ore cars starting on January 19, 1981. Renumbered from 27000–33000 series ore cars and renumbered into 60000 series, they were rebuilt with angle iron reinforcing to keep pellets from leaking out the bottom of the cars. The series would eventually expand to 1143 cars numbered 60000–61142. DMIR 60080, former DM&IR 29453, Forbes, October 2, 1999. (D.P. Holbrook)

mini-quads to handle the additional tonnage caused the 70-ton "regulars" to be pressed into taconite service in 1980. An immediate problem was that the doors on most of these cars did not close enough to prevent leakage of the taconite pellets. Prior to 1980, a number of low-side ore cars had been maintained for pellet service and called "clean and tights." These had "CT" stenciled on the side for easy identification but having random numbers became

Prior to 1972, ore cars and 30 Rogers ballast cars built in 1941, W1300–W1329, were used in ballast service. Four 70-ton ore cars were rebuilt with Morrison-Knudsen patented ballast doors in 1972. Renumbered in 1978 to 1400–1403, they were joined by an additional 100 cars 1404–1503 in 1978. DM&IR 1431 Proctor April 21, 1980 (S.D. Lorenz)

The first two Morrison-Knudsen patent ballast door conversions, 24100 and 24350, renumbered to 1400 and 1401 in 1977, had no side extensions when built. Later 9¾-inch side extensions were added. All additional ballast car conversions would have a standard 14-inch side extensions. Keenan, Minnesota, May 28, 1983. (D.P. Holbrook)

DM&IR 60932 Alborn, Minnesota, September 27, 2001 (D.P. Holbrook)

Repainting of ore cars, except taconite service cars, was discontinued in 1966. By September 22, 1999, these two cars are typical of paint on the remaining "regular" ore cars. (D.P. Holbrook)

Many ore cars reached the 50-year overage rule in the 1990s. To identify FRA mandated inspections DM&IR painted a white star on the ore cars and added a star each year when car was inspected beyond the 50-year rule. DM&IR 60036 is a former Class U26 29383 built in 1948. Keenan, Minnesota, October 2, 1999. (S.D. Lorenz)

a problem. Beginning January 19, 1981, 224 regular ore cars were shopped at Proctor and additional angle iron reinforcing was placed on the bottom doors to prevent this leakage making the cars clean and tight for pellet, flue dust or, if needed, natural ore loading. As cars were shopped they were renumbered into the 60000 series. By March 27, 1981, 465 cars had been renumbered: two Class U21, 130 Class U23, 172 Class U24, 79 Class U25 and 82 Class U26 to 60000–60464 series. Renumbering allowed the rebuilt cars to be more easily identified. This number series eventually expanded to 1143 cars numbered 60000–61142.

TACONITE AND T-BIRD CARS

Two World Wars and the Korean conflict caused large amounts of natural ore to be mined and shipped from the Missabe and Vermillion Ranges. By the early 1950s steel companies were looking for new sources—both foreign and domestic—to satisfy the need for raw materials. Taconite, a low-grade ore, was available in large amounts in Northeast Minnesota, but had always been passed over for the higher iron content of natural ore. Early attempts at commercial taconite processing took place at the Mesabi Iron Co. plant at Babbitt in 1922, which closed in 1924.

Not until some tax law changes in 1940 did any of the steel companies act. Pickands Mather formed Erie Mining Co. in 1940 and Oglebay Norton formed Reserve Mining in 1939 to make sure when the high-grade natural ore was depleted there would be another source of iron for steelmaking. Edward W. Davis, working at the University of Minnesota Mines Experiment Station, became the "Father of Taconite." Processes he developed would finally allow large-scale commercial development of taconite. Erie and Reserve would both build their own railroads to haul taconite. U.S. Steel built its experimental taconite processing plant in 1953 but it would be 1967 before large-scale taconite production would begin at the Minntac Plant in Mt. Iron.

DM&IR did not have equipment at the time to handle the processed taconite. The 70-ton ore cars could not be filled to capacity unless side extensions were applied to the cars. A non-sideboard equipped ore car could only handle 61.8 gross tons, but with sideboards this increased to 74 gross tons. The economic downturn in the early 1960s forced DM&IR to store many of the 70-ton ore cars. Because of the new nature of taconite it was felt rebuilding old equipment instead of buying new would be the most economical way of providing cars for taconite service. A number of heights of extensions were applied to ore cars including, 33-inch, 27½-inch, 19½-inch and eventually 9¾-inch became the standard.

During 1961 27½ inch sideboard extensions were applied to five 70 ton Class U-31 ore cars 33002, 33073, 33098, 33106 and 33143 for hauling agglomerates from the U.S. Steel Extaca pilot taconite plant located at Virginia. To make it easier to track these cars the number "5" was added to the original car number; 33002 became the 533002, and so on. During 1962 an economic study was done to determine the costs and benefits associated with the use of these cars after one year of utilization. Two different designs of high sideboards for 70-ton ore cars were designed to increase loading capacity for sintered and taconite pellet shipments. Based on this study the original cars were rebuilt again in 1964 by replacing the 27½-inch extensions with 33-inch extensions for hauling bulk material (limestone and coal) from Hallett Dock at West Duluth to UAC and AS&W at Steelton. The other proposed side board height was 20-inches for handling of the taconite pellets.

Once the issues with processing taconite were solved, it fell to DM&IR to handle the crude and processed taconite. Over 1,000 70-ton ore cars were in storage in 1964 account declining natural ore shipments. The total fleet of ore cars was more than sufficient to handle natural ore and rather than purchase new

Crude taconite and taconite pellets required side extensions. Beginning during May 1964 with 69 cars, 19½-inch sideboard extensions with shaker pockets were applied to crude taconite service. Taconite pellet service cars would receive the same style extensions without the shaker pockets. Overloading of taconite pellet cars caused the side extensions to be reduced to 9½ inches in 1971. The fleet of taconite cars would eventually surpass 1500 cars. T-Bird service DM&IR 50031 at Virginia, July 27, 1969 (Owen Leander)

equipment the choice was made to modify existing ore cars for both crude and processed taconite service. A plan was developed to equip 400 70-ton ore cars with both heights of sideboards to increase capacity, one for crude taconite and one for processed taconite pellets. The newest Class U31 ore cars 33000–33499 built

DM&IR 53400, a T-Bird crude taconite service ore car, at DM&IR Rainy Jct. Yard, Virginia, July 27, 1969. Built as 33400, it was renumbered 533400 in 1964 when high sides were added, to 53400 August 2, 1967, and finally 40226 on October 15, 1976, when repainted into T-Bird paint scheme. (Owen Leander)

DM&IR 53000, a T-Bird crude taconite service ore car, at DM&IR Rainy Jct. Yard, Virginia, July 27, 1969. The X on the side sill is the "stencil group" car was placed in to identify air brake testing years. Built as 33000, it was renumbered 533000 in 1964 when high sides were added, to 53000 August 2, 1967, and finally 40233 on October 18, 1976 when repainted into T-Bird paint scheme. (Owen Leander)

in 1957 were selected to be rebuilt with sideboards. While this was going on, the first taconite plant, Eveleth Taconite Co. was preparing plans for construction of their pellet plant at Forbes, Minnesota, and the crude taconite loading pocket, T-Bird, at Virginia.

AFE 9753-2 was approved in May 1964 for the purchase and application of high sideboards to 69 70-ton ore cars. Per letter of September 14, 1964, from E.J. Bosmeny, general supervisor revenue accounting, to R.M. Seitz, chief mechanical officer, all rebuilt side board-equipped ore cars were to be renumbered by adding the number "5" in front of the present car number. Work was programed for October through December 1964. By October 28, 1964, 19½-inch sideboards had been completed on 56 cars with the 13-car balance to be completed the following week. Cars were completed by November 1, 1964, and were to be used in initial crude ore service between Eveleth Taconites T-Bird mine and Fairlane pellet plant at Forbes. Cars selected were: 14 Class U27 29500–30499, 2 Class U28 30500–30999, 19 Class U30 32500–32999 and 34 Class U31 33000–33499. Because cars were completed before the first taconite plant was in operation, a study was made to determine benefits of using the sideboard-equipped car in Missabe Jct.-to-Steelton limestone service. On April 5, 1965, the 69 original high-side ore cars were placed in service for loading of limestone from Hallett Dock to AS&W Duluth Works and UAC plant. This released regular 70-ton ore cars for ore service. Cars remained in this service until Eveleth Taconite was opened on December 10, 1965, at which time they were placed in shuttle train service between the T-Bird mine and Fairlane taconite pellet plant at Forbes.

A new series of car numbers was adopted for these cars and the new numbers appeared in the next issue of the *Official Railway Equipment Register* with the new AAR designation of "HMAS" rather than the standard 70-ton ore car "HMA" designation, the "S" standing for "special equipment." The 1965 capital expenditure forecast included the application of an additional 465 car sets of sideboards scheduled for application beginning in March and continuing through October. In connection with this program, an investigation was being made of semi-automatic welding equipment to be used at Proctor shops to speed up the application process.

On Jan 6, 1965, a Major Methods Project (MMP) was completed on the economic benefits of application of sideboards to 216 Class U31 33000–33499 ore cars for the first phase of Minntac and Eveleth Taconite pellet movement in 100-car trains. AFE 9941 was approved on March 26, 1965, for the initial application of 108 cars for Minntac and 108 cars for Eveleth taconite service. This was later increased by two cars to allow nine additional cars in each service to protect bad orders and routine maintenance. Economic benefits of application of sideboards to ore cars for use in AS&W-UAC stone movement and benefits of roller bearings were also assessed. Sideboards for the cars were manufactured by Marine Iron and Ship Building Co., Duluth, and then hauled by truck to Proctor. Once the sideboards arrived it was anticipated that 15 cars per week could be done and installation was completed by November.

The 1965 application of high-sides to 216 Class U31 cars 33000–33499 was begun with a sample car on September 10th. Work on the remaining 215 cars began on September 27th, working 3 cars per day. On November 3rd, 91 of the 216 cars were completed and by December 19th, 184 cars had been completed. It was decided to complete the remaining 32 cars in the spring bcause of a ventilation problem at the Proctor shops. Poke holes (steaming holes) on these cars were tack-welded shut. Through December 14th, 109 high-side cars were cleaned of snow and ice, reweighed, restenciled and had ACI labels corrected to reflect new information. Before the rebuilding project was complete it was determined that the 4-inch channel used as a top chord member would create a potential hazard, allowing taconite pellets to accumulate and subsequently fall off cars while in motion. A modification from channel to angle iron top chords was approved on March 17, 1966. Angles were applied to the top sides of 102 high-side ore cars with the remaining channel equipped cars sent back through the shops for modification. Triangular holes near the top sides of high-side ore cars had to be closed by welding a triangular plate to prevent leakage of pellets. Welding of plates took place at Proctor rip track and roundhouse. Testing of modified high-side cars showed pellets were leaking out the car bottoms. Six high-side ore cars were modified late in 1965 to eliminate spillage onto dock during unloading, and once tested and approved the remaining cars were also modified. By the end of 1965, 260 70-ton ore cars were equipped with 19½-inch sideboards for taconite pellet service.

During the DM&IR board meeting on September 22, 1965, discussion took place about the original five high-side ore cars that were equipped with 33-inch sideboards. These cars had been found to be extremely difficult to load using front-end loaders. It was recommended that they be cut down to a 19½-inch height. Chairman D.B. Shank recommended instead of going to the expense of cutting the cars down, consideration should be given to consigning them to the NP for sugar beet movement along with other cars to fill their lease request for 1965. It was noted at the board meeting that as DM&IR handled more taconite, the car requirements would drop and the winter car repair program could perhaps be handled on a 12-month basis instead of seasonally. This arrangement was later adopted and had a significant impact: lower manpower needs and higher utilization of shop capacity.

During 1966 engineering work on taconite ore "super cars" of 140- to 150-long ton capacity were in progress. Various alternative car designs were considered including 3-axle trucks on each end of a car, articulated ore cars, and drawbar couplings of longer and larger capacity cars. Many new innovations in car design, air brake arrangement, and train operation were incorporated in the study. Further engineering and financial evaluations were conducted on the three most promising designs but no further action took place. DM&IR elected to continue to rebuild 70-ton ore cars.

Per letter from R.M. Seitz, chief mechanical officer, on May 4, 1966, all 293 taconite cars with 6 digit numbers 529615–529991(four cars) 530031–533499(289 cars) were to be renumbered to 5-digit numbers by moving the "5" in position 1 to be in position 2; 530031 would become 50031. Eighty of these cars were equipped with shaker pocket side extensions and were assigned to T-Bird loading. All cars were renumbered by January 1, 1967.

Approved in spring 1966, sideboards were applied to an additional 34 70-ton ore cars during summer 1966. Painting of additional high-side ore cars began on June 13th and by July 29th 127 cars had been completed. Painting requirement on crude ore high-side cars remained high because of the need to use flame throwers for unfreezing the crude ore. It was felt that car shakers could possibly eliminate the problem. Application of high-sides to 34 additional ore cars was started on July 13th and completed August 1st, with all cars being repainted maroon and gold. The remaining five 33-inch side board cars were shopped starting on July 13th and completed on August 9, 1966, to lower them to the standard 19½-inch side-boards. Cars for crude taconite service had shaker pockets and cars for taconite pellet service did not

August 21, 1967, found DM&IR with 293 high-side ore cars, with 70 cars assigned to Eveleth Taconite T-Bird crude ore service and the remaining 216 cars assigned equally to Minntac and Eveleth Taconite for taconite pellet service. At the time, unit pellet trains with 96 cars and two 1750/1800 hp locomotives were considered the standard operating requirement. Data from Minntac in August, 1967 indicated production would increase from 4.25 million tons to 5.1 million tons in 1968, which would require 104 additional ore cars to be modified with high-sides for taconite loading. AFE 10273 was approved August 23, 1967, and all 104 Class U31 cars 33000–33499 were modified during November and December 1967 and renumbered by changing first digit from 3 to 5. Eight of these cars, 53041, 53086, 53114, 53172, 53200, 53244, 53265 and 53371, had shaker pocket high-sides applied for use in T-Bird crude taconite loading. All of these cars were painted when rebuilt.

Cost Authority C-6005 was requested on August 2, 1967, and approved on October 20, 1967, for the application of "Orinoco-type" retainers to 220 high-side ore cars 50031–59991(various numbers) and associated control equipment on five RS Class SD9 locomotives. The Orinoco variable retainer control, also called "straight air" by DM&IR, was a secondary air line that required the train brakes to be applied first with the automatic brake. Once that was done the Orinoco brake valve would be set to maintain the approximate pressure the automatic application produced. Then the automatic brake would be released.

The Orinoco straight air pipe could then be raised or lowered a few pounds to maintain exact speed going down grades. This eliminated the need to stop a train at Proctor or Two Harbors to manually walk the entire train and set retaining valves on each car. This project was initiated because of the unit train operations agreement that allowed DM&IR to have train crews operate direct from the ore docks in Duluth to iron range taconite plants and return with a train to the docks. Normally this operation would have involved one crew to handle the train from the ore docks to Proctor, a road crew from Proctor to the taconite plants and return and another crew to handle the train from Proctor to the ore docks. The project was completed on August 22, 1968. The Orinoco braking system was originally developed for use on U.S. Steel's Orinoco iron ore operations

Starting in 1961 DM&IR began testing various heights of sideboard extensions on ore cars for loading taconite. A standard 19½-inch sideboard was established in 1964 and by early 1970, 388 cars had been converted. Overloading of these cars became a problem and in November, 1971 the sideboards began to be cut in half to 9¾-inch with the top half used to equip additional taconite cars. A decision was made in 1972 to drawbar-couple these cars in four-car sets called "mini-quads." DM&IR 51062, former Class U29 31062 is outside Proctor shops during July, 1976. Note the lack of stencil group. It will be stenciled with a mini-quad "set" number on the side sill for computer tracking of the car sets. (Bob's Photo collection)

in Venezuela which involved heavy grades and utilized all ABEL-equipped cars.

On December 31, 1968, AFE 10386 was approved for application of high-sides, drop door seals and Orinoco-type air brakes to 30 Class U31 33000–33499 ore cars to replaced 30 high-sides destroyed in the July 27, 1968, Mt. Iron wreck. Once rebuilt the first number was changed from 3 to 5. One car from this group, 53144 was equipped with shaker pockets on the high-sides. By August 1, 1970, a total of 388 high-side ore cars 50031–59991(various numbers) were in service. An additional 108 Class U31 cars 33000–33499 were to be equipped with high-sides during the 1970-71 winter shopping. This request was made to allow three taconite trains to be operated in an eight-hour period during the winter months and in theory would only require one dock switch crew for one shift to dump the trains.

Early in 1971 it was found that the 19½-inch high-side cars were a problem. The high sideboards caused taconite mine operators to overload the cars. Minntac was producing 6 million tons of pellets and expected this to expand to 12 million tons by mid-1972. Increased shipping of Minntac pellets caused DM&IR to temporarily place both regular and hi-side ore cars in taconite service between the plant and the ports of Duluth and Two Harbors. Solving this problem became a priority.

During June 1971, B.E. Lewis superintendent of motive power & cars, proposed, on Cost Authority C-6013. that the 19½-inch sideboard cars be cut down. This would be accomplished by cutting off the top half of the 268 current hi-side taconite cars and placing this half on 268 regular ore cars. This would increase the fleet of taconite cars to 536 cars with 9¾-inch sideboards, creating a "mini high-side." The conversion also involved applying straight air brake retainer equipment (Orinoco), type E couplers and removal of AC and application of AB brakes on all cars in

Class U31. Cutting off of the tops of existing cars was a significant cost savings because no new material was required and the only cost was labor to accomplish the rebuilding. Crude taconite service cars would continue to have 19½-inch side extensions.

Prior to the taconite plants being built, iron ore was shipped from April to November because of the problems with ore freezing in the cars. Year-round operation of the taconite trains caused the DM&IR to experience new problems in train operation. Snow, ice and cold temperatures created their own problems. The most severe was maintaining air pressure in the brake systems. The best way to solve this problem was to eliminate as many connections between cars as possible.

DM&IR 51184 one of four cars in mini-quad set 70-1 is shown at Proctor, October 4, 1998. Set 70-1 consists of 51184, 51800, 52361, 52380. (D.P. Holbrook)

DM&IR 51692, former 31692 at Forbes, October 2, 1999. Mini-quad set 105-2 includes: 51206, 51692, 51872, 52234. Note the "blaze orange" corner post indicating the location of couplers on mini-quad sets. (D.P. Holbrook)

By November 1971, the first two hi-side taconite cars were converted to "mini hi-sides." DMIR 53151 was one of the first two cars converted. Cars were equipped with a one-piece air hose, standard couplers between the two cars and released from Proctor Shops late in January 1972. The uncoupling levers were made inoperative to prevent tearing the air hose connection apart and all four corner posts on the ends of the pair of cars were painted blaze orange to call attention to the location of uncoupling levers. These cars had 9¾-inch sideboard heights, which would become the DM&IR standard for taconite pellet service cars. Crude taconite cars in T-Bird service would continue to have the 19½-inch sideboards with shaker pockets.

Starting in May 1971 until early 1972 DM&IR had hi-side, mini-hi-sides and regular ore cars in taconite pellet service. By January 1972, 28 hi-side taconite cars had been converted to mini hi-sides. During summer and fall of 1972, modifications to 268 hi-side taconite cars were completed.

Once testing of the first pair of cars was completed the mechanical department's innovate solution was to use three solid drawbars between four cars, creating a "mini-quad" set of the now-standard 9½-inch sideboard taconite cars. This eliminated three-fourths of the glad hands by connecting the cars with solid one-piece hoses between the cars. The standard E-type couplers on each end were replaced with F-type interlocking couplers to help cars stay in-line with track in case of a possible derailment. All of this was possible because of cars being in assigned taconite service so there was no need to switch individual cars. Threaded pipe fittings were eliminated and replaced with welded ones to eliminate another leakage point in the brake system. Added during the shopping was an additional straight air retain-

Eliminating air hose glad hands and couplers, DM&IR created "mini-quads" in 1972. These were 9½-inch sideboard equipped taconite cars with drawbars connecting the first through the fourth car and air hoses without glad hands between these cars. Stencil groups were eliminated and each four-car set was assigned a set number. Set 92-3 is shown at Allen Jct. Minnesota May 3, 1986. The smaller number was used to identify air brake testing years. Mini-quads had accelerated testing and maintenance account their high utilization and mileage. DM&IR 51494, part of a mini-quad set exhibits blaze orange corner posts so crews can identify locations of couplers instead of drawbars between cars. (Doug Buell)

er system (Orinoco) in addition to standard AB systems. When released from the shop, the cars were painted in DM&IR maroon with blaze orange corner posts on the ends of the cars that retained couplers. This allowed easy identification of the coupler ends of each set of cars.

The first 4-car set consisted of cars 52619, 52589, 52915 and 52856 which were released from Proctor Car Shop at 10:30 a.m., Friday July 14, 1972. Each car retained its individual car number but each set was assigned a "mini-quad" set number that was painted in the center of the side-sill (see photos). At the time the first "mini-quad" set was released a total of 356 high-side ore cars had been converted to Mini hi-sides. A total of 171 hi-side cars remained to be converted to mini hi-sides. Maintenance shopping on an annual basis was established with a 42-week shopping period. This allowed for shopping 12 cars per week. (See mini-quad roster for set and car numbers.) This was also done for computer tracking of the sets of cars. Before the project was completed both Minntac and Eveleth Taconite requested additional cars, Minntac for movement of pellets to Two Harbors for loading in newer wide-beam Great Lakes boats and Eveleth Taconite for additional crude cars. Authority was requested on AFE 10712 approved September 13, 1972 for purchasing mini-hi-side extensions for an additional 266 cars. Cars were rebuilt from various cars in classes U28, U29, U30 and U31

-continued on page 222

Duluth Missabe & Iron Range Railway – Mini-Quad set High-Sides by group, January 15, 1982

Group/ Unit	Car 1	Car 2	Car 3	Car 4
1-1	52517	52548	52779	52959
1-2	53013	53020	53026	53040
1-3	52558	52644	52851	52924
2-1	52519	52541	52643	52914
2-2	52601	52651	52804	52994
2-3	52703	52837	52894	52972
3-1	52770	52869	52925	52979
3-2	52534	52543	52838	52920
3-3	52821	52862	52890	52981
4-1	52572	52575	52616	52875
4-2	52598	52619	52856	52915
4-3	52743	52781	52917	52947
5-1	52556	52622	52807	52936
5-2	52586	52625	52887	52973
5-3	52621	52762	52763	52861
6-1	52501	52759	52963	52975
6-2	52505	52606	52700	52950
6-3	52515	52640	52736	52941
7-1	52551	52614	52784	52906
7-2	52661	52673	52693	52930
7-3	52655	52772	52727	52960
8-1	52646	52767	52867	52901
8-2	52522	52639	52749	52811
8-3	52734	52921	52955	52967
9-1	53006	53069	53079	53112
9-2	52612	52683	52750	52818
9-3	52532	52657	52933	52998
10-1	52537	52746	52777	52952
10-2	52504	52853	52857	52992
10-3	52752	52929	52953	52983
11-1	53011	53060	53090	53156
11-2	53017	53059	53130	53139
11-3	53028	53035	53045	53142
12-1	52597	52610	52653	52937
12-2	52538	52645	52713	52956
12-3	52630	52690	52738	52974
13-1	53066	53080	53083	53126
13-2	52502	52780	52940	52996
13-3	53027	53036	53100	53138
14-1	52500	52626	52712	52949
14-2	52628	52635	52694	52868
14-3	52629	52918	52971	52993
15-1	52632	52735	52910	52990
15-2	53014	53122	53335	53342
15-3	53075	53109	53116	53159
16-1	52724	52834	52902	52977
16-2	52523	52557	52714	52866
16-3	53033	53107	53150	53151
17-1	53029	53153	53199	53461
17-2	52595	52793	52886	52978
17-3	52528	52559	52765	52942
18-1	53121	53141	53149	53203
18-2	53008	53022	53060	53162
18-3	53051	53054	53125	53426
19-1	53009	53049	53084	53091
19-2	53046	53056	53089	53098
19-3	53023	53039	53044	53082
20-1	53015	53016	53101	53124
20-2	53047	53058	53076	53111
20-3	53005	53093	53094	53132
21-1	53002	53053	53087	53464
21-2	53074	53127	53140	53283
21-3	53032	53102	53123	53136
22-1	53004	53106	53117	53469
22-2	53034	53072	53096	53166
22-3	53007	53031	53042	53145
23-1	53071	53081	53135	53176
23-2	53024	53103	53118	53218
23-3	53158	53167	53219	53259
24-1	53070	53110	53298	53328
24-2	53078	53212	53266	53273
24-3	52544	52568	52658	52832
25-1	53038	53052	53184	53488
25-2	53168	53303	53322	53331
25-3	53189	53522	53315	53483
26-1	53179	53290	53415	53473
26-2	53133	53210	53268	53280
26-3	53063	53099	53161	53453
27-1	53227	53255	53330	53341
27-2	53195	53332	53413	53489
27-3	53205	53357	53378	53450
28-1	53282	53365	53456	53462
28-2	53143	53163	53183	53206
28-3	53204	53232	53237	53264
29-1	53231	53304	53310	53327
29-2	53010	53320	53460	53495
29-3	53170	53171	53197	53377
30-1	53152	53178	53207	53409
30-2	53271	53289	53308	53366
30-3	53105	53129	53246	53480
31-1	53030	53073	53454	53478
31-2	53085	53104	53414	53451
31-3	53165	53329	53395	53434
32-1	53177	53190	53360	53411
32-2	53217	53286	53291	53446
32-3	53088	53297	53314	53370
33-1	53182	53221	53245	53367
33-2	53001	53144	53316	53402
33-3	53252	53344	53422	53498
34-1	53336	53350	53369	53390
34-2	53326	53338	53410	53487
34-3	53192	53249	53375	53436
35-1	53241	53261	53263	53427
35-2	53119	53169	53346	53349
35-3	53269	53299	53359	53390
36-1	53154	53404	53412	53440
36-2	53389	53394	53401	53471
36-3	53260	53288	53393	53486
37-1	53223	53229	53447	53481
37-2	53134	53191	53239	53416
37-3	53230	53285	53379	53465
38-1	53194	53301	53312	53439
38-2	53185	53224	53381	53397
38-3	53155	53302	53339	53467
39-1	53202	53281	53318	53470
39-2	53146	53262	53323	53348
39-3	53358	53387	53405	53475
40-1	53256	53362	53398	53468
40-2	53198	53257	53324	53372
40-3	53147	53233	53384	53493
41-1	53226	53250	53292	53497
41-2	53215	53248	53356	53383
41-3	53213	53234	53364	53484
42-1	53275	53368	53391	53479
42-2	53238	53353	53472	53490
42-3	53340	53361	53431	53437
43-1	52692	53418	53449	53482
43-2	52585	52649	52709	52957
43-3	53012	53064	53067	53115
44-1	52810	52859	53050	53062
44-2	52520	52594	52613	52820
44-3	52516	52607	52636	52698
45-1	52545	52603	52909	52991
45-2	52536	52563	52659	52809
45-3	52615	52882	52898	52946
46-1	52521	52638	52823	52864
46-2	52581	52588	52873	52984
46-3	52835	52900	52982	52987
47-1	52695	52769	52803	52825
47-2	52510	52650	52696	52704
47-3	52674	52748	52904	52928
48-1	52573	52669	52961	52962
48-2	52539	52583	52608	52895
48-3	52554	52726	52794	52796
49-1	52691	52799	52816	52999
49-2	52513	52680	52728	52980
49-3	52578	52681	52776	52871
50-1	52634	52727	52802	52808
50-2	52577	52682	52831	52855
50-3	52701	52885	52939	52970
51-1	52689	52707	52732	52817
51-2	52672	52847	52879	52989
51-3	52506	52798	52908	52945
52-1	52511	52730	52756	52863
52-2	52593	52718	52905	52938
52-3	52555	52631	52708	52877
53-1	52623	52648	52676	52927
53-2	52571	52710	52829	52988
53-3	52641	52705	52711	52911

Group/ Unit	Car 1	Car 2	Car 3	Car 4
54-1	52525	52662	52846	52860
54-2	52590	52742	52844	52922
54-3	52560	52580	52751	52800
55-1	52514	52549	52550	52883
55-2	52582	52671	52854	52903
55-3	52535	52647	52715	52964
56-1	52664	52755	52897	52907
56-2	52524	52569	52944	52969
56-3	52686	52747	52789	52943
57-1	52642	52770	52815	52931
57-2	52605	52916	52926	52932
57-3	52530	52663	52678	52836
58-1	52546	52584	52741	52870
58-2	52706	52716	52865	52899
58-3	52733	52745	52775	52968
59-1	52540	52788	52850	52923
59-2	52526	52553	52627	52839
59-3	52518	52721	52822	52843
60-1	52589	52722	52888	52891
60-2	52564	52757	52761	52785
60-3	52687	52739	52884	52954
61-1	52529	52670	52801	52995
61-2	52509	52587	52768	52773
61-3	52624	52702	52819	52849
62-1	52566	52668	52684	52719
62-2	52527	52667	52677	52876
62-3	52604	52731	52827	52965
63-1	52666	52723	52826	52986
63-2	52533	52679	52764	52893
63-3	52574	52618	52727	52787
64-1	52565	52600	52754	52830
64-2	52652	52688	52842	52892
64-3	52602	52637	52654	52935
65-1	52542	52592	52786	52845
65-2	52717	52828	52848	52881
65-3	51359	51371	51615	52396
66-1	51136	51237	51482	52253
66-2	51140	51330	51674	52241
66-3	51339	51515	51614	51983
67-1	51028	51354	51690	52023
67-2	51602	51871	52103	52440
67-3	51131	51334	51992	52088
68-1	51199	51965	52076	52239
68-2	51035	51118	51783	51922
68-3	51595	51632	51671	51947
69-1	52111	52187	52385	52427
69-2	51076	51183	51419	52425
69-3	51296	51830	51934	52330
70-1	51184	51800	52361	52380
70-2	51217	51461	51708	52250
70-3	51320	52109	52130	52258
71-1	51589	51626	51711	51810
71-2	51308	51309	51797	52368
71-3	51196	51390	52080	52370
72-1	51043	51660	51808	52173
72-2	51201	51290	51480	51624

Group/ Unit	Car 1	Car 2	Car 3	Car 4
72-3	51122	51754	52081	52086
73-1	51127	51268	51409	51943
73-2	51503	51604	51611	51881
73-3	51216	51380	52424	52481
74-1	51267	51973	52094	52402
74-2	51071	51439	51672	52397
74-3	51211	51336	51942	52289
75-1	51210	51470	51568	52244
75-2	51067	51915	52443	52465
75-3	51130	51275	51430	52163
76-1	51640	51839	51862	52190
76-2	51763	51935	52018	52098
76-3	51178	52175	52208	52419
77-1	51616	51758	52123	52398
77-2	51047	51423	52193	52351
77-3	51081	51203	51891	52263
78-1	51542	51723	52340	52404
78-2	51191	51474	52041	52270
78-3	51001	51347	51500	52233
79-1	51246	51929	51968	52343
79-2	51162	51176	51625	52283
79-3	51020	51291	51823	52017
80-1	51346	52412	52417	52429
80-2	51564	51661	51953	51972
80-3	51391	51446	51623	52332
81-1	51059	51532	51951	51956
81-2	51225	51331	51796	51997
81-3	51212	51436	51588	51874
82-1	51032	51278	51876	51984
82-2	51567	51818	52468	52913
82-3	51049	51781	52512	52774
83-1	51539	52359	52508	52985
83-2	52415	52665	52795	52976
83-3	52562	52697	52766	52858
84-1	52570	52760	52912	52997
84-2	52503	52729	52792	52958
84-3	52620	52872	52919	52966
85-1	52507	52561	52576	52934
85-2	51689	51886	52056	52805
85-3	52725	52790	52806	52812
86-1	51161	51814	51920	52067
86-2	51180	51349	51597	51918
86-3	51314	51751	52096	52345
87-1	51398	51925	52093	52280
87-2	51725	51854	51998	52114
87-3	51107	51717	52309	52358
88-1	51030	51078	51204	51628
88-2	51522	51534	52129	52191
88-3	51413	51748	52144	52238
89-1	51062	51097	51442	52143
89-2	51167	51647	51963	52388
89-3	51288	51516	51770	52019
90-1	51186	51658	51855	52000
90-2	51618	51970	52003	52489
90-3	51010	51113	51190	52334
91-1	51529	51790	52247	52347

Group/ Unit	Car 1	Car 2	Car 3	Car 4
91-2	51282	51691	52180	52472
91-3	51437	51477	51630	52485
92-1	51449	51733	52375	52376
92-2	51994	52148	52273	52498
92-3	51466	51494	52185	52432
93-1	51202	51256	51417	52226
93-2	51362	51530	52242	52316
93-3	51641	51799	52466	52478
94-1	51051	51590	52186	52248
94-2	51213	51593	52268	52438
94-3	51133	51174	51540	52178
95-1	51088	51313	51945	51967
95-2	51518	52031	52454	52482
95-3	51104	51317	51558	52211
96-1	51188	51747	51849	52344
96-2	51337	52044	52194	52315
96-3	51207	51579	51897	52389
97-1	51447	51581	51620	52176
97-2	51479	52004	52049	52276
97-3	51065	51086	51322	51344
98-1	51092	51141	51753	52090
98-2	51124	51664	51938	52008
98-3	51096	51735	51950	52394
99-1	51310	51333	51767	52464
99-2	51073	51121	51166	52414
99-3	51223	51284	52042	52285
100-1	51396	51443	51562	52399
100-2	51415	51537	51652	51720
100-3	51138	51312	51520	51734
101-1	51018	51170	51250	52492
101-2	51137	51279	51729	51773
101-3	51048	51248	51659	51710
102-1	51077	51700	52128	52401
102-2	51502	52092	52390	52445
102-3	51185	51687	52267	52378
103-1	51000	51080	51493	52118
103-2	51378	51563	52084	52030
103-3	51027	51177	51631	52014
104-1	51205	51824	51826	52227
104-2	51569	51682	52192	52496
104-3	51134	51192	52007	52444
105-1	51050	51400	51572	52299
105-2	51206	51692	51872	52234
105-3	51095	51804	52021	52354
106-1	51111	51789	51868	52353
106-2	51052	52177	52254	52408
106-3	51646	52170	52297	52409
107-1	51082	51156	51663	52073
107-2	51713	51791	51894	52295
107-3	51241	51526	52221	52306
108-1	51085	51541	51730	51765
108-2	51977	52025	52095	52215
108-3	51471	51916	52433	52455
109-1	51505	51592	51727	52156
109-2	51820	51995	52009	52249
109-3	51332	51535	51657	51742

Group/ Unit	Car 1	Car 2	Car 3	Car 4
110-1	51070	51462	52050	52314
110-2	51752	52203	52228	52271
110-3	51153	51556	51812	52052
111-1	51142	51219	51286	52064
111-2	51233	51252	52304	52352
111-3	51712	51757	51846	52272
112-1	51108	52140	52174	52382
112-2	51024	51264	52125	52387
112-3	51003	51454	51582	51896
113-1	51831	52045	52282	52490
113-2	51521	51946	51948	52224
113-3	51817	52141	52293	52328
114-1	51787	51974	52122	52284
114-2	51608	51966	52037	52471
114-3	51060	51865	52386	52494
115-1	51327	51485	51804	52091
115-2	51144	52272	51274	52070
115-3	51022	51341	51376	52183
116-1	51011	51435	51921	52243
116-2	51129	51253	51318	52020
116-3	51236	51377	51798	51900
117-1	51101	51622	51906	52460
117-2	51004	51079	51639	52365
117-3	51305	51483	52071	52220
118-1	51263	51300	51679	52085
118-2	51383	51707	51964	52431
118-3	51132	51668	52197	52355
119-1	51165	51285	51425	52338
119-2	51570	51841	52002	52199
119-3	51147	51260	51832	52452
120-1	51014	51394	52055	52058
120-2	51198	52134	52168	52434
120-3	51103	51247	51651	51873
121-1	51019	51038	51355	51686
121-2	51084	51816	51988	52155

Group/ Unit	Car 1	Car 2	Car 3	Car 4
121-3	51021	51091	51301	51762
122-1	51238	51335	51352	52035
122-2	51524	51629	51649	52133
122-3	52016	52102	52252	52366
123-1	51635	51655	52060	52169
123-2	51289	51695	51709	52201
123-3	51266	51453	51619	52467
124-1	51621	51864	51923	52202
124-2	51696	52006	52318	52435
124-3	51100	51554	51853	52275
125-1	51847	51919	52259	52379
125-2	51168	51232	52210	52298
125-3	51158	51445	51653	52051
126-1	51361	51669	51699	52291
126-2	51110	51784	51910	51930
126-3	51087	51258	51574	52209
127-1	51187	51389	52360	52480
127-2	51061	51234	51822	52393
127-3	51157	51908	52022	52423
128-1	51221	51231	51739	52068
128-2	51802	52161	52350	52487
128-3	52024	52377	52442	52470
129-1	51777	51809	51884	51917
129-2	51933	52230	52367	52395
129-3	51776	51867	52311	52323
130-1	51801	51875	52113	52189
130-2	52151	52321	52348	52439
130-3	52124	52213	52216	52288
131-1	51179	51272	51356	51961
131-2	51455	51656	51878	52214
131-3	51328	51681	51936	52392
132-1	51547	51980	52046	52484
132-2	51002	51152	51522	51552
132-3	51869	52164	52232	52369
133-1	51350	51880	51982	52450

Group/ Unit	Car 1	Car 2	Car 3	Car 4
133-2	51834	52157	52374	52441
133-3	51772	52240	52264	52281
134-1	51150	51848	52115	52132
134-2	51099	51226	51888	52256
134-3	52034	52053	52322	52333
135-1	51576	51989	51993	52384
135-2	51228	51325	51702	52277
135-3	51093	51827	52158	52337
136-1	51410	51911	51960	52375
136-2	51885	51979	52106	52146
136-3	51432	51662	52391	52411
137-1	51249	51905	52078	52126
137-2	51218	51412	51893	52061
137-3	51293	51428	51636	52205
138-1	51283	51580	51882	52320
138-2	51261	51307	51675	52172
138-3	51075	51240	52040	52043
139-1	51220	51379	51486	52100
139-2	51245	51265	51342	52336
139-3	51251	51401	51987	52287
140-1	51007	51009	52329	52477
140-2	51005	51304	51550	51794
140-3	51026	51326	51949	52447
141-1	51214	51633	51786	51991
141-2	51135	51324	51429	51941
141-3	51587	52038	52217	52483
142-1	51732	51402	51666	52302
142-2	51069	51643	51858	52346
142-3	51478	52147	52279	52462
143-1	52084	52204	52265	52413
143-2	51427	51523	51738	52112
143-3	51543	51811	51852	51863

Cars are not necessarily in the order of the group. Car 1 may be the 2nd, 3rd or 4th car in the set or in any random order.

-continued from page 219
and renumbered into the 51001–59991 series. Work on the initial 508 cars was completed by December 31, 1972 and the remaining 266 cars done in early 1973. A small number, 1, 2 or 3, was added to the set numbers to further divide up the cars for air brake testing during specific years. This was done because unlike regular ore cars the mini-quads were in year-round service and instead of a stencil group test year, the small numbers were used and cars were air-brake tested on a more frequent basis.

During June 1973 Class U30 70-ton ore car DMIR 52931 was painted gold for the DM&IR Veteran Employees Association to celebrate their 50th Anniversary (1924 to 1974) annual meeting on May 10, 1974. Car was adorned with special lettering and the Scotchlite safety-first DM&IR monogram.

AFE 10893 of July 10, 1974 requested 244 additional regular ore cars be converted to mini hi-sides and drawbar-coupled into mini-quad sets to provide equipment for the new Minorca Taconite plant located at Virginia. All cars were rebuilt by November 2, 1976 but 124 cars had not been painted. Painting on these was completed July 14, 1977.

An additional 244 mini-quads were requested on AFE 10834 on January 9, 1975. Expansion of Eveleth Taconite in 1975 required 168 additional mini hi-sides, with the remaining 76 cars being rebuilt for service to other taconite plants and allow for sufficient cars to meet annual shopping of equipment. All were to be coupled in four-car mini-quad sets.

AFE 11006 was approved on December 1975 for the rebuilding of 340 additional mini-quads to be placed in the 51001–53498 series. Prefabricated steel extensions for the tops of the carbodies were purchased from Paper Calmenson of Duluth. These were completed on October 5, 1977, bringing the total ownership of mini-quads to 1,528 cars.

The speed of dumping operations at the Eveleth Taconite Co. plant (EVTAC) became a concern in 1974. EVTAC was going to increase production and they felt application of air-operated doors to the drop-door mechanism would speed up unloading. On June 5, 1974, Cost Authority C-6027 was approved for $5,400 to install air-operated doors on two shaker-pocket crude ore cars. Application was completed on January 10, 1975 and cars placed in service. This allowed cars to have drop doors operated either with air or a trapping machine. No additional cars were converted.

EVTAC, through equipment modifications and improvements, increased annual pellet production from 1.6 to 2.4 million tons of pellets, a 50 percent increase, during the period from November 1970 to January 1973. During the same period DM&IR increased their shaker pocket fleet of hi-side cars from 77 to 94 cars in an attempt to keep pace with the higher production. Operations experience during 1973 determined that the current 94 cars did not provide sufficient operating flexibility and cars for maintenance outages. Initially, crude ore trains from T-Bird to EVTAC averaged 17 trips per week. By 1973 they averaged 29 trips weekly. The ownership of 94 shaker pocket cars and an average train length of 89 cars showed that the excess five cars did not provide enough equipment to allow shop time. Additional cars were needed for T-Bird service and AFE 10852 was approved on March 13, 1975, for the conversion of ten additional shaker pocket-equipped hi-side cars. Cars were selected from Class U29

Class U30 52931 was repainted in this gold paint scheme in June 1973 to celebrate the 50th Anniversary of the DM&IR Veteran Employees Association. Proctor, June 1973 (MRHS collection)

Overhead view of 51362, part of mini-quad set 93-2 at Biwabik, September 22, 1986. Note the interior bracing for the side extensions. (D.P. Holbrook)

31000–32498 series with first number changed to 5 when rebuilt. Car numbers: 51025, 51386, 51460, 51600, 51638, 51782, 52074, 52099, 52116 and 52245 were all converted at Proctor during April 1975. Track modifications were made at EVTAC during 1974 to allow T-Bird crude taconite trains to be increased from 89 cars to 96 cars. The DM&IR rostered 104 T-Bird crude taconite cars and 776 mini-quad taconite cars at the end of 1975.

Eveleth Taconite would further expand their operations in 1975 and 1976 and add

The opening of the second T-Bird mine, "South" in 1976 caused DM&IR to rebuild 168 T-Bird crude ore service cars 40000–40167 during summer of 1976. Starting on October 6, 1976, the original T-Bird service cars were renumbered to 40168–40271 from 51000, 52000, 53000 series numbers. Placing all in the 40000 series allowed for easier tracking of cars by type of service. All were repainted with the T-Bird logo above the reporting marks. DM&IR 40255 Keenan, September 10, 1978. (R.F. Kucaba)

Expansion at Eveleth Taconite crude ore haulage created the need for 168 additional T-Bird service cars 40000–40167. Beginning on July 29, 1976, Proctor began rebuilding these cars and painting them in the special T-Bird service paint scheme. Starting on Ocobert 6, 1976, earlier T-Bird service cars scattered in the 51000–53000 series were renumbered to 40168–40271 and also painted in the T-Bird paint scheme. DM&IR 40030, former 31302, then renumbered 531302 when high sides were applied in 1964, and finally to 40030 on April 28, 1976, when repainted for T-Bird service. Virginia, September 6, 1976. (R.C. Anderson)

another crude taconite loadout, T-Bird South, which would ship its first crude taconite on July 13, 1976. The 104 cars were insufficient to handle production from the two T-Bird loadouts and Proctor Shops was called upon to build additional shaker-pocket hi-side cars for this service. AFE 10934 was approved Feb 21, 1975 for the rebuilding of 168 mini hi-side taconite cars, 168 additional T-Bird crude taconite cars and four cabooses to cover the increased carloadings.

During January 1976 ETCO approached the DM&IR for a solution to the continuing problem of frozen crude ore in the T-Bird cars. The problem was expected to become even more significant when pellet production was increased from 2.6 million tons to 6 million tons annually. ETCO request the DM&IR insulate the exterior of the end slope sheets on the fleet of 104 existing cars. Two cars had been equipped with foam insulation during the fall of 1974 and favorable results had been obtained.

Starting on July 29, 1976, 168 cars numbered 40000–40167 were rebuilt for T-Bird service at Proctor with all equipped with shaker pockets under AFE 10934. The decision to start the new 40000 series numbers was made to allow easier tracking of cars in T-Bird service. These were all completed by September 9, 1976. Cars were painted with a special T-Bird logo above the initial and number on the sides. All were sprayed with polyurethane foam insulation underneath the slope sheets to reduce freezing of the crude ore. Prior to these 168 cars being rebuilt, there were a total of 104 T-Bird cars with shaker pockets mixed in the 51000, 52000, 53000 series.

Starting on October 6, 1976, the original 104 T-Bird cars in the 51000, 52000, 53000 series were renumbered 40168–40271 to group them in the new 40000 series. Cars were returned to Proctor Shops and had the end slope sheet area covered with foam insulation and were renumbered at that time. All of these cars were also repainted with the special T-Bird logo when renumbered. Eight additional T-Bird cars numbered 40272–40279 were rebuilt on September 21, 1978 from 31000, 32000 series cars. No additional cars would receive the T-Bird paint job.

During November 1978 an additional U31 70-ton ore car, DMIR 52454, part of mini-quad set 95, was painted gold for the loading of the 100 millionth ton of pellets loaded at Minntac on November 15, 1978.

As of February 27, 1978 the roster included 1,560 mini-quads assigned: 1,040 to Minntac, 276 to EVTAC and 244 to Minorca. AFE 11205 was made out on February 8, 1978, for the conversion of 156 additional regular ore cars to mini-quads in 50001–59991 series. All were to be equipped with shaker pockets. Prior to these cars all other mini-quads were not equipped with shaker pockets. Project was started on November 15, 1979, and all were completed by April 12, 1980. The additional mini-quads were needed to serve increased taconite pellet production. Total mini-quad taconite car ownership would reach an all-time high of 1,716 cars on July 31, 1980.

By the winter of 1981, the existing 280 T-Bird cars were showing wear and tear from the constant pounding during the loading of crude taconite at T-Bird North and South mines. Cars were found to have excessive slope sheet wear, cracked side sheet stiffeners, weld seams failing and numerous other problems. AFE 11565 was approved for the retirement of the existing T-Bird cars numbered 40000–40279 and transfer of the top end extensions to other stored regular ore cars. These old 19½-inch extensions were cut in half horizontally and used on mini-quad cars that needed replacement extensions. The "new" T-Bird crude cars were numbered 40280–40460. Retirements and conversions began during March 1982 and were completed December 8, 1982.

AFE 11719 in 1984 was authorized for the retirement of 100 T-Bird crude ore cars and replacing them with 100 regular ore cars converted for crude ore service. Subsequent to the

DM&IR 40265 in T-Bird service paint scheme. Built as Class U31 33435, it was renumbered 533435 in 1964, to 53435 on August 2, 1967 and finally to 40265 on October 27, 1976. Proctor, October 13, 1977. (D.P. Holbrook collection)

The 100 millionth ton of taconite pellets at the U.S. Steel Minntac Plant was produced on November 15, 1978. To commemorate the occasion, DM&IR repainted Class U29 ore car 52454 in this gold paint scheme. Forbes, June 13, 1987. (R.C. Anderson)

The final group of 156 regular ore cars rebuilt to mini-quad sets in late 1979 were equipped with shaker pockets. These would be the only mini-quads so equipped. DM&IR 51587, Mt. Iron, May 12, 2005. (D.P. Holbrook)

Two cars, 52823, 52521 from mini-quad set 46-1 Proctor, July, 1975. The other two cars in the set are 52864 and 52638. (D.P. Holbrook collection)

AFE approval, Transportation Superintendent P.D. Stephenson requested that additional air brake modifications, specifically the application of straight air (Orinoco), be made to 60 of the 100 converted cars. Experience had shown from 20 to 25 occurrences of flat wheels on T-Bird crude ore trains. It was determined that this was being caused by the engineer reducing feed-valve air pressure during loading to achieve needed braking without a service application.

To eliminate this practice it was recommended that eight T-Bird cars be equipped with straight-air on each end of the three T-Bird train sets. An experiment during June 1984 using one set of mini-quads (four cars) on each end of T-Bird North trains and two sets of mini-quads (eight cars) on one end of the T-Bird South trains resulted in a solution to the flat wheel problem. During the winter of 1985 until the fall of 1987 a total of 60 mini-quads were assigned to T-Bird service to assist in eliminating the flat wheel problems. Normal operation at this time was to use two engines on T-Bird trains between December and April and one engine from April through November.

AFE 11912 and CA-6134 were requested in 1986 for authority to modify 60 crude ore cars with the application of Orinoco straight air. It was determined after the project started that 50 cars would be converted. These cars were numbered 45001–45050 to allow crews to distinguish between straight air and non-straight air equipped cars. Cars were converted between September 1986 and November. 1987. Once converted they were added to the ends of the T-Bird train sets.

Until 1986 260 of the T-Bird crude ore service cars continued to use plain-bearing trucks. CA-6147 was approved in August 1986 to convert these trucks to roller bearings. This eliminated the periodic need to rebuild plain-bearing trucks for these cars. Also beginning in 1986, with the assistance of jigs, gauges and technical help from Timken, a five-year project was begun to replace all 6x11-inch plain-bearing trucks with roller-bearing inserts on all 1,716 mini-quad ore cars.

The loss of Rhude & Fryberger natural ore business in 1987 and aging of the ore car fleet caused DM&IR to rationalize the ore car roster. Ownership in 1987 consisted of 3,961 active ore cars broken down as: 1,503 regulars, 1,716 mini-quads, 280 T-Bird crude service and 462 60000 series regulars qualified for pellet service. The inactive rosters consisted of: 989 stored cars, 19 authorized for disposal and 377 cars out on lease. It was determined that 700 of these cars were excess and needed to be disposed of. By January 1990 this reduction had been accomplished through sales and retirements.

T-Bird crude cars had an 8–10 year service life because during loading the unprocessed taconite rock caused damage and excessive wear

Thirty 40481–40510 of the 50 40461–40510 ore cars rebuilt in 1988 and 1989 for T-Bird service were equipped with rubber belt material on the slope sheets to minimize damage when loading crude taconite in the cars. DM&IR 40508 Alborn, September 27, 2001. (D.P. Holbrook)

DM&IR 40730 converted to T-Bird crude taconite service in 1999 was one of 70 added to T-Bird fleet in that year. Wolf, May 12, 2005. (D.P. Holbrook)

on side and slope sheets. AFE 12142 was for 50 "First Generation" crude cars, 40461–40510 regular T-Bird cars. The first 20 cars 40461–40480 were rebuilt during 1988 without authority retaining their original regular ore car numbers. Once AFE 12142 was approved they were renumbered. During mid-1989 it was decided to use secondhand 1-inch thick rubber conveyor belt material to line the slope sheets on T-Bird cars to minimize the damage caused by loading crude taconite. Testing of this theory began in late July 1989. Tests were successful and 30 cars 40481–40510 were equipped with the rubber belting. First Generation crude cars were cars without rubber belting material and the Second Generation were cars that were equipped with rubber belting on the slope sheet areas. All future T-Bird cars would be Second Generation cars.

AFE 12220 was approved in 1989 for 25 Second Generation T-Bird crude cars: two cars, 45051 and 45052, with straight air, eight cars 49001–49008 with permanent ETD bracket without side extensions, and 15 cars 40511–40525 as regular T-Bird cars. Project was terminated early and the last 3 cars, 40523–40525 were not rebuilt. Project resumed in May 1994 with 40523–40525 completed and eight additional cars added to the project, 40526–40533.

Retirements of damaged crude cars and the need to support up to ten trains per day consisting of two 96-car and one 84-car set to support increased pellet production at EVTAC required additional crude taconite cars to be rebuilt. AFE 12529 was approved in June 1994 for the rebuilding of 30 additional crude cars. Eighteen cars 40534–40550 and 45053 were completed in July 1994 and the remaining 12 cars 40551–40562 in October 1994.

Approved in October 1994 was AFE 12531 for rebuilding 25 additional T-Bird crude taconite cars 40563–40587 and AFE 12529 for one additional straight-air equipped T-Bird car 45053 for EVTAC service.

T-Bird cars were averaging 12-year service lives in 1995. An additional 25 T-Bird cars 40588–40612 were constructed under AFE 12590 approved July 26, 1996. Almost as soon as these were completed, an additional 100 cars, 40613–40712 were approved on AFE 12654 July 1, 1997. These 125 cars were used as replacements for previously rebuilt cars dating to 1946 that would be overage in 1996, exceeding the 50-year rule. Some 22 additional cars, 40713–40727 and replacement T-Bird idler cars 49001, 49003, 49005, 49007 and 49008 were rebuilt during 1998.

AFE 1031 was approved in March 1999 for the rebuilding of 100 additional T-Bird

Once freight cars reached 50 years of age they became non-interchange qualified. As such the DM&IR mini-quad fleet would be overage between 1998 and 2007. Looking to purchase replacement cars, Kingsport Railcar of Kingsport, Tennessee rebuilt this drawbar-coupled two-car set of demonstrator ore cars from old ACF Coalveyor bathtub bottom coal cars in 2001. Cars tested on DM&IR in 2002 but no orders were placed. Proctor, March 2, 2002. (D. Schauer)

crude taconite cars. Changes in operations at EVTAC in 1999 allowed for a reduction from three sets of equipment to two sets, reducing the need to 72 cars 40728–40797 and replacement idler cars 49004, 49006.

The final groups of T-Bird crude cars would be 10 cars 40798–40807 rebuilt in 2002, and 24 cars 40808–40831 rebuilt in 2004 under AFE 5423.

NEW CARS?

The turn of the century in 2000 brought about the need to consider replacing the taconite car fleet. The mini-quads had all been rebuilt from regular ore cars constructed between 1952 and 1957 and were approaching the 50-year overage rule which would cause the cars to need FRA waivers to continue operating.

Kingsport Railcar Co. of Kingsport, Tennessee, built an articulated 143-ton capacity two-car drawbar-coupled set of demonstrator ore cars during 2001. Kingsport rebuilt two ACF Coalveyor-type unit train coal gondolas by cutting one end off and shortening the car and installing new slope sheets on the cutoff end and repositioning the trucks. These cars, RCSX 1001 and 1002, arrived on DM&IR for testing in early 2002. After initial testing the slope sheets were modified by May 2002 to allow cars to have the proper capacity and prevent overloading of the cars. Normally an articulated car would carry one number, but builder elected to build cars with each having its own air and hand brake system and have a number on each car in the event DM&IR elected to remove the drawbar and replace it with standard Type E couplers. Cars were returned to builder in 2003.

DM&IR would not order any new cars to replace the older mini-quad fleet and elected to obtain waivers from FRA to operate the older cars. DM&IR would be acquired by Canadian National in May 2004. After the merger CN would purchase multiple orders of new cars to replace the old mini-quads.

DM&IR CABOOSES

At the time of the merger there were a total of 44 D&IR and 30 DM&N cabooses on the roster. DM&N cars retained their numbers and D&IR cabooses had the numeral "1" added to their number when relettered DM&IR. Beginning on May 11, 1938 the prefix letter "C" was added to the number on cabooses. After December 9, 1939 road numbers with the letter C

-continued on page 233

Duluth, Missabe & Iron Range Cabooses

No.	Builder	Cl	Built	Retired	O.L	Notes
C-5	Duluth Mfg Co	G	1910	11/8/1977	31'	SU applied 1940 - Donated to LSMT 11/8/77
C-8	Duluth Mfg Co	H	1893	8/12/1974	36'0"	Side door equipped - SU applied April. 1945. Trucks removed and car body placed alongside track at T-Bird North Mine January, 1966. Placed back in service in 1969. Retired 8/12/74
C-9	Duluth Mfg Co	H	1893	11/8/1977	36'0"	Side door equipped - Cupola in Center of Caboose 9 - SU applied April. 1945. Donated to LSMT 11/8/77
C-10	Duluth Mfg Co	H	1893	6/4/1975	36'0"	Side door equipped - Cupola removed June 1910 - SU applied 4/4/45. Cupola reinstalled ? - To work service W10 in 1/29/75 AFE 10935
C-12	Duluth Mfg Co	H	1894	3/3/1975	36'0"	SU applied in 1940 - Donated to LSMT 11/8/77
C-15	Duluth Mfg Co	G1	1894	5/22/1975	31'	Sold to D&NE 2/29 repurchased from D&NE 5/42 -SU applied 4/4/45 - Sold to Hyman Michaels 5/22/75
C-18	American Car & Foundry	G1	1899	Rebuilt 1948	31'	SU applied 1940 - Retired April 1947 - Rebuilt to DM&IR C191 12/48
C-22	American Car & Foundry	G1	1900	4/13/1973	31'	Sold to D&NE 2/29 repurchased from D&NE 5/42 - SU applied 4/45 - Sold to Hyman Michaels 5/22/75
C-25	American Car & Foundry	G1	1900	11/8/1977	31'	SU applied 1940 - Sold to LSRM 11/6/80, resold to Bud Bulgrin
C-26	American Car & Foundry	G1	1900	4/15/1969	31'	SU applied 1940 - Destroyed by fire on Soo Line 4/15/69
C-30	D&IR Two Harbors	G1	1911	11/8/1977	31'	SU applied 1940 - Sold to LSMT 11/6/80
C-38	D&IR Two Harbors	G1	1911	8/9/1977	31'	SU applied 1940 - Witte engine and Radio telephone equipped in 1948. Sold to L Tomporowski Proctor MN 8/9/77
C-39	D&IR Two Harbors	G1	1911	12/30/1974	31'	SU applied 1940 - Witte engine and radio telephone in 1948 Radio removed 1952 to C200 series. Sold to D&NE 12/30/74
C-40	D&IR Two Harbors	G1	1911	4/15/1969	31'	SU applied 1940 - Destroyed by fire on Soo Line 4/15/69
C-43	D&IR Two Harbors	G1	1911	10/8/1981	31'	SU applied 1940 - Witte engine and radio telephone equipped in 1947 - Radio removed 1952 to C200 series. Sold to LSMT 10/8/81
C-44	D&IR Two Harbors	G1	1911	7/30/1981	31'	SU applied 1940 - Witte engine and radio telephone equipped in 1947 - Radio removed 1952 to C200 series. Sold to LSMT 7/30/81 then resold to private individual 1981
C-50	D&IR Two Harbors	G1	1911	5/28/1975	31'	SU applied 1940 - Sold to Hyman Michaels 5/28/75
C-51	D&IR Two Harbors	G1	1911	Retired before May 1943	31'	Steel underframe
C-52	D&IR Two Harbors	G1	1911	7/30/1981	31'	SU applied 1940 - Witte engine and radio telephone equipped in 1948 - Radio removed 1952 to C200 series - Radio and engine reinstalled July, 1966. Sold to LSMT 7/30/81
C-55	D&IR Two Harbors	G1	1911	7/4/1976	31'	Steel underframe - Donated to Biwabik, MN 7/4/76 then to Museum of Mining Chisholm, MN
C-56	D&IR Two Harbors	G1	1911	10/8/1981	31'	SU applied 1940 - Witte engine and radio telephone equipped in 1948 - Radio removed 1952 to C200 series. Sold to LSMT 10/8/81
C-57	D&IR Two Harbors	G1	1911	10/16/1962	31'	Steel underframe - Witte engine and radio equipped in 1948. Radio removed 1952 to C200 series
C-58	D&IR Two Harbors	G1	1911	10/8/1981	31'	SU applied 1940 - Witte engine and radio telephone equipped in 1948 - Radio removed 1952 to C200 series. Sold to LSMT 10/8/81 then to private individual 1981, then to Iola Historical Society, Iola, WI
C-60	DM&N Proctor Shops	H1	1912	10/18/1960	36'5"	Center cupola - side door equipped - SU applied 4/45
C-61	DM&N Proctor Shops	H1	1913	4/1/1975	36'5"	Center cupola - side door equipped - SU applied 4/45 - To work service W-61 1/29/75. Then to private individual at Popular, WI in 2004
C-70	DM&N Proctor Shops	G2	1924	5/26/1966	33'2"	Equipped with Witte engine and radio telephone in 1947
C-71	DM&N Proctor Shops	G2	1924	4/??/1967	33'2"	Equipped with Witte engine and radio telephone in 1947
C-72	DM&N Proctor Shops	G2	1924	9/8/1959	33'2"	Equipped with Witte engine and radio telephone in 1947
C-73	DM&N Proctor Shops	G2	1924	3/6/1963	33'2"	Equipped with Witte engine and radio telephone in 1947. Sold unknown date for cabin at Lake Vermillion, MN then donated to Tower, MN in 2001
C-74	DM&N Proctor Shops	G2	1924	09/??/1965	33'2"	Equipped with Witte engine and radio telephone in 1947. Sold to Mid-Continent RR Museum North Freedom, WI September 1965
C-75	DM&N Proctor Shops	G2	1924	10/25/1963	33'2"	Equipped with Witte engine and radio telephone in 1947
C-76	DM&N Proctor Shops	G2	1924	Unknown	33'2"	Equipped with Witte engine and radio telephone in 1947
C-77	DM&N Proctor Shops	G2	1926	11/11/1963	33'2"	Equipped with Witte engine radio telephone in 1947. Barber trucks as of 8/15/58. Sold to LS&I, then to Waunakee, WI for display
C-90	EJE-Joliet Shops		Rcvd 1/49	11/8/1977	34'8"	Former EJE 5. SU when built. Sold to LSMT 8/1/80
C-91	EJE-Joliet Shops		Rcvd 1/49	5/22/1975	34'8"	Former EJE 88. SU when built. Sold Hyman Michaels 5/22/75
C-92	EJE-Joliet Shops		Rcvd 1/49	10/26/1965	34'8"	Former EJE 96. - SU when built. Sold to LS&I 10/26/66
C-93	EJE-Joliet Shops		Rcvd 1/49	2/27/1964	34'8"	Former EJE 101. SU when built. Sold to FJ Paripovich, Hibbing, MN 5/6/64

No.	Builder	Cl	Built	Retired	O.L	Notes
C-94	EJE-Joliet Shops		Rcvd 1/49	11/9/1976	34'8"	Former EJE 120. SU when built. Donated to LSMT 11/9/76
C-95	EJE-Joliet Shops		Rcvd 1/49	5/28/1975	34'8"	Former EJE 123. SU when built. Sold Hyman Michaels 5/28/75
C-96	EJE-Joliet Shops		Rcvd 1/49	11/8/1971	34'8"	Former EJE 144. SU when built. Sold to LSMT 11/8/71
C-97	EJE-Joliet Shops		Rcvd 1/49	10/8/1976	34'8"	Former EJE 145. SU when built. Sold to Oglebay Norton Co. Virginia, MN 10/8/76 then donated to City of Eveleth then moved to Iron World Chisholm, MN
C-98	EJE-Joliet Shops		Rcvd 1/49	1/22/1971	34'8"	Former EJE 155. SU when built. Destroyed Wilpen, MN Sept. 1970. Dismantled on site. Sold to Hyman Michaels 3/15/71
C-129	D&IR Two Harbors	K1	1909/10	10/3/1972	31'	SU applied 1940 - Equipped with Witte engine and radio telephone in 1947 - Radio removed 1952 to C200 series. Sold to Hyman Michaels 10/3/72
C-132	D&IR Two Harbors	K1	1909/10	5/14/1975	31'	SU applied 1940 - Equipped with Witte engine and radio telephone in 1948. Radio removed 1952 to C200 series. Wrecked Coons-Pacific Washing Plant 8/29/74. Sold to Hyman Michaels 5/14/75
C-135 (1st)	D&IR Two Harbors	K1	1909/10	6/15/1982	31'	SU applied 1940 - Equipped with Witte engine and radio telephone in 1948 -Radio removed 1952 to C200 series Rebuilt in 1967 to (2nd) C-135, cupola removed and replaced by winterized side window for T-Bird service
C-135 (2nd)	DM&IR Proctor Shops		1967	6/15/1982	31'	Sold to LSMT 6/15/82
C-136	D&IR Two Harbors	K1	1909/10	12/16/1975	31'	SU applied 1940 - Equipped with Witte engine and radio telephone in 1948 - Radio removed 1952 to C200 series - Sold to Hyman Michaels 12/16/75
C-137	D&IR Two Harbors	K1	1909/10	8/30/1967	31'	6-bunks - SCS 1927 - Equipped with radio telephone in 1948. Sold to LS&I 8/30/67
C-138	D&IR Two Harbors	K1	1909/10	11/8/1977	31'	SU applied 1940 - Equipped with Witte engine and radio telephone in 1948. Sold to LSMT 4/22/81
C-143	D&IR Two Harbors	K1	1909/10	9/24/1981	31'	Equipped with Witte engine and radio telephone in 1948 -Back-up horn added 7/26/74. Sold to LSRM 9/24/81
C-144	D&IR Two Harbors	K1	1909/10	8/13/1975	31'	Equipped with Witte engine and radio telephone in 1948 - Destroyed by fire Endion, MN 4/25/75. Sold to West End Iron & Metal 8/13/75
C-147	D&IR Two Harbors	K1	1909/10	8/23/1967	31'	6-bunks - SU applied 1927. To switch tenders' shanty at TH 8/23/67. Car body resold to JC Severson, Duluth, MN 8/29/79
C-151	D&IR Two Harbors	K1	1909/10	9/22/1965	31'	6-bunks - SU applied 1935 - Equipped with Witte engine and Radio telephone in 1948
C-152	D&IR Two Harbors	K1	1934	Retired before May 1943	31'	6-bunks - SU when built
C-153	D&IR Two Harbors	K1	1909/10	10/12/1946	31'	6-bunks - SU applied 1934 - Burned Tower Jct. 8/30/46
C-154	D&IR Two Harbors	K2	1909/10	9/22/1965	31'	6-bunks - SU applied 1934 - Equipped with Witte engine and radio telephone in 1948
C-156	D&IR Two Harbors	K2	1909/10	1/4/1949	31'	6-bunks - SCS 1927 - Destroyed by fire Toivola, MN 10/13/48
C-157	D&IR Two Harbors	K2	1909/10	8/23/1967	31'	6-bunks - SCS 1927 - To switch tenders' shanty Rainy Jct. Yard 8/23/67
C-158	D&IR Two Harbors	K2	1909/10	8/30/1967	31'	6-bunks - SCS 1927 - Sold to LS&I 8/30/67
C-159	D&IR Two Harbors	K2	1909/10	3/23/1959	31'	6-bunks - SCS 1927
C-160	D&IR Two Harbors	K2	1909/10	3/11/1965	31'	6-bunks - SCS 1927 - Sold to Laona & Northern 4/30/65
C-161	D&IR Two Harbors	K2	1909/10	4/5/1968	31'	6-bunks - SCS 1927 - Equipped with Witte engine and radio telephone in 1947 removed in 1951 - Radio reinstalled 6/66
C-162	D&IR Two Harbors	K2	1909/10	6/9/1969	31'	6-bunks - SCS 1927 - Sold, less trucks, to Dr. John Wandmaker, Duluth, MN 6/9/69
C-163	D&IR Two Harbors	K2	1909/10	2/19/1961	31'	6-bunks - SCS 1927 - Equipped with Witte engine and radio telephone in 1947 - Radio removed 1952 to C200 series
C-164	D&IR Two Harbors	K2	1909/10	4/5/1968	31'	6-bunks - SCS 1927
C-165	D&IR Two Harbors	K2	1909/10	8/30/1965	31'	6-bunks - SCS 1927
C-166	D&IR Two Harbors	K2	1909/10	11/??/1968	31'	6-bunks - SCS 1927
C-167	D&IR Two Harbors	K2	1909/10	6/17/1969	31'	6-bunks - SCS 1927 - Sold, less trucks to Mr. Gerald Brechlin, Duluth, MN 6/17/69
C-168	D&IR Two Harbors	K2	1909/10	6/16/1969	31'	6-bunks - SCS 1927 - Equipped with Witte engine and radio telephone in 1947 - Radio removed 1952 to C200 series. Sold, less trucks to Dr. John Wandmaker, Duluth, MN 6/16/69
C-169	D&IR Two Harbors	K2	1909/10	9/26/1967	31'	6-bunks - SCS 1927 - Sold to D&NE
C-170	D&IR Two Harbors	K2	1909/10	06/??/1960	31'	6-bunks - SCS 1927
C-172	D&IR Two Harbors	K2	1909/10	5/22/1953	31'	6-bunks - SCS 1927 - Equipped with Witte engine and radio telephone in 1947 - Radio removed 1952 to C200 series
C-173	D&IR Two Harbors	K2	1909/10	8/30/1967	31'	6-bunks - SCS 1927 - Sold to LS&I
C-174	D&IR Two Harbors	K2	1909/10	8/30/1967	31'	6-bunks - SCS 1927 - Sold to LS&I
C-175	D&IR Two Harbors	K2	1909/10	1/3/1967	31'	6-bunks - SCS 1927 - Equipped with Witte engine and radio telephone in 1947 - Radio removed 1952 to C200 series
C-176	D&IR Two Harbors	K2	1909/10	10/26/1965	31'	6-bunks - SCS 1927 - Sold to LS&I

No.	Builder	Cl	Built	Retired	O.L	Notes
C-177	D&IR Two Harbors	K2	1909/10	7/21/1969	31'	6-bunks - SCS 1927 - Donated to John V Wolak 7/21/69. Restored in 1971 at St. Stephen's Church St. Cloud, MN
C-179	D&IR Two Harbors	K2	1909/10	1/3/1967	31'	6-bunks - SCS 1927 - Sold to Ira Sack c/o Mid-South Society for Preservation Memphis, TN. Relettered ISC for movement
C-180	D&IR Two Harbors	K1	1909/10	11/8/1977	31'	Equipped with Witte engine and radio telephone in 1948 - Sold to LSRM 4/22/81 then to Pflueger Construction Co. Wabasha, MN late 1981.
C-181	D&IR Two Harbors	K3	1909/10	01/??/1956	40'4"	SCS 1935
C-182	D&IR Two Harbors	K4	1910	4/15/1969	34'	6-bunk - Bunks and Stove removed in April 1937 - SU 2/10/47. Destroyed by fire on Soo Line 4/15/69
C-183	D&IR Two Harbors	K4	1910	11/8/1977	34'	6-bunk - Bunks and Stove removed in April 1947 - SU 4/16/47. Sold to LSMT 8/1/80 then to private individual Elroy, WI late 1980 then to Mid-Continent RR Museum North Freedom, WI in 1985
C-184	Rebuilt D&IR Two Harbors	K3	1920	1/29/1975	40'4"	6-bunks - SCS 1921 - SU 9/14/46 - To work service W-184 1/29/75 AFE 10935. Donate to LS&M
C-185	Rebuilt D&IR Two Harbors	K3	1920	4/13/1942	40'4"	Wrecked at Waldo, MN 3/6/42 - Dismantled Two Harbors 4/13/42
C-186	Rebuilt D&IR Two Harbors	K5	1923	10/29/1967	40'7"	6-bunks - side doors - Center Cupola - SU 3/31/47 - Equipped with Witte engine and radio telephone in 1948. Radio telephone removed in 1951
C-187	Rebuilt D&IR Two Harbors	K3	1920	1/29/1975	40'4"	6-bunks - SCS 1921 - SU 7/10/46 - Equipped with Witte engine and radio telephone in 1948 - To work service W-187 1/29/75 AFE 10935. Donated to LS&M
C-188	Rebuilt D&IR Two Harbors	K3	1920	4/15/1975	40'4"	6-bunks - SU 1921 - Equipped with Witte engine and radio telephone in 1948. Sold to Hyman Michaels 4/15/75
C-189	Rebuilt D&IR Two Harbors	K3	1920	1/29/1975	40'4"	6-bunks - Steel underframe 1921 - Equipped with Witte engine and radio telephone in 1948 - To work service W-189 1/29/75 AFE 10935
C-190 (1st)	DM&IR Two Harbors		1942	1965	31'	Steel underframe when built - Equipped with Witte engine and radio telephone in 1947 - Constructed on AFE M4858 5/4/42. Rebuilt to (2nd)C-190 at Proctor by removing cupola and adding winterized side window for T-Bird service 1965
C-190 (2nd)	DM&IR Proctor		1965	7/25/1967	31'	Wrecked 1967
C-191 (1st)	DM&IR Two Harbors		1947	1965	31'	C18 retired 1947 and rebuilt to C191 in Dec 1948 - Equipped with Witte engine and radio telephone in 1948. Rebuilt to (2nd) C-191 at Proctor by removing cupola and adding winterized side window for T-Bird service 1965
C-191 (2nd)	DM&IR Proctor		1965	6/15/1982	31'	Sold to LSMT 6/15/82. Resold to private individual Pengilly, MN
C-192	DM&IR Two Harbors		1951	10/8/1981	31'	Built with SU - Equipped Witte engine and radio telephone when built. AFE 6951 1/30/50. Sold to LSRM 10/8/81. Sold to Burger King Kenosha, WI in late 81
C-193	DM&IR Two Harbors		1951	10/11/1979	31'	Built with SU - Equipped with Witte engine and radio telephone when built AFE 6951 1/30/50. Sold to Reserve Mining Co. 10/11/79
C-200	International Railway Car Co.		1952	4/9/1990	39'10"	Straight air equipped 1968 Donated to Proctor, MN in early 1990's. On display
C-201	International Railway Car Co.		1952	4/9/1990	39'10"	
C-202	International Railway Car Co.		1952	4/9/1990	39'10"	Straight air equipped 1968
C-203	International Railway Car Co.		1952	4/9/1990	39'10"	
C-204 (1st)	International Railway Car Co.		1952	1976	39'10"	Rebuilt to (2nd) C-204 2/6/76 - Cupola removed and replaced with bay window
C-204 (2nd)	DM&IR Proctor Shops		1976	4/9/1990	39'10"	
C-205 (1st)	International Railway Car Co.		1952	1976	39'10"	Rebuilt to (2nd) C-205 2/6/76 - Cupola removed and replaced with bay window
C-205 (2nd)	DM&IR Proctor Shops		1976	4/9/1990	39'10"	Sold to NSM 7/3/2000. Purchased from NSM by LSMT in 2003
C-206	International Railway Car Co.		1952	10/19/1973	39'10"	Destroyed in wreck at Highlands, MN 10/19/73
C-207	International Railway Car Co.		1952	4/9/1990	39'10"	
C-208	International Railway Car Co.		1952	4/9/1990	39'10"	Donated to Rose School Duluth, MN in late 1990's
C-209 (1st)	International Railway Car Co.		1952	1967	39'10"	Rebuilt to (2nd) C-209 in 1967 - Cupola removed and replaced with side winterized window
C-209 (2nd)	DM&IR Proctor Shops		1967	4/9/1990	39'10"	
C-210 (1st)	International Railway Car Co.		1952	1967	39'10"	Rebuilt to (2nd) C-210 in 1967 - Cupola removed and replaced with side winterized window
C-210 (2nd)	DM&IR Proctor Shops		1967	By 1995	39'10"	

No.	Builder	Cl	Built	Retired	O.L	Notes
C-211	International Railway Car Co.		1952	4/9/1990	39'10"	Straight air equipped 1968
C-212	International Railway Car Co.		1952	4/9/1990	39'10"	Straight air equipped 1968
C-213	International Railway Car Co.		1952	4/9/1990	39'10"	Straight air equipped 1968
C-214	International Railway Car Co.		1952	4/9/1990	39'10"	Straight air equipped 1968
C-215	International Railway Car Co.		1952	4/9/1990	39'10"	Straight air equipped 1968
C-216	International Railway Car Co.		1952	4/9/1990	39'10"	Straight air equipped 1968
C-217	International Railway Car Co.		1952	4/9/1990	39'10"	Straight air equipped 1968
C-218 (1st)	International Railway Car Co.		1952	2/6/1976	39'10"	Straight air equipped 1968 Rebuilt to (2nd) C-218 2/6/76. Cupola removed replaced by bay window.
C-218 (2nd)	DM&IR Proctor Shops		1976	By 1995	39'10"	
C-219 (1st)	International Car Co.		1952	2/6/1976	39'10"	Straight air equipped 1968 Rebuilt to (2nd) C-219 2/6/76. Cupola removed replaced by bay window
C-219 (2nd)	DM&IR Proctor Shops		1976	4/9/1990	39'10"	Rebuilt 2-6-76 - Cupola removed and replaced with bay window - radio equipped when rebuilt
C-220	International Car Co.		1974	4/9/1990	42'5"	Straight air and radio equipped when built
C-221	International Car Co.		1974	4/9/1990	42'5"	Straight air and radio equipped when built
C-222	International Car Co.		1974	Feb. 2005	42'5"	Straight air and radio equipped when built. Sold to Duluth Port Authority in Feb. 2005 for use with Schnabel car movements
C-223	International Car Co.		1974	4/9/1990	42'5"	Straight air and radio equipped when built
C-224	International Car Co.		1974	By 1995	42'5"	Straight air and radio equipped when built
C-225	International Car Co.		1974	4/9/1990	42'5"	Straight air and radio equipped when built
C-226	International Car Co.		1974	4/9/1990	42'5"	Straight air and radio equipped when built
C-227	International Car Co.		1974	2000	42'5"	Straight air and radio equipped when built - Sold to NSM 7/3/00
C-228	International Car Co.		1974	4/9/1990	42'5"	Straight air and radio equipped when built
C-229	International Car Co.		1974	4/9/1990	42'5"	Straight air and radio equipped when built
C-230	International Car Co.		1974	4/9/1990	42'5"	Straight air and radio equipped when built
C-231	International Car Co.		1974	4/9/1990	42'5"	Straight air and radio equipped when built
C-232	International Car Co.		1974	4/9/1990	42'5"	Straight air and radio equipped when built
C-233	International Car Co.		1974	4/9/1990	42'5"	Straight air and radio equipped when built
C-234	International Car Co.		1974	4/9/1990	42'5"	Straight air and radio equipped when built
C-235	International Car Co.		1974	4/9/1990	42'5"	Straight air and radio equipped when built
C-236	International Car Co.		1974	4/9/1990	42'5"	Straight air and radio equipped when built
C-237	International Car Co.		1974	4/9/1990	42'5"	Straight air and radio equipped when built
C-238	International Car Co.		1974	4/9/1990	42'5"	Straight air and radio equipped when built
C-239	International Car Co.		1974	4/9/1990	42'5"	Straight air and radio equipped when built

Abbreviations

LS&M Lake Superior & Mississippi Railroad, West Duluth, MN

LSMT Lake Superior Museum of Transportation, Duluth, MN

D&NE Duluth & North Eastern RR, Cloquet, MN

EJ&E Elgin, Joliet & Eastern

LS&I Lake Superior & Ispheming

NSM North Shore Mining

Rcvd Received

All cabooses are 8-wheel.

AFE 5882 Ten locomotives, 17 cabooses(C-43, C-44, C-70, C-71, C-72, C-73, C-74, C-75, C-76, C-77, C-129, C-161, C-163, C-168, C-172, C-175,C-190) and 7 wayside office equiped in 1947 with Aireon inductive radio. Each caboose equipped with Witte 4-HP one-cylinder diesel engine.

AFE 6180 19 cabooses (C-38, C-52, C-56, C-57, C-58, C-132, C-135, C-136, C-138, C-143, C-144, C-151, C-154, C-180, C-186, C-187, C-188, C-189, C-191) equipped with radio and Witte 4-HP engines in 1948.

Following cabooses had radio and Witte 4-HP engines installed unknown date: C-39, C-96, C-152, C-153, C-164, C-165, C-166, C-167, C-173, C-174.

Following cabooses plywood sheathed between 1964-70: C-39, C-44, C-50, C-58, C-135, C-136, C-138, C-143, C-144, C-180, C-191, C-192, C-193.

Following cabooses "Duroply" sheathed between 1964-70: C-15, C-38, C-43, C-50, C-56, C-91.

Following cabooses confirmed in "Arrowhead" paint scheme: C-43, C-44, C-52, C-56, C-58, C-138, C-143, C-144, C-180, C-192, C-193, C-204, C-205, C-210, C-218, C-219.

Cabooses C-55, C-96 and C-95 are on hand at Mitchell on July 28, 1969. C-96, former EJ&E 123, and C-96, former EJ&E 144, were purchased in January 1949. Note the roof profiles, arched on C-96 and peaked on C-95 and the different style letterboards. (D.P. Holbrook collection)

-continued from page 228
were added to the ends of cupolas. A hyphen was added between the letter C and the road number beginning on January 12, 1942.

Prior to merger, D&IR cabooses did not have an assigned Class and were assigned classes K1, K2, K3, K4 and K5 after the merger. See DM&IR caboose roster for details.

An inventory of cabooses and types of underframes on October 9, 1939 showed C-129, C-132, C-135, C-136, C-138, C-143, C-144, C-180, C-151, C-154, and C-182 through C-187 equipped with wood underframes that had been reinforced with steel rails. Cabooses C-137, C-153, C-156 to C-170, C-172 to C-177, C-179, C-181 were equipped with steel center sills. Cabooses C-147, C-188, C-189 were equipped with steel underframes. A total of 16 cabooses with wood underframes, 10 with steel-rail reinforced underframes, 25 with steel center sills and 3 with steel underframes, all former D&IR cabooses

Cabooses C-55, C-57, C-70 through C-77 were all equipped with steel underframes. Cabooses C-5, C-8 thru C-10, C-12 through C-58 were all equipped with wooden underframes. A total of 20 with wood and 10 with steel underframes, all former DM&N.

AFE 4138 was approved for $8,500 on October 27, 1939 for rebuilding eight class G1 cabooses with steel underframes. Chosen were C-5, C-12, C-26, C-30, C-39, C-40, C-43 and C-58. Eight additional cabooses, C-18, C-25, C-38, C-44, C-50, C-52, C-56 and C-132 were approved for steel underframes under AFE 4329, and four more, cabooses C-129, C-135, C-136, and C-138 under AFE 4344. All used Cardwell NY-11-E draft gears for the underframes, being more flexible than the draft gears which were applied to other former D&IR steel-underframe cabooses. The former G-1 series was chosen for rebuilding because all had the same 24-foot underframes. Project was started on November 7, 1939, and completed on March 18, 1940. When the project was completed only five 24-foot and 31-foot or longer cabooses were left with wooden underframes.

Increased shipments of ore with the start of World War II caused a severe caboose shortage. Two former DM&IR cabooses, D&NE 3 and 5, were purchased and delivered back to DM&IR at Saginaw on May 15, 1942. Both were taken to Two Harbors for repairs and restencilling, D&NE 3 to C-22 and D&NE 5 to C-13. To assist in the shortage in 1942, two cabooses were leased from NP. An unknown number of cabooses were leased from sister road EJ&E during early 1947 and returned the same year at the end of the ore

A caboose shortage during 1946 caused DM&IR to lease a number of cabooses from sister road EJ&E. EJ&E 109 is with Class K 2-8-0 DM&IR 1204 pulling interchange cars from NP Rice's Point Yard. (R.V. Nixon photo, Museum of Rockies collection)

season. Also leased were an unspecified number of Omaha Ry. bay-window cabooses.

Request was made in 1944 to equip seven wooden underframe cabooses, C-8, C-9, C-10, C-15, C-22, C-60, C-61 assigned to Proctor Hill ore transfers with steel underframes. AFE M5248 was approved April 14, 1944 and all work completed at Two Harbors by April 4, 1945.

After the war many railroads began to install radio communication in engines, cabooses, yard offices, dispatcher's offices and depots to allow for better communication and safer operations. AAR Communications Section had established a special Committee 7 on December 11, 1944, to consider the application of communication equipment to rolling stock. Discussion during these meetings determined that engines, both steam and diesel, had sufficient electrical capabilities but recommended an additional turbo-generator on steam engines to supply 117 volts, single-phase AC power for radios. The committee considered axle generators and storage batteries for cabooses. Both had disadvantages; axle generators were only effective if train speed was sufficient to charge on-board batteries. Storage batteries without an axle generator would require charging locations at each terminal, with cabooses switched to those locations. The third possibility was a small internal combustion engine connected to an AC or DC generator. In 1944, the only engine available was fueled by gasoline and considered a fire hazard.

DM&IR approved AFE 5882 on November 30, 1946, at a cost of $90,000 for the installation of inductive radios, Witte diesel engines for electrical power and fuel oil tanks on 17 cabooses: C-43, C-44, and C-70 through C-77, C-129, C-161, C-163, C-168, C-172, C-175, and C-190 on the Missabe Division. Also equipped with radios were steam engines 502, 513, 220, 223, 228-231, 235 and 237. Testing of this equipment was begun on June 30, 1947 using engine 235 and caboose C-172. Caboose radio installations were completed on April 10, 1947, at Two Harbors shop.

Radios were installed on additional engines 81, 190, 1207, 90–93, D&RGW 1550, D&RGW 1551 and D&RGW 1553 during October 1947 for use in Proctor yards per AFE 6176. During May 1948, radio equipment was moved from engine 81, 190 and 1207, to 506, 514, and 515. D&RGW engines 1550, 1551 and 1553 had radios removed in summer 1948 and reinstalled on DM&IR engines 508, 510 and 512.

Installation of inductive radio equipment for the Iron Range Division was approved on AFE 6180 on December 5, 1947, for installation on 15 cabooses, 14 engines and six wayside locations at a cost of $116,000. An additional 19 cars were approved under AFE 6180 in 1948: C-38, C-52, C-56, C-57, C-58, C-132, C-135, C-136, C-138, C-143, C-144, C-151, C-154, C-180, C-186 through C-189 and C-191. All 10 M4 Class and 4 M3 Class 2-8-8-4 Yellowstones were equipped with inductive radios during 1948.

DM&IR formed a caboose committee in 1947 consisting of eight labor and 14 management employees to design a new wood caboose. After many meetings a caboose design, similar to those constructed in 1937, was approved.

The design was to consist of:

- Wood caboose bodies with additional bracing and screws in place of nails.
- All metal objects, such as stoves, tables, coal boxes, lockers, window sashes, door jambs, handhold and shelving would have rounded corners.
- All glass would be safety glass.
- All aisleways coated with a non-skid surface.
- Ends platforms and steps were to be of the open grid type.
- Caboose steps would be lowered to a height of not over 14 inches above rail.
- Inside of all cupolas would be sealed so there are no protruding ribs or braces.
- Cupola seats and back rest would be cushioned by springs or foam rubber.
- Thermos jug would be installed.
- All to have steel underframes.
- Windows in cupola had to be moved as close to corner of cupola as possible.
- Conductor's desk would be raised to a height of 32 inches.
- Raise ladder from platforms to roof of caboose 30 inches above the latitudinal running board, but in no case, higher than cupola roof.
- Install in all cabooses, suitable wash basin, the same as in sample caboose C-18 which was rebuilt to caboose C-191.
- Standardize on end railings, the same as were then installed on Class G-2 cabooses.
- Reverse coal box by putting it against end of the long seat between seat and stove and padded.
- Provide larger containers for yellow fusees.
- Standardize on type of cupola on the Class G-2 cabooses, except end windows would be the same as in sample caboose C-18 (C-191). The height of windows and height of cupola were to be the same as Class G-2 cabooses.
- All cabooses rebuilt or purchased for use in road, ore or local way freight service to be as wide as widest ore car.

The caboose shortage was too great in 1948 and a decision was made to find suitable second-hand cabooses to tie DM&IR over until shop time was available at Two Harbors. Caboose C-18 was destroyed by fire and retired in April 1947. Parts were salvaged and new caboose C-191 was built from these in July, 1947 using the designs approved by the 1947 caboose committee.

EJ&E CABOOSE PURCHASE

On December 22, 1948, DM&IR approved the purchase for $14,000 of nine secondhand cabooses from EJ&E. These cars, EJE 5, 88, 96, 101, 120, 123, 144, 145, and 155 were shipped from EJE Joliet Yard on December 26, 1948, and delivered from Omaha Ry. at South Itasca to DM&IR on January 3, 1949. Cabooses were immediately placed in service, still with EJE orange paint. Cars were renumbered, in order, as DM&IR C90–C98. All cars had 12-inch channel steel center sills, 10-inch channel steel side and ends sills. A letter from P.M. Sullivan, superintedent motive power & cars, on January 19, 1949, requested that all nine cabooses be painted. The final two cars were painted on

Caboose shortages during 1946 caused DM&IR to lease cabooses from EJ&E. Still requiring cabooses DM&IR purchased nine EJ&E cabooses; 5, 88, 96, 101, 120, 123, 144, 145, 155 in December 1948 and placed them in service. It took until February 21, 1950, for all of them to be repainted and renumbered as DM&IR C-90 to C-98. Built to EJ&E designs, they took on their own appearance. Carbodies and underframes were identical, but cupola roofs and letterboards were different from car to car. DM&IR C-91 at Mitchell, May 11, 1960 (Bruce Meyer)

DM&IR installed plywood sheathing to 12 cabooses and "DuroPly" sheathing to six cabooses between 1964 and 1970. DM&IR C-91 received "DuroPly" sheathing and is at Virginia sporting new paint, Scotchlite monograms and lettering on July 27, 1969. C-91 was the only former EJ&E caboose that was sheathed. It was sold to Hyman Michaels scrap yard on May 22, 1975. (Owen Leander)

February 21, 1950. Cars required additional repairs which were approved on AFE 6498 for $4,050 on March 28, 1949. This included the installation of conductor's table and chair, folding table, small water tank, seat cushions and backs, dry hopper toilet and ventilator, marker brackets and Aladdin lamps. The nine wandered the rails while waiting to be painted with one being used out of Mitchell Yard at Hibbing in March and April, 1949.

BUILDING MORE CABOOSES

Still faced with a continuing caboose shortage, the construction of two new wood cabooses was approved on AFE 6951 on March 30, 1950. The Mechanical Department was instructed to submit a report on the possibility of constructing the caboose bodies on existing merchandise underframes. Two Harbors was completing the rebuild of six refrigerator cars with new underframes, ends and application of AB brakes, and notified management that once work was completed that an assembly line would be established for fabrication of the new caboose. Because of the difficulties and excessive man-hours involved in reworking underframes on suitable merchandise cars it was decide to build new underframes.

Cabooses C-192 and C-193 were approved for construction based on the 1947 approved design. Caboose C-153 had burned on the Missabe Division and was still awaiting repairs and it was recommended that this underframe and one new one be used. After inspection of C-153, it was decided to fabricate two complete underframes. If secondhand trucks were used it was estimated that each could be built for between $7,000 and $7,600. Secondhand trucks from box cars 5312 and 5334 were used. Radios and electric lighting were removed from C-186 and C-161 and used on C-192 and C-193. Management wanted the cabooses completed by May 1, 1951, in time for the 1951 season. This was insufficient time to complete engineering drawings and it was requested that caboose C-191 and a Class G-2 caboose be spotted at Two Harbors to be used a samples for checking dimensions while the cars were constructed. Cars were placed in service during July 1951. It is unclear if cars were plywood sheathed when built or if plywood sheathing was applied at a later date.

FIRST EXTENDED-VISION CABOOSES

After C-192 and C-193 were completed management made the decision to buy instead of build. International Car Co. of Kenton, Ohio, had designed a new cupola caboose with the cupola extending over the side sheets of the car. This new "extended-vision" caboose offered the advantages of a bay-window caboose but was safer by placing the crew members at a higher vantage point which allowed better inspection of trains and kept them away from possible shifted loads on flat cars and gondolas.

The caboose shortage during World War II caused DM&IR to form a caboose committee in 1947 to develop a new wooden caboose design. One prototype car C-191 was rebuilt from fire damaged C-18 in April 1947. Finally in 1950 shop time was available and two additional 1947-design cabooses C-192 and C-193 were constructed at Two Harbors. It is unclear if cars were plywood sheathed when built. C-193 at Steelton, June, 1972. (Ted Schnepf collection)

The demonstrator caboose completed by International Car was of all riveted construction; subsequent cars for other railroads would be all welded with a few minor exceptions. DM&IR placed an order with International Car on AFE 7073 September 17, 1951, to purchase twenty extended-vision cabooses for $291,421.76. Numbered C-200 through C-219, this included the riveted demonstrator caboose numbered as C-200. All of the cars were of riveted construction, but the rivet pattern on C-200 was different then the C-201 thru C-219.

Cars were delivered painted yellow with black lettering and roofs, 9-inch reporting marks on side and cupola ends, and a DM&IR stencil type safety-first monogram centered below the cupola windows. These were the first new-built extended vision cabooses in North America and the design would go on to become the most common caboose in North America. Cars were delivered with cupola roof-mounted marker lights. These marker lights began to be removed in late 1969 and replaced with red marker lights installed underneath the peak of the roof on each end. They were delivered dur-

By 1950 most railroads had started to replace their aging wood caboose fleets with steel cars. DM&IR decided in 1951 to place an order for 20 riveted steel cabooses C-200 through C-219 with International Car Co. International had developed an "extended vision" caboose that had a cupola that extended over the edges of the carbody allowing a high and wide vantage point for crew members to observe their trains. The first car, C-200, originally built as a demonstrator caboose, differed from the others by having a different rivet pattern. Not delivered with radio equipment, DM&IR quickly installed radios after delivery as shown by C-203 at Proctor, July 27, 1957. (Bob's Photo collection)

Delivered with DM&IR stencil monogram on cupola it has been painted over in this view of C-201 at Duluth, July 7, 1963. The riveted extended vision caboose would be unique to DM&IR with all others built for other railroads being of welded construction. (D. Repetsky, John C. La Rue, Jr. collection)

ing 1952 and placed in service in numeric order; 5 in April, 10 in May, 5 in June, one in July.

On February 27, 1952, instructions were issued to remove radios and Witte engines from 15 wooden cabooses in the poorest condition. These were: Missabe Division-assigned C-43, C-44, C-129, C-163, C-168, C-172 and C-175, and Iron Range Division-assigned C-38, C-52, C-56, C-57, C-58, C-132, C-135, and C-136. Radios were then installed in the C-200 to C-214 once delivered. Cabooses C-215 to C-219 received brand-new radios once delivered.

Cost savings on DM&IR became a driving factor in the 1960s with the declining ore business. When wood sheathing on cabooses deteriorated beyond economical repair it would be replaced by plywood sheathing. From 1964 until 1970 cabooses C-39, C-44, C-58, C-135, C-136, C-138, C-143, C-144, C180, C191, C-192, and C-193 were rebuilt with plywood sheathing. Cabooses C-15, C-38, C-43, C-50,

1960s standard extended vision caboose paint job is depicted by C-212 at Keenan, September 10, 1978 with maroon roof, lettering and early "Scotchlite" monogram and no red hash stripes on the side of the carbody. (R.F. Kucaba)

C-56 and C-91 were rebuilt with "Duroply" sheathing which looked much like plywood. The economic downturn during 1962 caused 21 cabooses to be placed in dead storage at the end of the 1962 shipping season and not to be used during 1963. Changes in environmental laws in Wisconsin forced cabooses C-44 and C-180 to be equipped with retention toilets by 1974, allowing them to be used on transfer runs on the Interstate Branch.

Extended vision caboose C-218 is in fresh paint at Duluth, on July 7, 1963. (D. Repetsky, John C. La Rue, Jr. collection)

T-BIRD AND MINNTAC BAY-WINDOW CABOOSES

Development of taconite created the need for specialized equipment to handle this new product. Taconite pellet and crude taconite ore cars were rebuilt from older 70-ton ore cars. Class Q covered hoppers were acquired to transport bentonite clay. Eveleth Taconite at Forbes, required DM&IR to operate crude taconite trains from the T-Bird North loadout in late 1965 and later the T-Bird South loadout in 1976. Both loadouts could accommodate ore cars but cupola-equipped cabooses would not fit through these facilities.

Operations of the T-Bird crude trains required a caboose at each end of the train because there was no loop track at the loadout and DM&IR wanted to avoid switching the cabooses at the loadout. Cabooses C-190 and C-191 were chosen to be rebuilt for T-bird crude taconite service on AFE 9924 June 30, 1965. A third caboose was considered for rebuilding to allow out-of-service time but R.H. Seitz, chief mechanical officer, indicated that it would be considered at a later date. Cupolas were removed, smokestack was shortened and roof access ladders on ends were cut down to roof level, creating a low-profile caboose that would fit in the loadout. A locomotive-style winterized window was installed on the side of the carbody below the original cupola location allowing crews a side view of their train. New locomotive-type revolving seats were installed inside next to the winterized windows. Rebuilding was completed September 30, 1965.

DM&IR Management Services Department in 1964 was looking at ways to improve loading time at Minntac. It indicated having bay window cabooses on Minntac pellet trains would help the loading process at an annual cost savings of $9,950. Frank King, industrial engineer for DM&IR, submitted project MS-

The crude taconite loadouts at T-Bird North and South required non-cupola cabooses to clear loading equipment. Three wood cabooses were modified for this service, two, C-190 and C-191, in 1965 and after C-190 was wrecked in 1967, C-135 was rebuilt in 1967. All were identical and equipped with a winterized window to replace the cupola. C-191 is seen at Rainy Jct. Yard Virginia, July 27, 1969. It was sold June 20, 1977, to Lake Superior Railroad Museum on June 15, 1982 for fund-raising purposes. (Owen Leander)

The T-Bird logo was unique to the three wood rebuilt cabooses and to the first two rebuilt groups of T-Bird ore service cars. C-135 Virginia, July 27, 1969. (Owen Leander, John C. La Rue, Jr. collection)

C-209 Keenan, June 20, 1977. Note access doors and exhaust stack for Witte engine and standard marker lights in the right corner and. (R.F. Kucaba)

One of two Minntac service extended-vision cabooses rebuilt in 1967 with side windows was C-210 at Virginia, on July 4, 1971. Before rebuilding was complete the plan to use them in Minntac service was eliminated and both cabooses were placed in regular service. (Owen Leander, John C. La Rue, Jr. collection)

Above: The opening of T-Bird South loadout for crude taconite on July 21, 1976, caused four cabooses, C-204, C-205, C-218, C-219, to be rebuilt in February 1976 for T-Bird mine service. Unlike earlier side-window conversions, DM&IR purchased bay-window kits from International Car to modify the extended-vision cabooses. C-218 at Keenan, September 10, 1978.

Right: Winter always found DM&IR with excess equipment available for lease. C-205 is at Dolton, Illinois, on lease to Louisville & Nashville R.R. on January 3, 1978. (Two photos, R.F. Kucaba)

1110 to cancel this rebuilding on August 15, 1967, but it was tabled because rebuilding had already begun. Three cabooses were chosen to be rebuilt in 1967; C-135 for T-Bird service and C-209, C-210 for Minntac service. T-Bird service caboose C-190 was wrecked in June 1967, retired on July 25, 1967, and C-135 was rebuilt as its replacement. All three were rebuilt in late 1967 to identical specifications as C-190 and C-191. C-135 was a wood caboose, but C-209 and C-210 were originally built as steel extended-vision cabooses in 1952. By 1969, C-135 and C-191 had received a maroon placard with "T-Bird" logo in yellow. By the mid-1970s this logo had been removed. It is unknown if C-209 or C-210 were ever used on Minntac pellet trains as assigned cabooses.

Eveleth Taconite started to expand operations again in 1975 and on July 21, 1976, the first crude taconite train was loaded at the new T-Bird South loadout. Having two loadouts, South and North, required DM&IR to rebuild four additional cabooses for this service. AFE 10834 was approved on January 9, 1975, for extended0vision cabooses, C-204, C-205, C-218, C-219 with all completed on February 6, 1976. Rebuilding involved the same modifications as previous T-bird service caboose with the exception of the side windows. DM&IR had taken notice of other carrier's bay-window cabooses and ordered bay-window conversion kits from International Car Co. Rebuilt cabooses retained their riveted carbodies and were repainted with the road number placed at bottom of the bay window and a small "Arrowhead" logo above the number. Once

International Car supplied DM&IR with the second order of extended-vision cabooses C-220 through C-239 in 1974. Light on cupola roof was used as a locator much like beacons on locomotives. Each end had four small red marker lights, two in each corner until FRA standards mandated effective July 1, 1978. that it be located on center rear of cabooses and be a larger size. C-233 Keenan, June 20, 1977 (R.F. Kucaba)

DM&IR C-220 at Rainy Jct. Yard in Virginia, on July 28, 1975. Note that the two red marker lights are in each corner of the roof above the end platform. Light centered on rear was used by crews to inspect track and passing trains. (R.F. Kucaba)

Overhead view of C-223 at Saunders, Wisconson, May 27, 1983. (D.P. Holbrook)

these four cabooses were completed, bay windows were assigned as follows: two to T-Bird North trains, one for T-Bird South, one spare and two in general service. C-135 and C-191 would eventually be removed from T-Bird service and placed in general caboose service and be retired in 1982. Remaining bay windows were retired in the 1990s.

NEWER EXTENDED-VISION CABOOSES

October 29, 1973 found the DM&IR owning 59 cabooses. Twenty of these were newer extended-vision cabooses built in 1952, with the remaining 39 being of wood construction and ranging in age from 1884 to 1951, many in deteriorated condition. Accounting records from September 21, 1973, indicate 53 cabooses assigned.

Desiring to retire most of the remaining wood cabooses, the DM&IR ordered a total of 20 extended-vision cabooses from International Car Co. in 1973. Cabooses C-220 through C-223 were purchased under a separate AFE 10820 of August 3, 1973 for $158,400. This was approved on November 7, 1973. These four cars were purchased separately by using the funds from the settlement on 55 ore cars destroyed on the CNW. Tax laws allowed funds to be converted into similar property within two years of destruction and avoid tax on the financial gain.

Cabooses C-224 through C-239 were purchased under AFE 10821 of Oct 21, 1973. This was approved on November 7, 1973. Purchase of C-220 through C-239 allowed the retirement of 24 wooden cabooses.

Prior to 1978, DM&IR used portable electric markers on cabooses C-200 to C-219. Cabooses C-220 to C-239 were equipped with four permanent electric markers mounted in the platform roof corners. FRA mandated effective July 1, 1978, under Part 22, Title 49 of Code of Federal Regulations that all freight trains shall have highly visible markers during periods of darkness or whenever weather conditions restrict clear visibility and will be located centered under the roof eave. Permanent rear end marker

C-226 at Keenan in May 1980. The 20 extended-vision cabooses C-220 to C-239 were built in 1974 to replace the majority of the remaining wooden cabooses. (D.P. Holbrook collection)

lights began to be installed on 39 steel cabooses C-200 through C-239 in February 1979 and all were completed by October 15, 1980.

On April 9, 1990, on AFE 12290 the decision was made to retire 34 of the remaining 39 cabooses becuse of the successful implementation of end-of-train (EOT) device operation on the DM&IR. The only ongoing need was for two T-Bird crude cabooses (no cupola), one to service the Louisiana-Pacific waferboard plant near Two Harbors, and two spares. The five retained cabooses were: C-210, C-218, C-222, C-224 and C-227. C-205 was retired April 9, 1990, but retained as a warming facility. Cabooses C-205 and C227 were sold to Northshore Mining on July 3, 2000. Caboose C-222 was the last car on roster and was sold to Duluth Port Authority for use with their Schnable flat car.

CABOOSE ASSIGNMENTS SEPTEMBER 1973

Proctor Pool
Proctor Hill & Transfer service – two Commercial and three ore service cabooses
Minntac unit trains – two cabooses
Taconite Jct. ore trains – one caboose
Fraser ore trains – two cabooses
Eveleth taconite unit trains – one caboose
Rainy Jct. ore trains – one caboose
Biwabik ore trains – two cabooses
Coleraine Local – one caboose
Rainy Jct. Local – one caboose
Steelton Ore transfers – three cabooses

Rainy Jct. Pool
Rainy Jct. Yard – two commercial service, two T-Bird crude and five mine run cabooses
Minntac Pellet service – three cabooses

Two Harbors Pool
Minntac unit trains – three cabooses
Rainy Jct. ore trains – one caboose
Virginia Local – two cabooses
Wales Branch local – one caboose.

Endion Pool
Ely Local – two cabooses
Endion Yard & Transfer – one caboose

Biwabik Pool – one commercial service and three for mine run cabooses
Taconite Pool – three mine-run cabooses
Mitchell Pool – one commercial service and three mine run cabooses
Steelton Pool – one commercial service caboose

All passenger cars used by DM&IR were conveyed from D&IR and DM&N except for one. Coach *Minnesota II,* former NP 517, was purchased from BN on June, 6, 1975. The car was donated to Lake Superior Museum of Transportation on July 31, 1998, Superior, Wisconsin. (D.P. Holbrook)

Duluth, Missabe & Iron Range Passenger Cars

Kind	Truck	Name/ No	Cl	O.L	Builder	Built	Notes
Coach	4-axle	***Minnesota II***		85′	Pullman	1946	Purchased from BN 06/06/75 former NP 517. Donated to LSMT 7/31/98

DM&IR PASSENGER CARS

After the merger of D&IR and DM&N, DM&IR continued to operate passenger service until July 15, 1961, with all remaining D&IR/DM&N cars repainted DM&IR but not renumbered. Once passenger service was discontinued the remaining passenger cars were retired or placed in MofW service. Retained for management and customer service purposes were combination car W24, coach W-33, and business car *Northland.*

From time to time these three cars were not sufficient to support the needs of management and a decision was made in 1974 to purchase a coach for the business car fleet. Purchased from Northern Pacific was coach 517 that was originally built by Pullman in 1946. Car was 85 feet long, with a seating capacity of 62, two washrooms, and a lounge area, and rode on two 8-foot 6-inch GSC 2-axle trucks. On arrival on DM&IR the car was reconditioned and named *Minnesota II.* As needed, this car was added to the business car consists and the annual American Institute of Mining Engineers special. DM&IR General Manager D.B. Shank was instrumental in forming the Lake Superior Museum of Transportation and felt that, as needed, this car could be used for fund-raising activities for the new museum.

When *Minnesota II* was purchased it was intended to name two additional cars. Car W-24 would become the "Missabe II" and car W-33 would become the "Vermillion II." This naming never took place. ■

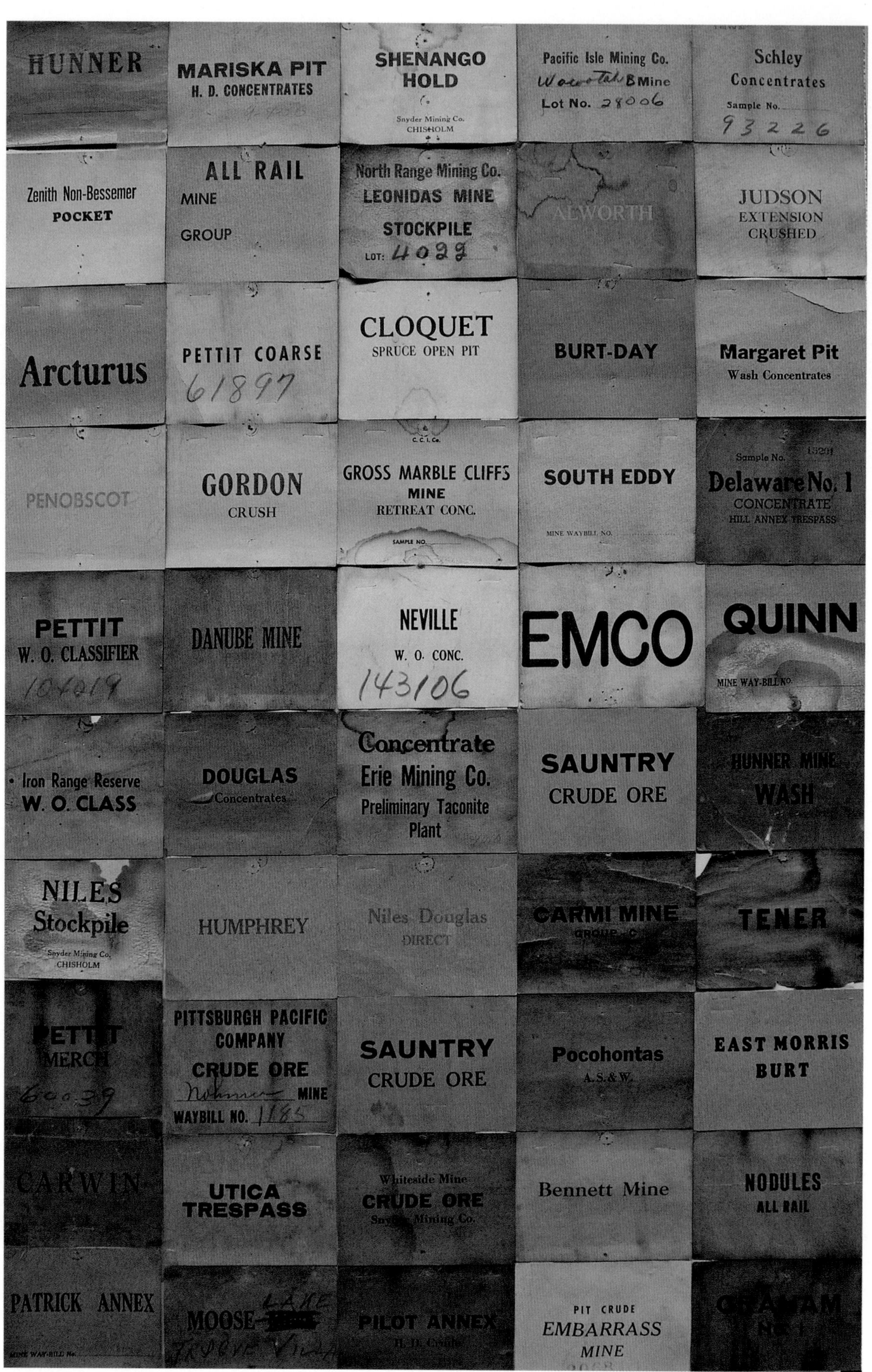

Mine Tags at Lake Superior Museum of Transportation. (D. Schauer)

CHAPTER FIVE

OPERATIONS

Coal mines in the eastern U.S. established about 1900 a method of placing a cardstock "tag" on coal hoppers to "tag" the shipper and allow movement from the mine before car was waybilled, a destination supplied, or a customer identified. Norfolk & Western was an early user of this process. Tags were removed at major yards when cars were weighed, kept in order by a yard clerk or weighmaster, and then given to clerical staff to make out waybills. During this same time period, mines on the Mesabi and Vermillion Ranges adopted much the same practice.

MINE TAGS

Beginning in 1900, steel ore cars were delivered with a "bad order card" holder mounted on the right end of the side sill adjacent to the grab irons. Mounted higher on the left-hand bottom corner of the side sheet, specified on Pressed Steel Car drawings of 50-ton steel ore cars, was a "ore card holder" slightly larger than the bad order card holder. All D&IR cars would have one high and one low holder. Beginning in 1913, DM&N 50-ton steel ore cars had the smaller bad order card holder replaced with the larger "ore card holder" and moved to the right hand corner of the side sheet. Ore card holders were universally used for both "ore cards," later called "mine tags" and bad order cards. DM&IR would retain the DM&N practice until the final ore cars were ordered.

Internal Oliver Iron Mining documents indicate that cardstock mine waybills were called "Mine Tags" or "Sample Tags" and placed in the "Mine Tag" holders by mine loading personal. Two tags were used. One was placed on the ore car and the other tag was placed in the "can" holding the sample of ore. While the car was in transit from mine to docks, the sample would be analyzed for chemical content. Some tags had additional lines at the bottom identified as sample number, waybill number or lot number to allow the mining companies to identify the loads. In some cases a grease pencil was used and the track number written on the bottom of the tag. Tags were used to allow cars to be picked up and moved before the revenue

Mine Tags at Lake Superior Museum of Transportation. (D. Schauer)

waybills were made out. This kept the flow of cars moving without delaying them for billing. On arrival at Proctor and Two Harbors, tags allowed for easy identification of which mine had loaded which cars and this information was combined with car weights after weighing to produce revenue billing. This method was used by all other mining companies on the iron range. DM&IR limestone dock personnel also used tags for coal unloaded at the docks to indicate the coal company name for the coal being shipped.

Computers would gradually put an end to the mine tag practice. Hanna Mining by 1975, and U.S. Steel by 1977 had discontinued using the tags, but Cleveland Cliffs was still using them in 1980.

ACI – AUTOMATIC CAR IDENTIFICATION

As computers became useful to the railroad industry, many progressive uses were found for their application. The process of checking trains and recording car numbers by hand had become antiquated. During 1962, a study was commissioned to develop a system to scan and track freight cars on the DM&IR. An optical scanning system was jointly developed with Sylvania Electronic Division of General Telephone and Electronics (GTE) and the Association of American Railroads (AAR). Initially called Kartrak, the system became known as Automatic Car Identification (ACI)

Test labels were applied to 25 of the 70-ton ore cars. The ACI labels were developed by Minnesota Mining and Manufacturing Co. (3M). Each label element was 6-inches wide and 1-inch high and the panel consisted of 13 of these elements. The labels were attached to an ACI plate which was 10½-inch wide by 22-inches tall. They contained a start element and the ten digits 0 through 9. Each label element had two positions that contained the color, orange, blue or white in the first position and orange, blue or black in the second position. This allowed for 13 color combina-

Car shop employee installing ACI plate on ore car in 1964. Note the "ore card holder" to the right of the 10-64 date, used for "ore cards" to identify the mine and type of iron ore once car was loaded. "Ore cards" also were called "mine tags" and "sample tags". Starting in 1900 with the first steel ore cars, D&IR and DM&N equipped the cars with a "ore card holder" in the left corner of the car body (facing the car) and a smaller "bad order card holder" in the right corner of the side sill. Starting in 1913 with the delivery of DM&N Class U9 50-ton ore cars 14500–15499, the smaller bad order card holder was replaced by the larger "ore card holder," with both located on the corners of the side sheets. (MRHS collection)

Above: ACI reader at the Proctor, Minnesota, scales in 1965. (D.P. Holbrook collection)

Right: Starting in 1962, DM&IR became the first road to install Automatic Car Identification (ACI) which had been developed in conjunction with the Association of American Railroads and Sylvania Electronics Division of General Telephone and Electronics. This development was driven by the computerization of railroad records and the need for a suitable system that could track car movements. DM&IR was chosen for the initial installation because of the controlled environment with ore cars moving, for the most part, only over DM&IR trackage. Pictured is the scalehouse at Highlands just outside Two Harbors, Minnesota, with ACI reader on the left. (D. Schauer)

tions and permited the coding of 12 individual pieces of information.

The ACI label information was: car initial and number, light weight, and a variable information field. The label was scanned by the trackside scanner developed by Sylvania which scanned the train at speeds up to 80 mph. The scanner was a simple hooded steel box mounted on a post at about 15 degrees from vertical to eliminate reflections and sunlight.

During 1964 the installation of the ACI system at key points in Proctor Yard was begun. The DM&IR was the first American railroad with ACI when installation of three scanners at Adolph, Proctor Scales and South Proctor, was completed during late October 1964. By 1968 there were a total of four ACI scanners on DM&IR. The Culver ACI scanner identified and entered into the computer system the car numbers in train order. This scanner was located on the west side of track. Scanning at this point provided verification of manual data entered at the Operations Center at Iron Junction, Minnesota, and also supplied car numbers for pellet movements. Errors detected by computer edit were sent to the Operations Center IBM 1050 computer for resolution.

Proctor Scales ACI scanner was located with two scanners, one on each scale track to verify inbound trains and interface weigh-in-motion scale weights with cars read by the ACI reader. These scanners were located on the east side of the track. In conjunction with the Culver ACI scanner, this allowed verification of both ACI labels on either side of a car in case of a no-read or misread of the ACI label. South Proctor ACI scanner identified and entered into the system the car numbers in train order for hill trains destined for the docks at Duluth. Computer response at the Duluth docks displaying disposition for each car controlled proper cargo loading in boats. The Adolph ACI scanner provided a record of northward empty car movement to yield better car turnaround time by mine loading point, and also recorded movement of ore loads and empties between Proctor and Steelton.

During its first year of operation in 1965, the new ACI system, with associated automatic weighing, contributed materially to the efficient movement of iron ore. The system reduced labor requirements, provided more timely car identification information, and enabled faster weighing. This was the first such equipment to be installed for daily operations on any railroad. A number of tests were conducted in cooperation with the AAR Research Center.

AFE 9739-1 was approved on May 15, 1964, for the application of ACI labels to 9,000 70-ton ore cars and 88 Class road switcher diesel locomotives. As of January 6, 1965, ACI labels had been applied to 8,858 70-ton ore cars and 48 locomotives. The balance were applied prior to the start of the ore-shipping season. Dirt-repellent material developed by 3M Company had by this time been applied to about 90 per cent of the labels.

By May 20, 1965, ACI was working satisfactorily, but the wet condition of ore at some washing plants was causing the ore to run over car sides, partially obliterating the labels. Labels were being cleaned by car inspectors at Coleraine, but not at other locations. 3M was contacted and they began applying a test solution to prevent buildup on the labels to avoid having to install washers for the labels. Application of dirt-repellent material to ACI labels had been completed by September 22, 1965, on approximately 7,700 ore cars.

Approximately 8,250 of the 9,180 cars bearing ACI labels had been treated with dirt-repellent material by November 3, 1965. In addition, a new type of label, designed to have improved dirt repellent characteristics, had been applied to 100 70-ton ore cars for test purposes.

Continued refinement of the ACI system during 1966 and installation of a SPADAR weighing and programing unit at Proctor ore scales further improved system accuracy and reliability, permitting cars to be weighed at approximately twice the previous rate. Additional scanners were installed at Highland and Allen Junction.

National application of ACI was begun in 1967 with a number of major railroads installing the necessary scanners for test and evaluation purposes. The first two scanners outside of the DM&IR installations were installed in 1966.

During the winter shopping program of 1968-69 the DM&IR started application of new ACI plates to the entire roster. The new ACI plate was a formed U-channel that was offset from the car side by one inch to prevent obliteration of the label when fine ores washed over the sides of the car.

In September 1971, the Federal Railroad Administration (FRA) had the Transportation Systems Center (TSC) evaluate ACI on a national level to recommend possible actions to improve the system. By this date, 92 percent of the American freight car fleet was labeled and 98 percent of all Canadian freight cars. ACI had been installed on 26 railroads and 145 scanners were in place. Many railroads were reluctant to install the ACI scanners because benefits were difficult to measure. The DM&IR participated in accuracy tests from April to July 1969 on 1,500 cars which showed in excess of 99 percent accuracy. Another test involving 9,000 car scans during the same period also showed 99 percent accuracy.

When ACI was initially developed it was revolutionary; however, technology was advancing quickly. Cameras for checking freight car reporting marks from yard offices and remote locations and the installation of computer car-reporting systems on many railroads became commonplace. ACI continued to be plagued by cars with dirty ACI labels and no labels. Each railroad developed their own system for car reporting and tracking and by 1978 ACI was offically phased out.

Automatic Equipment Identification (AEI) tags became commonplace on the national freight car fleet in 1992. Much like ACI, these passive transponders on the freight cars were read by trackside scanners. These AEI tags are now, in 2014, the preferred way of keeping track of equipment. Canadian National discontinued using the former DM&IR ACI readers in June 2007.

THE MINES, MILLS AND FACTORIES

D&IR ore docks at Two Harbors in July 1953 with FM Trainmaster demonstrator TM-4 pulling empty 50- and 70-ton cars down to the yard to assembly into a northbound empty train. (Bob's Photo collection)

DM&IR Iron Ore Traffic: 1955 to 1965

Ore Tonnage Shipments to	1955		1956		1957		1958		1959	
	U.S. Steel	Independent	U.S. Steel	Independent	U.S. Steel	Independent	U.S. Steel	Independent	U.S. Steel	Independent
Duluth Docks	16,359,553	862,607	11,823,479	773,851	16,136,881	971,947	7,952,893	587,463	7,614,870	636,926
Two Harbors Docks	14,653,868	2,507,119	12,447,652	1,807,618	15,749,566	1,399,030	11,318,625	486,777	8,356,637	800,906
Duluth Docks from GN, NP & SOO	2,112	170,187		102,670		336,413		168,132		58,891
Two Harbors Docks from GN, NP & SOO										
Total to DM&IR Docks	31,005,533	3,539,913	24,271,131	2,684,139	31,886,447	2,707,390	19,271,518	1,242,372	15,971,507	1,496,723
To NP Docks, Superior	8,326	123,263	1,252	346,966		471,300	78,994	797,274	19,505	918,040
To GN Docks, Superior	115,356	3,097,214	114,956	2,105,326	90,157	2,362,774	463,305	1,559,188	195,332	1,656,260
To AS&W Division, Steelton, MN	1,186,963		2,070,223		502,390		454,973		480,945	2,968
To other Duluth industries		160,035		148,824		164,445	3,053	171,021	2,043	125,570
All-Rail ore to Eastern Furnaces			244,865	120,205	1,088	146,212	95,488		151,102	15,058
All-Rail ore - Miscellaneous	795	75,820	3,042	6,245	4,107	65,393	4,060	16,031	3,027	7,337
Granite City (All-Rail)										
Grand Total All Ore	32,316,973	6,996,245	26,705,469	5,411,705	32,484,189	5,917,514	20,371,391	3,785,886	16,823,461	4,221,956
Flue Dust over docks		30,294		11,025						
Grand Total All Ore & Flue Dust	32,316,973	7,026,539	26,705,469	5,422,730		Note 1				

DM&IR Coal & Limestone Docks

Received-Tonnage	1955		1956		1957		1958		1959	1960	1961	1962	1963	1964	1965
	Duluth	T.H.	Duluth	T.H.	Duluth	T.H.	Duluth	T.H.	Duluth	Duluth	Duluth	Duluth	Duluth	Duluth	Duluth
Coal - Coal Docks - DMIR Company Use	100,383	98,504	83,749	105,547	152,327	156,878	None	None	None	21,810	None		Note 2	Note 3	Note 3
Coal - Coal Docks for AS&W			397,748												
Coal - Limestone Docks	924,279		532,606		883,174		455,032		357,435	484,836	382,612	384,362	Note 2		
Limestone - Limestone Dock	497,633		452,283		504,230		362,249		385,000	381,389	261,112	79,300	Note 2		
Scrap Iron & Steel - Limestone Dock	53,002		45,686		19,055		22,217		26,000	35,323	8,948		Note 2		
Ferro-Manganese - Limestone Dock	3,650		6,278		6,697		3,180		7,500	4,100	5,642		Note 2		
Steel Products - Limestone Dock	22,623		20,387		24,766		18,516		14,310	23,532			Note 2		
Cement Clinker - Limestone Dock					22,449		45,329		75,271				Note 2		
Iron Ore - Limestone Dock - Note 1	11,299								6,091	14,764		6,846	Note 2		
Grand Total Receipts	1,612,869	98,504	1,538,737	105,547	1,612,698	156,878	906,523		871,607	965,754	658,314	470,548	Note 2		
Shipments - Tonnage															
Steel Products	78,099		13,531		89,441		59,989		3,733	67,479	5,019	None	Note 2		

Merchandise Traffic - Number of cars

	1955	1956	1957	1958	1959	1960	1961	1962	1963	1964	1965
Coal	40,258	32,581	41,445	21,369	18,892	27,135	15,894	20,153	14,259	19,306	22,181
Pulpwood	10,639	10,979	10,931	10,459	10,903	10,666	9,900	10,436	10,824	10,130	7,408
Rock	4,067	7.095	3,617	2,032	814	3,510	3,097	1,642	898	n/a	n/a
Sand & Gravel	1,765	1,119	18	45	67	209	720	647	339	n/a	n/a
Crude iron ore						70,830	54,919	37,647	57,584	68,339	52,691
All other	96,327	88,073	86,071	63,194	70,188	35,735	31,249	30,782	30,626	37,092	54,123
Total Cars Misc. Traffic	153,056	139,847	142,082	97,099	100,864	148,085	115,779	101,307	114,530	125,706	136,403

T.H. = Two Harbors

Note 1 - Iron ore from Venezuela shipped to OIM Virginia, Minnesota for further processing before steel mill use.

Note 2 - Limestone dock Duluth not operated in 1963. Approximately 560,000 tons of coal, stone and bulk commodities handled by Hallett, Duluth Dock and Transport and Interlake Iron docks for movement to AS&W and UAC Plants.

Note 3 - Duluth Coal & Limestone Dock closed 1964. Previous traffic handled by other docks in Duluth.

1960		1961	1962	1963	1964		1965	
U.S. Steel	Independent	All Ore	All Ore	All Ore	U.S. Steel	Independent	U.S. Steel	Independent
10,676,065	1,167,556	3,526,345	3,996,256	13,166,016	14,060,036	400,133	15,282,000	339,000
14,090,873	1,348,838	11,764,208	10,547,370	Note 2	Note 2	Note 2	Note 2	Note 2
	341,366	254,897	212,031	127,126		218,286		145,000
		7,756	24,305					
24,766,938	2,857,760	15,553,206	14,689,962	13,293,142	14,060,036	618,419	15,282,000	484,000
139,555	542,012	275,628	479,670	507,647		481,119	92,000	551,000
140,487	1,901,238	2,272,142	2,664,553	2,128,941		2,385,090		2,306,000
683,876		562,601	574,992	631,942	849,095		685,000	
628	65,635							
90,172	138,533							
5,576	14,325			2,956				
				100,424	100,602		100,000	
25,827,232	5,519,503	18,663,577	18,409,177	16,665,052	15,009,733	3,484,628	16,159,000	3,341,000

1. Not included in 1957 totals are 20,174 tons of Anaconda MT ore. 14,746 tons over Duluth Docks and 5,428 tons over Two Harbors Docks

2. Two Harbors dock shut down at end of 1962 shipping season. All ore shipments in 1963, 1964, 1965, used Duluth ore docks. Two Harbors docks reopened May 1966.

Ore Docks Two Harbors

	Built	Retired	Wood/Steel	Number of Pockets	Storage Capacity Tons	Length of Dock
Dock 1	1884/1888/1899	1911	Wood	162(224)	16,000 (40,400)	1,076 (1,376)
Dock 1 (2nd)	1912	In Service	Steel	224	56,000	1,376
Dock 2	1885/1889/1898	1915	Wood	141(208)	18,000 (41,600)	1,006 (1,280)
Dock 2 (2nd)	1916	In Service	Steel	228	68,400	1,400
Dock 3	1893	1917	Wood	90(170)	16,000 (34,000)	540 (1,054)
Dock 4	1892/1893	1917	Wood	168	36,960	1,042
Dock 5	1895	1923	Wood	168	33,600	1,042
Dock 6	1909/1916/1924	Not in use	Steel	148	37,000	920

Two Harbors Dock 1, 312 feet of single pockets and 1,076 feet of double pockets.
When multiple built dates are shown, the first is the built date and other years are dates dock was rebuilt and/or lengthened

Ore Docks Duluth

	Built	Retired	Wood/ Steel	Number of Pockets	Capacity Tons	Length of Dock
Dock 1	1893	1913	Wood	384	57,600	2,236
Dock 2	1896	1915	Wood	384	69,120	2,336
Dock 3	1900	1919	Wood	384	80,640	2,304
Dock 4	1906	1927	Wood	384	119,274	2,304
Dock 5	1914	In service	Steel	384	115,200	2,304
Dock 6	1918	Out of service	Steel	384	153,600	2,304

Dock 1 was dormant from 1905 until 1913 when it was torn down.
Dock 4 was dormant from 1927 until 1929 when it was torn down.

Ore docks were constructed on Lake Superior at Two Harbors, Minnesota, by D&IR and (shown) at Duluth, Minnesota, circa 1912 by DM&N. Left to right, Dock 1 is being dismantled and Docks 2, 3 and 4 are in service. Note the mix of wood and steel ore cars. (University of Minnesota Duluth, Kathryn A. Martin Library Archives and Special Collections)

NATURAL ORE MINING

The main reason for the D&IR, DM&N and DM&IR to be in business was iron ore, also called "natural ore." It occurred in narrow strips of great vertical depth with some in pools and basins. It was covered in various depths by glacial sand, gravel, boulders and dirt, called overburden. Much of the mining on the Vermillion Range was underground with most of the Mesabi Range being open-pit mining. The ore varied from hard with many large lumps to fine as sand. Some of this ore was loaded directly from mining shovels into ore cars without further processing called "direct-shipping" ore. As mining developed some of the ore deposits had increasing amounts of detrimental material that made it necessary to wash or separate this waste material. Processed iron ore was loaded at "pockets."

By the mid-1950s more and more ore was imbedded in hard taconite and other rocky material which required crushing before shipping. Crushers were built at multiple mines across the range with the output defined as coarse or fine ore. Loading pockets at these plants had coarse and fine pockets, depending on the size of the ore being loaded. Washed fine ore had a tendency to leak out the bottom of ore cars. Most fine loading pockets were equipped with platforms to allow mining employees to place hay into the car to create a seal at the bottom of the cars to prevent leakage. During the late 1950s and early 1960s tree bark was also used,

The 647-foot boat *J.L. Mauthe* built in 1952 and operated by Interlake Steamship Co. of Cleveland, Ohio, is being loaded with iron ore at DM&IR Dock 5 in Duluth. In the foreground are two tracks of DM&IR Class Q triple hoppers that had just been loaded with coal at the DM&IR Duluth limestone docks. (Wes Harkins)

but was not as effective.

Many smaller open pit operations loaded direct-shipping ore into trucks in the late 1950s and 1960s, with the ore then dumped directly from trucks into ore cars. On arrival at steel mills, coarse ore was charged directly in the blast furnace, but fines required sintering to create larger pieces to charge. Processing could involve crushing, screening, jigging, heavy media, washing or a combination of these which were developed to process ore. Ore processed with these techniques was called "beneficiated ore" and grew from 12 percent of total ore mined in 1929 to over 27 percent in 1952.

The Trout Lake washing plant built by Oliver Iron Mining in 1909 at Coleraine, Minnesota, was the first. By 1917 there were more than 15 washing plants operating on the Mesabi Range west of Hibbing. In 1924 this amounted to only 12 percent of all the ore shipped, but by 1925 28 percent was being processed. The year 1957 would see 78 beneficiation plants of various types operating on the Mesabi and Vermillion iron ranges.

Natural ore was primarily shipped to Duluth and Two Harbors docks for loading into Great Lakes boats. Ore was also shipped to AS&W steel mill at Steelton, interchanged to

The Vermillion iron range ore was located deeper below ground and the majority of mines were operated as underground workings. Here we see the Pioneer B shaft and powerhouse at Ely, Minnesota, about 1915. Ore cars from right to left: a Class D 3450–3899 built by Pressed Steel Car Co. in 1930, three Class E 10500–11249 built by Standard Steel Car Co. in 1910, one more Class D 3450–3899 followed by two Class D 11400–11499 built by American Car & Foundry in 1910. Note in the far left background the D&IR skeleton log flats loaded with mining timbers. (University of Minnesota Duluth, Kathryn A. Martin Library Archives and Special Collections)

Besides direct-shipping iron ore being loaded in the open pit mines, many mines had loading pockets such as the Stephens Mine, Aurora, Minnesota, seen here on September 17, 1989. Ore was moved from the pits to the loading pockets by truck, rail haulage and conveyors and further processed by washing and beneficiation. Barely visible behind the pocket is a loading platform that was used for haying the cars. Washed iron ore leaked out the bottom of ore cars and in an attempt to plug the leakage hay would be placed in the bottom of the ore cars to seal up the bottom doors. (S.D. Lorenz)

Open-pit mining is depicted at Mt. Iron, Minnesota, May 13, 1943. Direct-shipping ore was scooped direct from the ground and loaded into ore cars without further processing. (University of Minnesota Duluth, Kathryn A. Martin Library Archives and Special Collections S3021)

Much of the Mesabi iron range ores were located close to the surface and had high enough iron content to be considered "direct-shipping" ore. Direct-shipping ores could be loaded directly from the ground into ore cars as shown at the Hull-Rust mine in Hibbing, Minnesota, about 1940. Oliver Mining Co. was the mining arm of U.S. Steel. Beginning in 1952 it became Oliver Iron Mining Division of U.S. Steel Corp. or OIM. OIM operated their own fleet of locomotives, steam and diesel, air dumps, flat cars and support equipment for pit operations. OIM 707, an 0-10-0, is spotting DM&IR 50-ton ore cars that, once loaded, will be moved to the DM&IR Hull-Rust yard to be assembled into a road train. (University of Minnesota Duluth, Kathryn A. Martin Library Archives and Special Collections)

Washed iron ore in later years had large amounts of water once loaded, creating a jelly-like consistence once loaded. DM&IR 26390 is at Proctor on July 21, 1974. These loads were very difficult to unload (R.F. Kucaba)

NP, Soo and GN for mixing purposes, all-rail to Granite City Steel, and during ore shortages and winter season, all-rail to eastern steel mills. Vermillion iron range ore was used by Zenith Furnace, later Interlake Iron at West Duluth for processing into pig iron.

As mining developed and blast furnace demands changed, much mixing of iron ore began to take place. Three- to five-car cuts of ore were sampled for chemical grade and shipped as a group. These blocks of cars would arrive at Proctor or Two Harbors and then be switched out based on blast furnace's requests. Once the correct number and order of cars was created it would be brought from Proctor to Duluth or direct from the yard at Two Harbors, dumped on the ore docks and loaded into lake boats.

Major shipping years were from 1942 until an all-time high of 49,317,625 tons in 1953. Af-

Loadouts ranged from huge washing plants to very simple wood-constructed dumping platforms shown here at the Sellers Mine in June 1939. The location where cars were loaded was called a pocket. A simple pull-behind scraper was used to drag ore up into the body and then shove it out over the opening in the platform, and a hydraulic cylinder would operate the dump mechanism. The Class E8 ore car in the foreground was one of 800 D&IR 11500–12299 built by Standard Steel Car Co. in 1913. Renumbered to DM&IR 41500–42299 with DM&IR merger, it displays the short-lived, 1937-1942, DM&IR monogram without the Safety First slogan.(University of Minnesota Duluth, Kathryn A. Martin Library Archives and Special Collections, S2386, Box 27 Folder 5)

Truck haulage was used at many open pits and if the ore needed no additional processing it was dumped directly from the truck into ore cars. Shown is Euclid Truck #82 operated by Snyder Mining at the Whiteside Mine in Buhl, Minnesota. Note the flexible hosing running alongside the track attached to the air hose on the end of the car. Gravity dropping of cars to loading pockets was used at most mines with brakes being applied by either handbrakes or by air, as shown here, supplied by a stationary air compressor and flexible hose. (University of Minnesota Duluth Kathryn A. Martin Library Archives and Special Collections S3742, Box 15 Folder 44 D. Barclay photo)

ter the Korean War, natural ore was in competition with imported ore from Brazil, Venezuela, and Canada. Imported ore was 2,754,216 tons in 1946 and grew to 23,443,220 tons in 1955, a growth from 4 percent to 21.8 percent of total iron ore mined. The Vermillion Range closed in 1967 with the closing of the Pioneer Mine at Ely, Minnesota, and Mesabi Range deposits were being depleted. From 1961 to 1965 DM&IR averaged 18 million tons of ore shipped each year. Lower grade taconite began to be processed into pellets that became preferred by blast furnace operators. Gradually as taconite was developed tonnages shipped began to rise and by 1969 23 million tons were shipped.

Natural ore shipments continued to decline through the 1970s as additional taconite plants were brought on-line.

The last natural ore shipments from the giant Hull-Rust pit were made by Rhude and Fryberger during 1983. The last shipments from the Canisteo District (far west end of Mesabi Range) were made in 1986. The Trout Lake washing plant was dismantled starting in 1975. The U.S. Steel Sherman Mine ran until 1981, with the last shipments being made from a reprocessing dump in 1986. The large Rouchleau plant, at Virginia, Minnesota, discontinued operations in the early 1970s. Other Eastern District ore continued to be processed by Pittsburg Pacific's Julia plant until 1986. The Stephens and Donora plants near Aurora were purchased by LTV Steel in the 1980s with shipping of natural ore discontinued in 1991.

The west end of the Mesabi iron range had large amounts of impurities imbedded in the ore. Processing involved large washing plants to remove the impurities. Pictured in 1918 the largest of these plants, the U.S. Steel Trout Lake washing plant at Coleraine was constructed in 1909 and dismantled starting in 1975. (University of Minnesota)

The end of the natural ore shipments from Mesabi Range mines was completed when the Auburn Mine at Virginia shipped its last ore on September 21, 2002. Red ore was no more.

OLIVER IRON MINING - MINNESOTA ORE OPERATIONS

Oliver Iron Mining Company became the Oliver Iron Mining Division of United States Steel (USS) in 1953. Oliver Iron Mining Division became USS Minnesota Ore Operations in 1963. The internal name for Minnesota Ore Operations was MOO, but was rarely used outside of the Administration building in Mountain Iron. In the field, the mines and plants were referred to individually, i.e. Plummer, Sherman, Sentry, Julia, Minntac, etc. Oliver used rail haulage of ore from pits to processing plants at Hull-Rust at Hibbing, Monroe Group at Chisholm, Sherman Group at Chisholm, Rouchleau Group at Virginia, Trout Lake Washer and associated mines at Coleraine, Pilotac at Mt. Iron and the Minntac taconite plant at Mt. Iron. Haulage was handled by, 0-4-0, 0-6-0 and 0-8-0 type steam engines, which were replaced beginning in the 1940s with diesels from Alco, Baldwin, GE, Whitcomb and EMD. Direct shipping ores were loaded directly into DM&IR ore cars but ore requiring further processing was loaded into 20-, 30- and 40-yard air-dump cars for movement to the processing plant. DM&IR provided ore car spotting and pulling at all the plants except Sherman, where OIM did the spotting and pulling of DM&IR ore cars.

Moving ore from the load-outs to assembly yards was accomplished by mine runs. East of Mitchell, Minnesota, on May 11, 1960, we see a mine run moving loads of ore to the assembly yard at Wilpen, Minnesota, where it would be picked up by a road crew out of Proctor. (Bruce Meyer)

TACONITE PLANTS

Until 1950, natural ore was plentiful and easy to mine. Two world wars and increasing competition from other sources around the world caused mining companies to look for other competitive ways of producing ore for consumption. Taconite was a plentiful low-grade rock, but low in iron content. U.S. Steel began testing taconite processing methods with the experimental Pilotac plant near Mountain Iron, Minnesota, during 1953. Pilotac was constructed for the concentration of low grade taconite. Initially 2 million tons of 25 percent to 30 percent iron-bearing taconite was processed into 500,000 tons of concentrate. The concentrate was shipped in DM&IR ore cars to the Extaca plant at Virginia, Minnesota, for agglomeration into egg-shaped nodules and sintered for movement to docks at Duluth and Two Harbors. Twelve years would elapse before large-scale production would be started and it would take three tons of crude taconite to produce one ton of pellets. Unlike natural ore, taconite was an extremely hard rock and required blasting before it could be hauled to the processing plants.

Tax laws in Minnesota, until 1964, taxed the mining companies with an "ad valorem" tax that taxed unmined natural ore. Taconite would fall under this law, and an intensive lobbying effort caused the Taconite Amendment"to be passed on November 3, 1964, shifting the law to a production-based tax. Investment in large-scale taconite plants was now possible.

Because taconite is handled on a year-round basis land storage facilities were built at Duluth in 1966 (Lakehead storage facility) and Two Harbors in 1977. These facilities allowed the stacking of ore pellets on the ground during the winter months when the Great Lakes were frozen over. These pellets were then reclaimed and shipped during the shipping season.

Besides the on-line taconite plants, the DM&IR during the 1970s received all-rail taconite trains loaded in DM&IR ore cars from Erie Mining during the winter months for movement to Bethlehem Steel plants at Burns Harbor, Indiana, and Johnstown, Pennsylvania. Prior to completion of Reserve Mining's own harbor facilities, taconite pellets were shipped by Reserve from Silver Bay, Minnesota to DM&IR Two Harbors ore docks.

MINNTAC

U.S. Steel constructed their taconite pellet plant, called Minntac, at Mt. Iron, Minnesota, in 1966 and 1967. The first train of taconite pellets departed Minntac for the DM&IR Duluth ore docks on October 25, 1967. By 1976, two trains per day were being operated from Two Harbors to Minntac and return, and one train from Proctor to Minntac and return. In 1978, after two expansion projects, Minntac produced 18.5 million tons of taconite pellets.

When an extra train was required it normally operated out of Proctor. Crude haulage of taconite at Minntac was accomplished by rail haulage in air-dump cars using a variety of EMD SW7, SW9, TR6, SW1200, SW1500 and Alco RS2, RS3, engines. After natural ore played out, Minntac became the largest shipper on DM&IR. Operating 24/7, during winter months pellets were stockpiled at Two Harbors for shipment in the spring. Pellets were also shipped all-rail to various other U.S. Steel mills.

During 1976, Minntac received 14 MP-15DC engines from EMD and several hundred new Difco 50-yard side-dump cars. All were built for the expanding pit operations hauling crude taconite from pit to crusher. Minntac continued to operate their EMD switcher fleet and a small number of RS3 Alco locomotives until rail haulage of taconite at Minntac was discontinued in late 1999.

Beginning in the 1980s Minntac shipped pellets all-rail to Ohio, Pennsylvania, and the Chicago area during the winter months and Birmingham, Alabama, on a year-round basis. The longest of these all-rail moves would begin in 1994 when pellets moved all-rail to Geneva Steel in Geneva, Utah via a DMIR-WC-SP routing.

Minntac began experimenting with making

U.S. Steel would build their first full-size taconite plant on the Mesabi iron range in 1967, with the first train of taconite pellets loaded on October 25, 1967. SD18s 176 and 174 are departing the plant with the first train. The pine tree atop the first car was a traditional indication of the opening of a new mine on the iron range. The last shipment when a mine closed had a vertically mounted broom to indicate a clean sweep. (Basgen photo, D.P. Holbrook collection)

fluxed pellets in 1985. Fluxed pellets contained a 50/50 mixture of dolomite and limestone that assisted blast furnace operators in making steel. Full-scale production of fluxed pellets began in July 1988. This 50/50 mix was loaded adjacent to the Duluth ore docks and brought to Minntac in twice daily 50- to 55-car trains using ore cars between April and October. Fluxstone, susceptible to freezing in ore cars, was hauled to the plant in two 35-car trains using air dumps between November and March.

EVELETH TACONITE CO.

Eveleth Taconite Co. was incorporated by a consortium of Ford Motor Co. and Oglebay Norton & Co. in 1964 and 1965. On December 10, 1965, Eveleth Taconite Co. began operations of the Fairlane Pellet plant at Forbes, Minnesota and the T-Bird mine at Eveleth, Minnesota. The first shipment of crude ore from T-Bird was made on November 1st and the first shipment of pellets was loaded on December 16 and 17. It was no coincidence that the pellet plant, Fairlane, and the mine T-Bird were named for automobiles manufactured by Ford Motor Co. In 1976, the T-Bird South mine was added and the old T-Bird mine renamed T-Bird North. During 1976 the Thunderbird (T-Bird) north and south mines produced 8,015,386 tons of crude taconite and the Fairlane pellet plant produced 2,010,846 tons of taconite pellets. Crude taconite was rail hauled by the DM&IR in 85-car trains from both the T-Bird North Mine located between Shelton Junction and Rainy Junction yard and the T-Bird South Mine located at Spruce to the Fairlane Pellet Plant at Forbes. The first pit blast at T-Bird South was on April 29, 1976, and the initial crude taconite was processed on

Taconite plants were large industrial complexes that required crushing, grinding, kilns and huge amounts of property to construct. Shown is Eveleth Taconite Fairlane Pellet Plant at Forbes, MN on October 25, 1990, constructed in 1964 and 1965. Ford Motor Co. was one of the owners so it was not a coincidence when automobile monikers were used, with the pellet plant was named "Fairlane" and the crude taconite mine loadout "T-Bird." To the far left of the photo are the taconite pellet train loadout silos. (S.D. Lorenz)

T-Bird North loading pocket on September 28, 2001. Crude taconite would be moved from open pits to crusher and then via the conveyor moved into the dome silo. Below the silo was a concrete-lined tunnel with a loading chute. Crude taconite was shipped via rail from T-Bird North and T-Bird South loading pockets to the Fairlane pellet plant located at Forbes, MN. Standard cupola cabooses would not clear loading chutes, requiring DM&IR to specially equip a number of cabooses with side or bay-windows for use on T-Bird trains. (D.P. Holbrook)

Inland Steel Minorca Pellet Plant at Virginia, Minnesota, October 1990. Limestone in center of photo is used for making "fluxed" pellets. The taconite pellet loading pocket is in the right center of photo. (E.M. DeRouin photo)

May 13th. The first regular T-Bird train was operated from T-Bird South to the Fairlane plant on July 13, 1976.

Beginning in 1974 development and construction was begun to expand Eveleth Taconite operations. The expansion was owned 40 percent by Armco Steel Corp., 23.5 percent by Steel Co. of Canada, 16 percent by Dominion Foundries and Steel Ltd. and 20.5 percent by Oglebay Norton. Construction was completed in 1977 and annual production increased to six million tons. All-time high pellet production of 508 million tons was reached in 1981 but would decline to 2.8 million tons in 1985 after Dominion Foundry sold off its share in 1983. The T-Bird South mine would close in October 1992 after crude taconite in this area played out. The majority of Eveleth Taconite pellets were handled by DM&IR trains operating from Proctor to Forbes and return and shipped through the port of Duluth. The economic downturn in the steel industry caused the plant to be shut down in 2003 with the final pellet train departing on May 15th. However, after the CN takeover the plant would reopen under new ownership.

MINORCA

On October 17, 1974, ground was broken on the Inland Steel Co.'s property north of Virginia, Minnesota, for the Minorca Taconite plant which when completed in 1977 had an annual capacity of 2,600,000 tons of pellets. Once completed, DM&IR trains operated from Two Harbors to Minorca and return. Six miles of this trip was over DW&P trackage rights from Shelton Junction to Minorca Junction. The first train of 112 cars was loaded at Minorca on June 9, 1977.

Minorca began making fluxed pellets on August 30, 1987. Pellet trains destined for loading at Minorca would have a 40- to 48-ore car

-continued on page 273

Mesabi Range Mines Serviced By GN/DMIR, 1951

Name	Natural Ore Shipped		Tons	Location	Railroad	Active-Inactive / Type - Ships To:	(Last) Operator	Notes
Aad	1904	1928	1,026,890	Eveleth	DMIR	INACTIVE-UG-TH	Hjalmer Jacobson	
Adriatic	1906	1918	1,167,731	Mesaba	DMIR	INACTIVE-UG-TH	OIM	
Agnew	1902	1967	7,022,665	Hibbing	GN	UG-ALL	CCI	
Agenw #2	1951	1974	4,918,050	Hibbing	GN	OP-ALL	Hanna	
Ajax	1899	1953	357,624	Biwabik	DMIR	INACTIVE-OP-DUL	Skubic Brothers	
Albany	1903	1965	15,474,906	Hibbing	DMIR/GN	OP-TH/DUL/ALL	PM	S(P)-WP(P)
Alberta	1907	1913	138,361	Virginia	DMIR	INACTIVE-DUL	OIM	
Alexander	1905	1910	233,351	Keewatin	GN	INACTIVE-OP-ALL	PM	
Alexandria	1917	1955	5,231,418	Hibbing	GN	INACTIVE-UG-ALL	Hanna/OIM	
Alice	1928	1957	1,097,144	Eveleth	DMIR	OP-DUL	W. S. Moore	
Allan	1913	1914	74,946	Calumet	DMIR	INACTIVE-UG-DUL	PM	
Alpena East	1900	1968	12,162,435	Virginia	DMIR	INACTIVE-OP-DUL	Hanna	
Alpena West	1955	1960	104,567	Virginia	DMIR	NOT OPEN YET-OP-DUL		
Alworth	1950	1957	10,005,109	Chisholm	DMIR	OP-TH/DUL	OIM	S(P)
Alworth Land	1948	1978	1,780,757	Hibbing	DMIR	OP-TH/DUL	CCI	
Arcturus	1917	1981	14,329,337	Marble	DMIR	OP-DUL	OIM	
Argonne	1941	1959	4,033,450	Nashwauk	GN	OP-ALL	Hanna	WP
Arne	1965	1969	558,838	Aurora	DMIR	OP-TH	Pittsburgh Pacific	
Aromac	1942	1958	1,394,195	Keewatin	GN	INACTIVE-OP-ALL	Butler Brothers	WP
Atkins	1943	1954	3,202,814	Kinney	GN	OP-ALL	CCI	
Auburn	1894	1997	8,083,513	Virginia	DMIR	OP-TH/DUL	OIM	
Bangor	1910	1918	1,274,669	Biwabik	DMIR	INACTIVE-UG-TH	OIM	
Barbara	1942	1943	473,075	Calumet	GN	INACTIVE-OP-ALL	Hanna	WP-HD
Belgrade	1908	1923	2,101,406	Biwabik	DMIR	INACTIVE-UG-TH	PM	
Bennett	1913	1963	20,360,734	Keewatin	GN	UG/OP-ALL	PM	WP(P)
Bennett Annex	1951	1964	1,406,297	Keewatin	GN	UG/OP-ALL	PM	
Billings	1919	1925	576,514	Fraser	GN	INACTIVE-UG-ALL	Hanna	
Bingham	1940	1970	3,808,238	Taconite		INACTIVE-OP		
Biwabik	1893	1955	24,484,116	Biwabik	DMIR	OP-TH/DUL	PM	
Boeing	1922	1962	4,878,794	Hibbing	GN	INACTIVE-UG/OP	CCI	
Bradford	1902	1957	492,324	Hibbing	GN	INACTIVE-OP-ALL	Pacific Isle Mining	
Bray	1909	1962	6,485,313	Keewatin	GN	OP-ALL	Hanna	C-S-WP(P)
Bruce	1927	1950	1,837,545	Chisholm	GN	INACTIVE-UG-ALL	J&L	
Bruce Annex	1929	1937	97,243	Chisholm	GN	INACTIVE-UG	International Harvester	
Brunt	1905	1964	1,960,446	Mt. Iron	DMIR	INACTIVE-OP-DUL	Hanna	
Buckeye	1943	1957	5,977,317	Coleraine	GN	OP-ALL	Hanna	HD-WP
Burns	1918	1952	747,347	Eveleth	DMIR	INACTIVE-OP-DUL	OIM	
Burt	1895	1951	16,118,075	Hibbing	DMIR/GN	OP-DUL/TH/ALL	OIM	
Campbell A	1932	1934	70,493	Hibbing	GN	INACTIVE-OP/UG-ALL	PM	
Campbell D	1955	1964	316,092	Hibbing	GN	INACTIVE-OP/UG-ALL	PM	
Canisteo	1907	1984	40,601,825	Coleraine	DMIR/GN	OP-DUL/ALL	CCI	WP
Canton	1893	1954	7,573,411	Biwabik	DMIR	OP-DUL/TH	OIM	
Carlz #2	1952	1957	1,683,279	Keewatin	GN	OPENING	Hanna	
Carmi	1952	1977	3,013,594	Hibbing	DMIR	OP-DUL	PM	
Carol	1937	1948	1,679,744	Nashwauk	GN	OP-ALL	Hanna	WP
Carson Lake	1919	1923	5,156	Leetonia	GN	OP-DUL	PM	
Cass	1903	1955	453,201	Biwabik	DMIR	INACTIVE-UG-TH	PM	
Cavour	1911	1916	177,964	Kinney	GN	INACTIVE-UG-ALL	W. S. Moore	
Chataco	1937	1960	648,702	Chisholm	GN	INACTIVE-OP-ALL	Pacific Isle Mining	HD-WP
Chester	1915	1972	1,348,543	Chisholm	DMIR	INACTIVE-UG-DUL	OIM	
Chisholm	1901	1967	9,022,013	Chisholm	DMIR	INACTIVE-UG-DUL	OIM	
Clark	1900	1925	7,030,357	Chisholm	DMIR	INACTIVE-UG-DUL	OIM	
Cloquet Annex	1943	1953	49,975	Eveleth	DMIR	INACTIVE		
Columbia	1901	1974	9,305,627	Virginia	DMIR/GN	OP-DUL/ALL(WP)	J&L	C(P)-S(P)-WP(P)
Commodore	1893	1960	8,081,382	Franklin	DMIR	OP-DUL/ALL(WP)	E. W. Coons	

Name	Natural Ore Shipped		Tons	Location	Railroad	Active-Inactive / Type - Ships To:	(Last) Operator	Notes
Coons	1940	1970	616,255	Virginia	DMIR	INACTIVE-OP-DUL	OIM	
Corsica	1901	1967	14,372,047	McKinley	DMIR	OP-TH/DUL	PM	C-WP(P)
Croxton-Syme	1902	1958	1,600,879	Chisholm	DMIR/GN	OP-DUL/ALL	Hedman Mining	WP(P)
Culver	1920	1979	3,233,305	Buhl		INACTIVE		
Cyprus	1903	1957	1,973,400	Hibbing	GN	OP-ALL(WP)	Pacific Isle Mining	S-WP
Dale	1911	1969	3,636,619	Hibbing	GN	OP-ALL(WP)	Pacific Isle Mining	WP
Danube	1919	1977	8,868,681	Bovey Coleraine	GN	OP-ALL(WP)	PM	J-WP
Day	1898	1976	9,560,645	Hibbing	DMIR/GN	OP-DUL/TH-ALL	OIM	
Deacon	1914	1919	348,912	Kinney	DMIR	INACTIVE-UG-DUL	OIM	
Dean	1915	1966	5,454,904	Kinney	GN	INACTIVE-OP/UG-ALL	Inland Steel	
Delware #1	1913	1976	8,921,974	Marble	DMIR	OP-DUL	OIM	
Delware #2	1942	1978	8,396,512	Marble	DMIR	OP-DUL	OIM	WP
Diamond	1907	1984	13,959,345	Taconite	DMIR	INACTIVE	OIM	
Dormer	1927	1968	2,064,754	Kinney	DMIR	INACTIVE-OP-DUL	OIM	
Douglas	1942	1977	5,842,079	Chisholm	DMIR/GN	OP-DUL/ALL	Hanna	WP(P)
Draper	1919	1950	304,054	Calumet	GN	INACTIVE-ALL	Hanna	WP
Draper Annex	1943	1947	2,065,157	Calumet	GN	INACTIVE-OP-ALL	Hanna	WP
Drew-Syme	1930	1957	355,017	Chisholm	DMIR/GN	OP-DUL/ALL	Hedman Mining	HD(P)-WP
Duluth	1893	1951	2,370,763	Biwabik	DMIR	OP-DUL/TH	OIM	
Duncan	1914	1970	6,610,392	Chisholm	DMIR/GN	OP-DUL/ALL	Hanna	WP(P)
Dunwoody	1917	1977	16,981,625	Chisholm	DMIR	OP-DUL	OIM	WP(P)
Eaton	1913	1913	3,548	Buhl	DMIR	INACTIVE-UG-DUL	PM	
Elba	1898	1926	3,481,872	Gilbert	DMIR	INACTIVE-UG-TH	OIM	
Elbern	1948	1960	1,257,473	Chisholm	GN	OP-ALL	Yaley-Young Mining	
Elizabeth	1910	1918	100,990	Chisholm	GN	INACTIVE-UG-ALL	Elizabeth Mining Co.	
Embarrass	1944	1977	18,624,258	Biwabik	DMIR	OP-DU/TH	PM	C
Emmett	1897	1956	539,314	McKinley	DMIR	OP-TH	Hedman Mining Co.	
Enterprise	1952	1961	6,912,744	Virginia		OPENING	Hanna	
Fay	1911	1924	1,266,676	Virginia	GN	INACTIVE-UG-ALL	OIM	
Fayal	1895	1965	37,055,594	Eveleth	DMIR	OP/UG-TH/DUL	OIM	
Fayal #1	1895	1948	7,435,924	Eveleth	DMIR	INACTIVE-UG-DUL/TH	OIM	
Fletcher	1927	1984	2,282,483	Bovey Coleraine		INACTIVE		
Forest	1904	1910	248,924	Keewatin	GN	INACTIVE-OP-ALL	Harrison Mining Co.	
Forster	1949	1979	17,200,190	Chisholm	DMIR/GN	OP-DLU/TH/ALL	OIM	S(P)-C(P)
Forsyth	1948	1956	301,245	Kinney	DMIR/GN	OP-DUL/ALL	W.S. Moore	
Fowler	1907	1922	926,196	Aurora	DMIR	INACTIVE-UG-TH	CCI	
Franklin	1893	1919	1,909,203	Franklin	DMIR	INACTIVE-UG-TH	Republic Steel Co.	
Frantz	1904	1928	744,475	Buhl	GN	INACTIVE-OP-ALL	Hanna	
Fraser	1928	1978	30,147,361	Fraser	DMIR/GN	OP/UG-DUL/TH/ALL	OIM	C(P)-WP(P)
Galbraith	1941	1955	4,755,794	Nashwauk	GN	OP-ALL	Hanna	C(P)-WP(P)
Galbraith Annex	1951	1952	153,294	Nashwauk	GN	OP-ALL	Hanna	
Genoa Fee	1896	1951	1,170,752	Gilbert	DMIR	OP-DUL/TH	E. W. Coons Co.	
Genoa-Sparta	1896	1956	10,081,500	Gilbert	DMIR	OP-DUL/TH	E. W. Coons Co.	
Gilbert	1907	1981	23,433,787	Gilbert	DMIR	OP-DUL/TH	OIM	C
Gilbert Sliver	1954	1957	77,130	Gilbert	DMIR	OP-DUL/TH	W.S. Moore	
Glen	1902	1957	13,575,115	Chisholm	DMIR	OP-DUL/TH	OIM	
Godfrey	1926	1963	12,227,759	Chisholm Hibbing	DMIR	OP/UG-DUL/TH	OIM	
Gordon	1952	1965	3,772,669	Keewatin	GN		Hanna	
Grace	1912	1919	608,491	Hibbing	GN	INACTIVE-UG-ALL	PM	
Graham	1913	1957	1,823,951	Mesaba	DMIR	INACTIVE-OP-DH	J&L	WP(P)
Grant	1902	1958	10,893,481	Buhl	GN	OP-ALL	J&L	S-C-HD-WP
Greenway	1940	1944	868,449	Grand Rapids	GN	OP-ALL	J&L	HD-WP(P) 4
Great Western	1951	1969	6,715,615	Virginia	DMIR	OP-DUL	OIM	
Gross Marble	1942	1976	8,383,055	Marble	DMIR	OP-DUL	OIM	
Halobe	1930	1959	4,139,727	Nashwauk	GN	OP-ALL	Hanna	HD

Name	Natural Ore Shipped		Tons	Location	Railroad	Active-Inactive / Type - Ships To:	(Last) Operator	Notes
Hanna-A	1909	1955	2,178,549	Mt. Iron	GN	OP-ALL	W. S. Moore Co.	
Harold	1910	1939	5,004,816	Hibbing	GN	INACTIVE-UG-ALL	PM	
Harrison	1914	1968	6,543,730	Cooley	GN	OP-ALL	Hanna	C-HD
Harrison Annex	1923	1952	267,566	Nashwauk	GN	INACTIVE-OP		
Hartley	1907	1978	8,031,087	Fraser	DMIR	OP-DUL/TH	OIM	C(P)-S(P)
Hartley-Burt	1910	1969	18,667,249	Fraser	DMIR	OP/UG-DUL/TH	OIM	C(P)-S(P)
Hawkins	1902	1968	25,224,621	Nashwauk	GN	OP-ALL	CCI	WP
Hector	1893	1953	568,280	Biwabik	DMIR	INACTIVE-UG-TH	Inter-State Iron Co.	
Helen	1924	1943	179,775	Cooley	GN	INACTIVE-OP-ALL	Hanna	
Helmer	1913	1950	1,376,507	Kinney	DMIR	INACTIVE-OP-DUL	PM	
Higgins #1	1914	1957	7,082,454	Biwabik	DMIR	INACTIVE-OP-TH	OIM	
Higgins #2	1904	1976	2,850,086	Franklin	DMIR	INACTIVE-UG-DUL	OIM	
Hill Annex	1917	1979	63,682,773	Marble Calumet	GN	OP-ALL	J&L	C(P)-S(P)-HD-WP
Hill Annex	1910	1976	15,376,668	Calumet	GN	OP-APP		
Hill-Trumbull	1910	1969	218,537	Marble	GN	OP-ALL	CCI	HD-WP
Hoadley	1906	1951	3,249,447	Nashwauk	GN	INACTIVE-OP-ALL	Hanna	J-WP
Hobart	1906	1969	1,433,041	Gilbert	GN	INACTIVE-UG-ALL	J&L	
Holland	1905	1957	420,139	Biwabik	DMIR	INACTIVE-UG-TH	Holland Mining	
Homan-Cliffs	1907	1970	24,732,085	Taconite	DMIR/GN	OP-DUL/ALL	CCI	HD-WP
Hudson	1910	1918	1,299,587	Aurora	DMIR	INACTIVE-OP-TH	PM	
Hull-Nelson	1901	1978	18,693,538	Eveleth	DMIR	OP-DUL/TH	OIM	
Hull-Rust(Hull)	1896	1983	122,252,529	Hibbing	DMIR/GN	OP-DUL/TH/ALL	OIM	
Hull-Rust(Rust)	1896	1983	88,889,991	Hibbing	DMIR/GN			
Humphrey	1909	1978	8,934,089	Chisholm Fraser	DMIR/GN	OP-DUL/TH/ALL	OIM	C(P)-S(P)
Impro A	1945	1950	5,613,946	Hibbing	DMIR/GN	OP-DUL/TH/ALL	OIM	C(P)-S(P)
Impro B Reserve	1944	1950	529,309	Hibbing	DMIR/GN	OP-DUL/TH/ALL	Hanna	C-S
Iron Chief	1942	1979	5,351,645	Buhl	GN	OP-ALL	OIM	
Iroquois	1903	1964	4,070,908	Mt. Iron	GN	INACTIVE-UG/OG-ALL	Inland Steel	
Itasca	1915	1965	1,242,611	Buhl	GN	INACTIVE-OP/UG-ALL	Inland Steel	
Jean	1916	1919	116,259	Eveleth	DMIR	INACTIVE-OP-TH	PM	
Jennings	1906	1909	213,317	Kinney	DMIR	INACTIVE-OP-TH	Hanna	
Jennison	1951	1953	866,368	Coleraine		OP-?	Hanna	
Jessie #1	1951	1962	1,391,649	Grand Rapids		OP-?	Jessie H. Mining	
Jordan	1902	1910	945,644	Chisholm	GN	INACTIVE-?-ALL	Hanna	
Judd	1913	1942	1,440,779	Bovey Coleraine	DMIR/GN	INACTIVE-OP-DUL/ALL	OIM	J-WP
Judson	1951	1961	1,459,952	Buhl		OPENING-OP	W. S. Moore Co.	
Julia	1895	1955	2,427,320	Franklin	DMIR/GN	OP-DUL/ALL	E. W. Coons Co.	WP(P)
Kerr	1916	1957	10,343,261	Hibbing	DMIR/GN	OP-DUL/ALL	Pacific Isle Mining	HD-WP 5
Kevin	1916	1968	9,496,664	Cooley	GN	OP-ALL	Hanna	
King	1952	1984	7,609,237	Coleraine	DMIR	OP-DUL/TH	OIM	7
Kinney	1903	1937	6,793,776	Kinney	GN	INACTIVE-OP-ALL	E. W. Coons Co.	
Kinney Scotch	1914	1936	55,634	Kinney		INACTIVE		
Knox	1909	1986	53,511	Aurora	DMIR	OP-TH	W. S. Moore Co.	
Knox Ext	1951	1985	1,090,142	Aurora	DMIR	OP-DUL/TH		
Kosmerl	1951	1978	7,057,874	Buhl		OPENING	OIM	
La Rue	1903	1943	7,786,598	Nashwauk	GN	INACTIVE-UG-ALL	Butler Brothers Co.	WP
Labelle	1902	1920	714,209	Gilbert	DMIR	INACTIVE-UG-TH	PM	
Lamberton	1917	1970	1,618,067	Hibbing	GN	OP-ALL	Pacific Isle Mining Co.	HD-WP
Lamberton Annex	1922	1970	28,557	Hibbing	GN	INACTIVE-OP-ALL	Pacific Isle Mining Co.	
Langdon	1929	1940	1,409,118	Cooley	GN	INACTIVE-OP-ALL	Hanna	J-WP
Langdon & Warren	1916	1956	137,885	Hibbing	GN	INACTIVE-UG-ALL	Hanna	
Larkin	1906	1948	264,948	Franklin	DMIR	OP-TH	Pacific Isle Mining Co.	WP 6
Laura	1902	1962	4,486,373	Hibbing	DMIR/GN	INACTIVE-UG-DUL/ALL	OIM	
Leach	1943	1961	1,288,603	Keewatin	GN	INACTIVE-OP-ALL		
Leetonia	1902	1938	299,440	Hibbing	GN	INACTIVE-OP-ALL	PM	
Leetonia	1902	1959	8,686,880	Hibbing	GN	INACTIVE-UG-ALL	J&L	

Name	Natural Ore Shipped		Tons	Location	Railroad	Active-Inactive / Type - Ships To:	(Last) Operator	Notes
Leonard	1903	1974	14,537,490	Chisholm	DMIR/GN	OP-DUL/ALL	OIM	
Leonard-Burt	1909	1974	6,925,913	Chisholm	DMIR	OP-DUL/TH	OIM	
Leonidas	1908	1980	23,961,466	Leonidas	DMIR	OP/UG-DUL/TH	OIM	WP(P)
Lincoln	1902	1973	8,415,859	Franklin	DMIR	INACTIVE-G-DUL	J&L	
Lind-Greenway	1940	1976	7,240,646	Grand Rapids	GN	UNDER DEVELOPMENT	J&L	
Lone Jack	1895	1962	7,742,224	Franklin	DMIR	OP-DUL/TH	OIM	S-C
Longyear	1902	1963	10,525,262	Hibbing	DMIR/GN	OP-DUL/ALL	J&L	C-S-WP
Mace #1	1910	1921	1,109,620	Keewatin	GN	INACTIVE-UG-ALL	Mace Iron Mining	
Mace #2	1916	1959	2,116,036	Nashwauk	GN	INACTIVE-OP-ALL	Hanna	WP
Mackillican	1953	1956	666,561	Nashwauk	GN	OP-ALL	CCI	
Maderia	1910	1917	196,421	Hibbing	GN	INACTIVE-UG-ALL	PM	
Madrid	1912	1922	218,919	Virginia	DMIR	INACTIVE-UG-DUL	OIM	WP(P)
Mahoning	1895	1973	108,818,262	Hibbing	DMIR/GN	OP-DUL/TH/ALL	PM	
Mahoning Grp 3	1952	1973	6,264,591	Hibbing	DMIR/GN	OP-DUL/TH/ALL		
Mahoning Grp 4	1949	1970	14,673,015	Hibbing	DMIR/GN	OP-DUL/TH/ALL	PM	
Mahoning Grp 6	1951	1960	147,007	Hibbing	DMIR/GN	OP-DUL/TH/ALL	PM	
Majorca	1912	1943	1,840,566	Calumet	GN	INACTIVE-OP-ALL	Hanna	J-WP
Malta & Malta Annex	1899	1937	1,417,824	Gilbert	DMIR/GN	INACTIVE/OP/UG-DUL/ALL	Malta Mining Co.	
Margaret	1918	1957	1,577,366	Buhl	DMIR/BN	OP-DUL/ALL	W. S. Moore Co.	WP
Mariska	1907	1963	1,445,596	Gilbert	DMIR	INACTIVE-UG-TH	W. S. Moore Co.	WP(P)
Mary Ellen	1924	1962	4,574,973	Biwabik	DMIR	OP-DUL/TH/ALL	Stanley Mining Co.	HD
Mayas	1906	1938	243,198	Mesaba	DMIR	INACTIVE-OP-TH	PM	
Mcewen	1905	1974	1,464,202	Franklin	DMIR	INACTIVE-UG-DUL/TH	OIM	
Mckinley	1907	1986	14,754,464	Biwabik	DMIR	INACTIVE-OP/UG-TH	OIM	
Meadow	1910	1963	898,239	Aurora	DMIR	INACTIVE-UG-TH	CCI	
Mesabi Chief	1913	1968	11,371,785	Keewatin	GN	OP-ALL	Hanna	C-S-WP(P)
Midget	1917	1955	152,529	Hibbing	DMIR/GN	OP-DUL/ALL	Pacific Isle Mining Co.	8
Midway	1915	1950	616,388	Kinney	DMIR	OP-DUL	OIM	
Midway #1 & #2	1953	1960	104,527	Kinney	DMIR	OP-DUL/ALL	OIM	
Miller-Mohawk	1905	1953	8,802,149	Aurora	DMIR	INACTIVE-OP-TH	PM	9
Minnewas	1893	9166	13,364,440	Virginia	DMIR	OP/UG-DUL/TH	E. A. Young	
Minorca	1902	1953	1,526,126	Franklin	DMIR/GN	OP-DUL/ALL	Pacific Isle Mining Co.	
Missabe Mountain	1893	1976	784,159,050	Franklin	DMIR/GN	OP-DUL/TH/ALL	Pacific Isle Mining Co.	10
Mississippi	1910	1930	2,779,878	Keewatin	GN	INACTIVE-UG-ALL	Hanna	
Mississippi #1	1942	1968	3,721,104	Keewatin	GN	INACTIVE-OP-ALL	Hanna	
Mississippi #2	1933	1955	1,047,625	Keewatin	GN	UG-ALL	Hanna	
Mississippi #3	1944	1965	3,740,578	Keewatin	GN	OP-ALL	Hanna	
Monica	1909	1915	469,723	Biwabik	DMIR	INACTIVE-UG-TH	Republic Iron & Steel	
Monroe-Tener	1905	1981	40,397,553	Chisholm	DMIR/GN	OP-DUL/ALL	OIM	
Moose	1926	1967	21,612,119	Franklin	DMIR	OP-DUL/TH/ALL	OIM	
Morris	1903	1965	46,088,908	Hibbing	DMIR/GN	OP/UG-DUL/TH/ALL	OIM	
Morrison	1926	1984	15,864,760	Coleraine	DMIR	OP-DUL	OIM	WP
Morrow	1913	1938	44,474	Eveleth	DMIR	INACTIVE-OP/UG-TH	W. S. Moore Co.	
Morrow Extension	1927	1929	37,649	Eveleth	DMIR	INACTIVE-OP-TH	OIM	
Morton	1912	1967	6,473,506	Hibbing	GN	OP-ALL	Hanna	
Mott	1949	1951	1,193,486	Mt. Iron	DMIR	INACTIVE-OP-DUL/TH	OIM	
Mountain Iron	1892	1956	48,664,453	Mt. Iron	DMIR	OP-DUL/TH(WP)	OIM	C-S-WP(P)
Myers	1905	1918	1,610,155	Chisholm	DMIR	INACTIVE-UG-DUL	OIM	
Nassau Mine	1907	1963	263,448	Hibbing	DMIR/GN	INACTIVE-UG-DUL/ALL	J&L	
Neville	1961	1966	380,122	Chisholm	DMIR		Hanna	
Neville R	1947	1978	1,036,392	Chisholm	DMIR		Pittsburgh Pacific	
Nordine	1951	1956	248,425	Hibbing	DMIR/GN	OP-DUL/ALL	Pacific Isle Mining	
Norman	1894	1963	7,339,064	Franklin	DMIR	OP-DUL/TH	W. S. Moore	
North Eddy	1915	1958	715,445	Hibbing	GN	INACTIVE-OP-ALL	Hanna	
North Harrison	1915	1960	8,370,887	Nashwauk	GN	OP-ALL	Hanna	HD
N Harrison Annex	1925	1943	686,047	Nashwauk	GN	INACTIVE-OP-ALL	Hanna	HD-WP

Name	Natural Ore Shipped		Tons	Location	Railroad	Active-Inactive / Type - Ships To:	(Last) Operator	Notes
North Shiras	1951	1956	196,664	Buhl		OPENED 1951-UG	Pacific Isle Mining	
North Star	1910	1970	4,386,517	Taconite				
North Uno	1910	1947	1,275,665	Hibbing	GN	INACTIVE-OP-ALL	Pacific Isle Mining	WP(P)
North Uno	1910	1963	2,339,491	Hibbing	GN	OP-ALL	Pacific Isle Mining	
Ohio	1895	1942	7,995,789	Franklin	DMIR	OP-DUL/TH	OIM	S(P)
Olson	1945	1968	5,647,942	Cooley	GN	OP-ALL	Hanna	HD
Onondaga	1907	1913	228,127	Franklin	GN	INACTIVE-UG-ALL	Malta Mining Co.	
Ordean	1916	1919	965,914	Virginia	DMIR	INACTIVE-OP-DUL	OIM	
Orwell	1929	1985	14,717,751	Taconite	DMIR	INACTIVE-OP-DUL		
Pacific	1937	1958	479,299	Aurora	DMIR	INACTIVE-OP-DUL	W. S. Moore Co.	
Palmer	1918	1921	908,390	Chisholm	DMIR	INACTIVE-OP-DUL	OIM	
Park Lots #1 & #3	1899	1917	278,718	Eveleth	DMIR	INACTIVE-UG-TH	OIM	
Patrick-Ann	1917	1968	14,147,388	Cooley	GN	OP-ALL	Hanna	C-HD(P)
Patrick Annex	1935	9168	2,353,287	Cooley	GN	OP-ALL	Hanna	HD
Pearce	1902	1960	466,817	Chisholm	GN	INACTIVE-UG-ALL	Meriden Iron Co.	
Penobscot	1897	1968	20,433,935	Hibbing	DMIR/GN	OP-DUL/TH/ALL	OIM	
Perkins	1909	1919	627,092	Aurora	DMIR	INACTIVE-OP-TH	W. S. Moore Co.	
Perkins Annex	1941	1975	149,181	Aurora	DMIR	INACTIVE-OP-TH	W. S. Moore Co.	
Perry	1949	1957	3,068,013	Keewatin	GN	OP-ALL	Hanna	
Petit	1902	1969	5,209,583	Gilbert	DMIR	OP/UG-DUL	J&L	
Pierce Group	1942	1977	32,216,577	Hibbing	DMIR/GN	OP-DUL/ALL		
Pillsbury	1898	1969	8,156,752	Hibbing	DMIR	OP-DUL/TH	OIM	
Pillsbury-Brown	1951	1978	216,776	Hibbing	DMIR/GN	OP-DUL/ALL	Pittsburgh Pacific	
Pilot	1919	1955	307,012	Mt. Iron	GN	OP-ALL	W. S. Moore Co.	
Pilot Annex	1951	1956	577,732	Mt. Iron	GN	OPENING	W. S. Moore Co.	
Plummer	1954	1966	19,477,150	Taconite	DMIR	NOT OPEN	OIM	
Pool	1903	1941	3,949,983	Hibbing	DMIR	INACTIVE-OP-DUL	PM	
Prindle	1914	1960	3,852,442	Virginia	DMIR/GN	OP-DUL/ALL	W. S. Moore Co.	HD-WP
Quinn	1914	1965	2,334,629	Nashwauk	GN	OP-ALL	Hanna	HD-WP
Rana	1915	1987	2,605,512	Kinney	DMIR	OP-DUL/TH		
Reed	1940	1941	38,617	Virginia	DMIR	INACTIVE-OP-DUL	E. W. Coons	
Rouchleau	1920	1976	52,003,748	Virginia	DMIR	OP-DUL/TH	OIM	C-S
Rouchleau Annex	1977	1986	150,366	Virginia	DMIR	OP-DUL/TH	OIM	
Roy	1944	1954	54,327	Hibbing	GN	OP-ALL	Republic Steel	
Ruddy	1911	1955	361,163	Biwabik	DMIR	INACTIVE-UG-TH	OIM	
Russell	1942	1964	3,602,011	Keewatin	GN	OP-ALL	PM	
Rust Reserve	1930	1942	357,109	Duluth	DMIR	OP-DUL	OIM	
St. Clair	1903	1978	3,906,740	Chisholm	DMIR	OP-DUL/TH	OIM	C(P)-S(P)-WP(P)
St. James	1916	1963	5,796,976	Aurora	DMIR	OP-TH	O/N	
St. Paul	1906	1964	6,825,641	Keewatin	GN	OP/UG-ALL	Republic Steel Co.	
St. Paul-Day	1923	1957	2,526,141	Keewatin	GN	OP-ALL	Republic Steel Co.	
Sally	1913	1970	3,770,829	Coleraine	DMIR/GN	INACTIVE-OP-DUL/ALL	Charleson Iron Mining Co.	
Sargent	1919	1968	9,919,437	Keewatin	GN	UG-ALL	CCI	
Sauntry	1899	1986	42,831,526	Virginia	DMIR/GN	INACTIVE-OP-DUL/TH/ALL	OIM	
Schley	1910	1969	6,214,704	Gilbert	DMIR	OP/UG-DUL	J&L	
Scott	1943	1943	35,316	Calumet	GN	OP-ALL	Hanna	
Scranton	1904	1960	24,137,096	Hibbing	DMIR/GN	OP-DUL/ALL	PM	C-S
Section 17	1912	1913	20,849	Buhl		INACTIVE-UG-?	OIM	
Section 18	1948	1955	4,103,215	Keewatin	GN	OP-ALL	Hanna	
Security	1961	1965	218,411	Eveleth	DMIR	NOT OPEN-OP-DUL	Rhude & Fryberger	
Sellers	1895	1979	91,146,628	Hibbing	DMIR/GN	OP-DUL/TH/ALL	OIM	C(P)-S(P)-WP(P)
Seville	1909	1953	271,008	Kinney	DMIR	OP-DUL/TH/ALL	Rhude & Fryberger	
Shada	1909	1957	1,585,273	Nashwauk	GN	INACTIVE-OP-ALL	Clement K. Quinn	11
Sharon	1901	1985	13,806,420	Buhl	GN	INACTIVE-UG-ALL	OIM	
Shaw	1917	1978	8,416,627	Franklin	DMIR	OP-DUL/TH/ALL	OIM	C-S
Shenango	1904	1978	17,453,926	Chisholm	DMIR/GN	OP/UG-DUL/ALL	Snyder Mining Co.	

Name	Natural Ore Shipped		Tons	Location	Railroad	Active-Inactive / Type - Ships To:	(Last) Operator	Notes
Sherman	1948	1980	32,897,376	Chisholm	DMIR	OP-DUL/TH/ALL	OIM	
Shiras	1914	1960	1,100,121	Buhl	GN	INACTIVE-OP-ALL	Hanna	
Sidney	1937	1960	3,839,657	Virginia	DMIR	OP-TH	E. W. Coons Co.	
Siphon	1907	1929	358,388	Mesaba	DMIR	INACTIVE-OP-TH	Hjalmer Jacobson	
Sliver	1908	1981	2,782,756	Virginia	DMIR	INACTIVE-OP-DUL	OIM	
Sliver	1909	1917	174,814	Virginia	DMIR	INACTIVE-OP-DUL	Hanna	
Smith	1917	1951	984,236	Hibbing	GN	OP-ALL	Pacific Isle Mining Co.	HD(P)-WP(P) 12
Snively	1950	1955	2,966,370	Mt. Iron	DMIR	OP-DUL	OIM	
Snyder	1935	1959	650,259	Cooley	GN	OP-ALL	Hanna	J-WP
South Agnew	1920	1975	13,124,080	Hibbing	GN	OP-ALL	Hanna	
South Eddy	1915	1966	3,960,456	Hibbing	GN	OP-ALL	Hanna	
South Longyear	1943	1976	3,407,462	Hibbing	GN	OP-ALL	J&L	C-D-WP
South Rust	1945	1964	10,449,915	Hibbing	DMIR/GN	OP-DUL/ALL	OIM	C(P)-S(P)-WP(P)
South Stevenson	1948	1965	343,837	Hibbing	GN	OP-ALL	Republic Steel Co.	
South Tener	1928	1981	915,344	Chisholm		OP	Snyder Mining Co.	
South Uno (Gn)	1911	1970	3,407,164	Hibbing	GN	INACTIVE-OP-ALL	Pacific Isle Mining	
South Uno (Np)	1911	1941	1,720,824	Hibbing	GN	OP-ALL	Pacific Isle Mining	12
Sparta	1897	1906	1,244,197	Gilbert	DMIR	INACTIVE-UG-TH	OIM	
Spruce	1895	1965	79,439,561	Eveleth	DMIR	OP/UG-DUL/TH	OIM	Mostly crushed
Stein	1940	1957	3,075,953	Keewatin	GN	OP-ALL	Hanna	WP
Stephens	1903	1991	51,429,490	Aurora	DMIR	INACTIVE-OP-TH	OIM	
Stevenson	1900	1957	15,425,654	Hibbing	GN	OP/UG-ALL	Republic Steel Corp	
Stevenson	1900	1912	1,317,118	Hibbing	GN	INACTIVE-OP-ALL	Corrigan McKinney Steel	
Stevenson Reserve	1944	1966	170,021	Hibbing	GN			
Stubler	1953	1957	153,464	Buhl		NOT OPEN-OP	W. S. Moore Co.	
Sullivan	1915	1948	313,249	Virginia	DMIR	INACTIVE-OP-DUL	Zontelli Bros. & Leach	
Sullivan #2	1947	1951	246,162	Cooley	GN	OP-ALL	J&L	13
Susquehanna	1906	1976	33,057,078	Hibbing	GN	OP-ALL	Republic Steel Corp	WP(P)
Sweeney	1920	1923	1,414,707	Hibbing	DMIR	INACTIVE-OP-DUL	PM	
Syracuse	1966	1968	145,554	Aurora	DMIR	OP-DUL/TH	Pittsburgh Pacific	
Thorne	1915	1948	460,448	Buhl	GN	INACTIVE-UG-ALL	Hanna	
Tioga	1916	1922	441,285	Chisholm	DMIR	INACTIVE-UG-DUL	Shenango Furnace Co.	
Tioga #2	1955	1961	3,742,671	Grand Rapids	GN	NOT OPEN YET-OP-ALL		
Troy	1903	1962	2,091,584	Eveleth	DMIR	OP-DUL	Rhude & Fryberger	
Trumbull	1910	1978	13,779,608	Marble	GN	OP-ALL	CCI	
Twin City	1942	1986	12,037,937	Chisholm	DMIR	ACTIVE-OP-DUL/TH	OIM	
Union	1900	1957	4,093,817	Franklin	DMIR	INACTIVE-OP-TH	OIM	
Utica	1902	1950	7,250,989	Hibbing	GN	OP/UG-ALL	PM	
Vernon	1924	1926	17,929	Cooley	GN	INACTIVE-OP-ALL	Hanna	
Victoria	1906	1957	713,931	Franklin	DMIR	INACTIVE-OP-TH	OIM	
Virginia	1910	1950	2,902,816	Eveleth	DMIR	INACTIVE-OP-DUL	Skubic Brothers	HM
Virginia Extension	1944	1950	1,401,775	Eveleth	DMIR	INACTIVE-OP-DUL	Snyder Mining Co.	
Vivian	1913	1947	363,611	Mesaba	DMIR	INACTIVE-OP-TH	E. W. Coons	
Wabigon	1925	1983	6,222,393	Buhl	GN	OP-ALL	Hanna	WP
Wabigon #2	1918	1977	1,846,190	Buhl	GN	INACTIVE-OP-ALL	Hanna	
Wacootah	1906	1964	8,816,741	Mt. Iron	DMIR/GN	OP-DUL/ALL	Pacific Isle Mining Co.	WP
Wade	1918	1968	3,615,479	Kinney	GN	INACTIVE-UG/OP-ALL	PM	
Walker	1909	1981	19,474,220	Bovey Coleraine	DMIR	OP-DUL	OIM	WP
Walker Ext	1955	1981	934,003	Bovey Coleraine		OP	CCI	
Walker-Hill #4	1941	1981	4,646,629	Marble	DMIR	OP-DUL	OIM	
Walker-Hill #5	1952	9178	1,349,997	Marble	DMIR			
Walker-Hill #6	1953	1969	1,911,163	Marble	DMIR			
Wanless	1914	1971	3,247,629	Buhl	DMIR	OP-DUL	CCI	
Warren	1917	1957	2,945,783	Hibbing	GN	INACTIVE-OP-ALL	PM	
Webb	1905	1964	18,099,205	Hibbing	DMIR/GN	OP/UG-DUL/ALL	Snyder Mining Co.	WP(P)
Weed	1915	1918	320,575	Aurora	DMIR	INACTIVE-UG-TH	OIM	

Name	Natural Ore Shipped		Tons	Location	Railroad	Active-Inactive / Type - Ships To:	(Last) Operator	Notes
Weggum	1915	1976	6,150,390	Hibbing	GN	OP-ALL	Hanna	C-D-WP
Wheeling	1931	1956	517,641	Mt. Iron	GN	INACTIVE-OP-ALL	Wheeling Steel Corp.	
White Iron	1909	1915	252,427	Chisholm	GN	INACTIVE-OP/UG-ALL	White Iron Lake Iron Co.	
Whiteside	1911	1979	5,051,295	Buhl	DMIR	OP-DUL	Snyder Mining Co.	
Whitney	1948	1983	2,835,176	Hibbing	DMIR	OP-DUL	Hanna	
Williams	1895	1955	530,723	Biwabik	DMIR	INACTIVE-UG/OP-DUL/TH	Hanna	
Wills	1902	1918	64,522	McKinley	DMIR	INACTIVE-UG-TH	Republic Iron & Steel	
Wisstar	1918	1960	179,778	McKinley	DMIR	INACTIVE-OP-TH	W. S. Moore	
Woodbridge	1912	1969	2,064,816	Buhl	DMIR	OP-DUL	CCI	
Wyman	1949	1957	1,401,656	Keewatin	GN	OP-ALL	Hanna	WP
Wyoming	1943	1986	2,648,287	Franklin	DMIR	INACTIVE-OP-DUL/TH	OIM	
Yates	1904	1909	679,038	Kinney	GN	INACTIVE-OP/UG-ALL	Hanna	
Yawkey	1907	1963	363,452	Franklin	DMIR	OP-TH	W. S. Moore Co.	
York	1911	1954	3,434,537	Nashwauk	GN	OP-ALL	Pacific Isle Mining Co.	WP

Vermillion Iron Range Mines Serviced By DMIR, 1951

Name	Natural Ore Shipped		Tons	Location	Railroad	Active-Inactive - Type - Ships To:	Operator	Notes
Armstrong Bay	1923	1923	4,748	Tower	DMIR	INACTIVE-UG-TH	OIM	
Chandler North	1891	1942	9,504,759	Ely	DMIR	INACTIVE-UG/OP-TH	Evergreen Mines	J(P)-WP(P)
Chandler South	1888	1957	2,396,154	Ely	DMIR	INACTIVE-UG/OP-TH	Evergreen Mines	J(P)-WP(P)
Consolidated Vermilion And Extension	1916	1920	22,893	Tower	DMIR	INACTIVE-UG-TH	Consolidated Vermilion & Extension Co.	
Mccomber	1917	1919	8,386	Tower	DMIR	INACTIVE-UG	Mutual Iron Mining Co.	
Pioneer	1888	1967	41,112,587	Ely	DMIR	UG-TH	OIM	14
Savoy	1899	1941	1,866,378	Ely	DMIR	INACTIVE-UG-TH	Evergreen Mines	
Section 30	1910	1923	1,457,295	Fall Lake	DMIR	INACTIVE-UG/OP-TH	Section Thirty Mining	
Sibley	1899	1954	9,808,202	Ely	DMIR	UG-TH	OIM	
Soudan	1894	1963	16,010,044	Tower	DMIR	UG-TH	OIM	C 15
Zenith	1892	1964	21,561,128	Ely	DMIR	UG-TH	PM	C-S

Early Taconite Operations

Name	Location	Railroad	Details	Operator
Argo (East Missabe)	Argo	DMIR	OP-TH	Reserve Mining
Erie Pilot Plant	Aurora	DMIR	OP-DUL/TH	PM

Taconite Plants, DMIR

Name	Natural Ore Shipped		Location	Railroad	Details	Operator	Notes
Extaca Plant	1951		Virginia	DMIR	NOD/SIN-DUL/TH	OIM	
Pilotac Mine	1952	1966	Mt. Iron	DMIR	OP-GRINDING	OIM	
Pilotac Plant	1952		Mt. Iron	DMIR	CONCENTRATE	OIM	1
Minntac	1967	Current	Mt. Iron	DMIR	OP-PELLETS	U.S. Steel Corp.	
Eveleth Taconite	1965	Current	Forbes	DMIR	OP-PELLETS	Oglebay Norton Co.	
Minorca	1977	Current	Virginia	DMIR	OP-PELLETS	Inland Steel Mining Co.	

Taconite Plants, GN/BN

Name	Natural Ore Shipped		Location	Railroad	Details	Operator	Notes
Hibbing Taconite	1976	Current	Hibbing	GN/BN	OP-PELLETS	Pickands Mather & Co.	
Butler Brothers	1967	1985	Nashwauk	GN/BN	OP-PELLETS	M.A. Hanna Co.	
National Taconite	1967	Current	Keewatin	GN/BN	OP-PELLETS	M.A. Hanna Co.	

Processing Plants, DMIR/GN

Name	Natural Ore Shipped		Location	Railroad	Details	Operator	Notes
Charleson Concentrator Plant				DMIR	JIGGING-DUL	Charleson Iron Mining	2
Harrison And Patrick Concentration Plants				GN	JIGGING-ALL	Hanna	3

Notes

TONNAGE SHOWN IS GRAND TOTAL AMOUNT SHIPPED DURING YEARS OF OPERATION UP TO AND INCLUDING 1986

Note 1 Ships to Extaca at Virginia, MN

Note 2 Treats stockpile ore from Missabe Mountain and Minnewas Mines

Note 3 Treats Harrison and Patrick Tailings basins

Note 4 Being developed for operation again in 1954

Note 5 HD and Washing at North Uno plant

Note 6 Low grade ore being treated at Coons Pacific plant

Note 7 Crude ore hauled by rail haulage to Trout Lake concentrating plant

Note 8 Treatment by washing and high density at North Uno plant

Note 9 Operated as Miller mine 1905-1927, Mohawk mine 1906-1923 and Miller-Mohawk 1950 onward

Note 10 Low grade ores treated at Coons Pacific plant

Note 11 Operated as Shada Mine No 1 and Shada Mine No 2 during various years

Note 12 Low grade ores treated at North Uno plant

Note 13 Crude ore trucked, stockpiled and then washed at Hill Annex Concentrator

Note 14 Lump ore produced by screening

Note 15 Entire production is "lump grade" iron ore

ABBREVIATIONS

C Crushed
D Dried
HD High density
HM Heavy Media
J Jigged
S Screened
WP Washing Plant
UG Underground
OP Open Pit
ALL Allouez, WI
DUL DMIR Duluth Dock
TH DMIR Two Harbors Dock
NOD Nodules
SIN Sinter
OIM Oliver Iron Mining
PM Pickands Mather & Co.
CCI Cliffs Iron Co.
Hanna M.A. Hanna
J&L Jones & Laughlin Steel Corp.
O/N Oglrbay Norton & Co..

-continued from page 265

block using mini-quads of fluxstone added to the head end of trains at Proctor. These would be set out at Minorca and the crew would pic up the previous day's empty fluxstone cars, add them to head end of their train and proceed to load a train of pellets for Proctor and eventual dumping at Duluth ore docks. Not equipped to handle air dumps Minorca received all their fluxstone during April to November and stockpiled on the ground what was needed for operations during the winter months. During 2001 Minorca-loaded pellet trains were shifted from Duluth ore docks to Two Harbors docks.

OTHER INDUSTRIES

MINNESOTA STEEL AT STEELTON

Completed in 1916, the Minnesota Steel Co. of U.S. Steel was located at Steelton, Minnesota on 1,832 acres 10 miles south of Duluth. Minnesota Steel Co. was leased to the USS subsidiary American Steel & Wire in 1932. U.S. Steel appears to have used the AS&W subsidiary as a place for troubled properties. It had the distinction of being the most northerly operated steel mill in the central western states. The plant was built with two blast furnaces and ten open-hearth furnaces for making steel. The first shipments of coal from boats and Duluth docks took place in August 1915. A 96-oven coke battery was located at the plant with the wire mill opening on July 5, 1922. It was reported in 1917 that it took 650,000 tons of iron ore, 500,000 tons of limestone (and 600,000 tons of coal to make 360,000 tons of coke) to make 500,000 tons of steel ingots that were equal to 360,000 tons of finished steel.

During 1931, the Great Depression caused both blast furnaces and half the open-hearth furnaces to be shut down. The Great Depression further impacted the mill when the southern blast furnace was torn down.

World War II brought about the need for increased steel production. The federal government's Defense Plant Corporation (DPC) dismantled and transported the Illinois Steel Co., Joliet, Illinois, blast furnace in 1942 and used it to re-established the second blast furnace at Steelton. The war took its toll on the furnaces and in 1949 both blast furnaces were demolished and replaced by two higher ca-

Minnesota Steel Co. constructed this large steel mill in 1916. It was leased to American Steel & Wire (AS&W), a U.S. Steel subsidiary in 1932. This aerial view was taken in the early 1950s and shows in the foreground the large ore bridge cranes straddling the limestone, coal and iron ore storage piles. Much of the barbed wire, steel fence posts and nails used in the western U.S. were manufactured here. Transportation of raw materials and finished products contributed greatly to DM&IR's bottom line. (University of Minnesota Duluth, Kathryn A. Martin Library Archives and Special Collections, S2386, Box 30 Folder 7)

pacity furnaces.

Inbound traffic was delivered by DM&IR. Iron ore came down from the iron range shipping points and was delivered from Proctor, 11 miles away, down Steelton Hill on transfer runs. Coal and limestone were unloaded from lake boats in Duluth and brought to Steelton by the Missabe Junction Hauler in both open hoppers and ore cars. A large material-handling yard was used to stockpile coal and limestone for use during the winter season when Lake Superior was frozen. A six-track iron ore thawing shed allowed ore to be shipped year-round.

DM&IR provided switching service for the raw material yard, blast furnace highline, thawing house and the four rolling mills. The remaining interplant switching was done by AS&W's own fleet of 0-6-0, 0-8-0 and 0-4-0T steam locomotives, which were later replaced by Baldwin VO-660s, VO-1000s, EMD SW-900s and a few smaller Atlas, Porter, and Plymouth 20- to 35-ton type engines. AS&W also had a fleet of flat cars, gondolas, ore cars, hopper cars and a few box cars for interplant traffic.

Outbound traffic consisted of billets, bars, steel fence posts, wire, woven wire fencing, nails and, after 1955, welded steel mats for highway construction. Some of the billets were shipped back to Duluth docks and loaded in lake boats. Beginning in the 1960s the coke plant shipped outbound coke to other U.S. Steel plants by rail. Most of this was shipped via DM&IR to Itasca, Wisconsin, for interchange to the C&NW. AS&W shipped 6,577 carloads of steel by rail in 1955. Also shipped in 1955 were 89,441 tons of steel products to Duluth for loading on lake boats. The AS&W plant began shipping 50 percent of their finished product in 1961 by truck, further reducing the shipping of finished product by rail.

Throughout its history AS&W received limestone and coal and shipped steel billets over the DM&IR limestone dock located just east of the Duluth ore docks. This movement was discontinued during mid-1963 when docks became obsolete. After the limestone dock closed, coal and limestone moved over the Duluth Dock & Transport docks and later, Hallett Docks 5 and 6. The cranes at the DM&IR-owned limestone docks were removed in April 1972.

During 1964, U.S. Steel merged seven operating divisions, this plant being renamed U.S. Steel, Duluth Works. The plant was the sole producer of steel wool and fence posts with the primary activity being wire fabrication.

The AS&W hot side was closed on November 13, 1971, and the finishing mills closed on October 1, 1973. A total of 9,022 loads of coke were anticipated in 1979 with DM&IR assigning 99 hopper cars to this service. Remaining carloads were handled in foreign ownership hoppers. The coke plant was shut down in May 1979 with final shipments of coke taking place during late 1979.

ATLAS CEMENT AT STEELTON

Atlas Cement Co. constructed a cement plant adjacent to Minnesota Steel in 1918 to use the slag by-product from the steel company to make Portland cement. Initial capacity was 4,000 barrels of cement per day. Bagged and bulk cement was loaded in box cars. One cement company reported that bulk cement shipments nationwide increased from 15,485 barrels in 1912 to 180,258 barrels in 1914. Atlas Cement had entered the cement market at a most opportune time. U.S. Steel acquired both Atlas Cement and Universal Cement Co. in 1930 and merged them into Universal Atlas Cement Co. (UAC) in 1930.

By the 1930s a number of car builders had developed covered hoppers for bulk shipment of dry bulk commodities. Between 1941 and 1956 the DM&IR acquired 40 70-ton covered hoppers to modernize bulk shipping of cement. Once the DM&IR was dieselized the UAC high-line trestle, where limestone, slag, flue dust and sand was unloaded, was restricted to SW9s until the fall of 1963. This restriction kept four SW9s on the roster until the high-line was rebuilt and SD9s were allowed on the high-line trestle.

During 1956, UAC began to ship cement in an interplant switch move from the plant to an unloading location inside the wye for transfer to truck. Facilities were completed in June 1961 for loading of trucks at UAC without the interplant move. This immediately resulted in a heavy diversion of cement traffic from rail to truck. UAC had shipped 34,600 tons of cement in 6,795 cars in 1955.

An annual gross profit statement from March 1968 indicated the following movement of cement on DM&IR. From UAC at Steelton: 49 cars to Hibbing (Panama location), 14 to Ely and 14 to Aurora. Outbound to foreign carriers: 1,432 to NP at Pokegama, 937 to GN at Saunders, 50 to Soo at Saunders, 102 to Soo at Ambridge and 433 to CNW at South Itasca. Cement received from foreign

Photographed from Ely's peak in late 1950's is the Universal Atlas Cement (UAC) plant in the foreground and in the background is American Steel & Wire (AS&W) steel mill at Steelton, Minnesota. Both of these U.S. Steel-owned operations contributed a significant amount of traffic to DM&IR's bottom line. UAC received limestone and coal from DM&IR Duluth limestone dock. Outbound shipments of bagged cement were made in box cars, and bulk cement moved in box cars and covered hoppers. Opened in 1918 the plant was closed on January 1, 1976. (D.P. Holbrook collection)

carriers: 12 from Soo at Saunders to UAC, 13 from NP at Missabe Jct. to Hibbing, 13 from NP at Endion to Aurora, Minnesota, and 15 from NP at Missabe Jct. to Minntac.

The UAC plant was closed on January 1, 1976.

LIMESTONE

Besides hauling limestone for AS&W and UAC, changes in the processing of taconite created a new need for limestone in 1988. Steel companies began to want a specialty type of taconite pellet and one of these was a "fluxed" pellet containing a small amount of a 50/50 mix of limestone and dolomite called "fluxstone." The fluxstone mixture in the pellets combined chemically with unwanted impurities at the blast furnace and carried them from the blast furnace as slag. Initially all the dolomite and limestone was unloaded at Hallett Dock 5 adjacent to DM&IR Dock 6 in Duluth. Mixing of the two was accomplished by alternating end-loader bucket loads from the piles of each material. Fluxstone had a tendency to freeze in the winter which caused the need to use side-dump cars during the winter months but allowed use of ore cars for summer shipments. During July 1995 both Minntac and Minorca approached DM&IR about handling it directly through their Lakehead Storage Area. This was ac-

Next to the ore docks in Duluth was the limestone dock shown in the foreground and the company coal docks in the far background. Steel billets shipped from American Steel & Wire at Steelton are being loaded into the 414-foot long boat *Clifford F. Hood* built in 1902 and operated by American Steel & Wire Division of U.S. Steel during the mid-1950s. Behind it is the 580-foot long boat *Eugene J. Buffington,* built in 1909 and operated by the Pittsburgh Steamship Division of U.S. Steel, unloading a load of limestone. The bottom left foreground shows limestone loaded in ore cars on the center track and the rear track is spotted with Class Q triple hoppers for coal or limestone loading. (Wes Harkins)

complished by installing an unloading hopper on the south side of Dock 5 that lake boats could self-unload the aggregates which were transferred to the storage area by conveyor. DM&IR did not win the contract for handling the fluxstone for Minorca until 2001. Quickly, fluxstone, with almost 1.5 million tons handled each year, became second only to pellets in tonnage hauled by the Missabe.

LOGGING

Logging railroads spread across Northeastern Minnesota from 1886 through 1916. Most of these railroads used 20-ton capacity 20-foot skeleton log flat cars. D&IR connected with a number of private logging railroads including: Duluth & Northern Minnesota (Knife River, Britton and York), N.B. Shank Co.(Biwabik),

Duluth & Northeastern (Hornby), Tower Lumber Co. (Murray), and Swallow & Hopkins (Ely). Many of these railroads had their own mills, but D&NM interchanged a significant part of their logs with D&IR at Knife River. A large number of finishing mills were located in Duluth, including: Scott-Graff, Red Cliff, Virginia and Rainy Lake, Mitchell & McClure (later Alger-Smith).

The importance of logging is represented in the following D&IR statistics.

Logs handled in regular freight trains:
Dec. 1901 – 397 empties and 623 loads
Feb. 1902 – 1264 empties and 895 loads
Mar. 1902 – 1292 empties and 1246 loads
Logs handled on log trains
Dec. 1901 – 6525 empties and 6277 loads
Feb. 1902 – 9712 empties and 9349 loads
Mar. 1902 – 9030 empties and 8690 loads.

The logging industry in Minnesota reached its peak in 1905.

During 1910 Duluth's six large sawmills cut almost 1.5 million board feet of lumber. By 1915, this number had fallen below 100 million board feet and in 1925, with the closing of the Scott-Graff mill, the lumber logging and milling era was all but over.

Loading logs on D&IR skeleton log flats with a McGiffert loader. Photo circa 1910. (D.P. Holbrook collection)

Sister road DM&N also hauled lumber from the vast timberland of Northern Minnesota. During the winter season from December 1 to April 30, 1904 lumber car loads amounted to: 5,626 of logs, 2,581 of mining timbers, 3,188 of pulpwood, 119 of cordwood, 18 of ship knees, 345 of ties and 161 of posts and pilings.

Much of the mining timbers shipped by mining companies were used by the same company at their underground mines.

PULPWOOD

Old-growth timber harvesting began to decline and by 1925 the last of the big sawmills were closing down. A secondary source of lumber business was found in hauling pulp-

Logs Shipped on D&IR
November 1901 through April 30, 1902

Location	Company	Board Feet
Knife River	Alger Smith	43,875,000
Britton	Scott Graff & Co	10,583,000
Drummond	Scott Graff & Co	785,000
Mile 37	Scott Graff & Co	5,890,000
York	Scott Graff & Co	11,054,000
Mile 39	Mitchel McClure	15,319,000
Highland	Red Cliff Co.	18,125,000
Cloquet River	LeSure Co.	13,000,000
Mile 61	LeSure Co.	1,800,000
Rivers	LeSure Co	2,100,000
Highland	Duncan Brewer Co.	6,675,000
Mile 82	Duncan Brewer Co.	24,000
Mile 86	Duncan Brewer Co.	300,000
Mile 88	Duncan Brewer Co.	3,627,000
Mile 90	Duncan Brewer Co.	55,000
Colby	Duncan Brewer Co.	169,000
Mile 44	Duluth Log Co.	211,000
Mile 63	Morgan Co.	587,000
Bassett	J. O'Brian	273,000
Biwabik	Alger Smith Co.	169,000
Various Points	Lunz's Mill	18,000
Various Points	Nolan Bros. Lumber	1,882,000
Murray	Tower Lumber	13,063,000
Total		137,884,000

wood. Pulpwood was initially obtained from hemlock and spruce which later included poplar and aspen-type trees. The wood fiber from these trees was made into the pulp, called "wood pulp," used in paper manufacturing. All of these were fast-growing trees. World War II brought about a rapid growth in the need for paper products, especially cardboard cartons to replace wood containers. The decline of sawed logs and sawmills also saw the creation of paper plants in northern Wisconsin and Minnesota that required the pulpwood.

Initially pulpwood was loaded at almost every station and on old logging spurs. To streamline the operation DM&IR began to develop log landings at towns that allowed multiple pulpwood cutters to pile their pulpwood on the ground adjacent to the sidings and if needed, store it on DM&IR-owned land until cars were available for loading.

The Wales Branch, constructed in 1917, was built into the rich timberland north of Two Harbors from a junction at Wales with the main line, 18 miles to Whyte. During 1947-48 a 32-mile extension was built from Whyte to Forest Center. The logging village at Forest Center was operated by Tomahawk Timber Co. which had secured timber rights on 150,000 acres of timber in the Superior National Forest during the 1940s. Log landings were located at Jordan, Whyte, Jaycee, Isabella, Sawbill Landing and Forest Center to allow loading of the pulpwood. Much of the pulpwood loaded here was sent to pa-

ogs Shipped on DM&N
ipped in 1903 up to April 1, 1903

rom	To	By	Cars	Board Feet
logan	Alborn	Northern Lumber Co.	3245	18,544,410
rand ake	Adolph	Books Scanlon Lbr. Co.	52	279,964
latch	Adolph	Brook Scanlon Lbr. Co.	26	127,400
dolph	Duluth	Merrill Ring Lbr. Co.	34	174,927
.Clarke	Duluth	Alger-Smith Lbr. Co	15	62,055
parta	Duluth	John O'Brien & Co.	109	559,745
ine	Duluth	D.B. Barber	34	160,100
e Suer	Duluth	Le Suer Lumber Co	80	395,584
latch	Duluth	J.W. Fee	2	9,345
.S. Road	Duluth	Duluth Log Co.	5	27,727
aginaw	Duluth	John O'Brien & Co.	107	630,018
lorrell	Duluth	Union Match Co.	41	198,827
.S. Road	Duluth	Price Lumber Co.	2	10,636
ilian	Duluth	Duluth Log Co.	5	24,664
ine	Duluth	John O'Brien & Co.	8	38,609
.S. Road	Duluth	John O'Brien & Co.	2	9,391
ayne	Duluth	Alger-Smith & Co.	5	26,409
lborn	Duluth	Duluth Log Co.	9	42,045
lborn	Adolph	Brooks Scanlon Lbr.Co.	15	70,409

Mining Timbers shipped on DM&N in 1903 up until April 1, 1903

From	To	By	Cars	Board Feet
Maxwell	Eveleth	Minnesota Iron Co.	454	1,816,000
Maxwell	Hibbing	Minnesota Iron Co.	39	156,000
Matthews	Hibbing	Minnesota Iron Co.	263	1,052,000
Morrell	Virginia	Interstate Iron Co.	40	166,865
Morrell	Hibbing	Minnesota Iron Co.	1015	4,060,000
Morrell	Eveleth	Minnesota Iron Co.	796	2,824,000
Morrell	Mt. Iron	Oliver Iron Mining Co.	2	8,000
Eveleth	Parmelee	Fayall Mine	15	60,000
Eveleth	Hibbing	Minnesota Iron Co.	172	688,000
Eveleth	Virginia	Republic Iron & Steel	11	44,000
Le Suer	Biwabik	Pitt Mine	6	28,000
Morrell	Parmelee	Fayal Mine	63	252,000
Iron Jct.	Virginia	Interstate Iron Co.	7	31,030
Jones	Eveleth	Troy Mine	50	250,000
Jones	Parmelee	Minnesota Iron Co.	3	10,000
Zim	Virginia	Interstate Iron Co.	3	12,000
Wolf	Virginia	Interstate Iron Co.	1	4,000
Virginia	Eveleth	Minnesota Iron Co	6	24,000
Payne	Eveleth	Minnesota Iron Co.	7	28,000
TOTAL			6649	32,926,180

Second to iron ore in tonnage, lumber and later pulpwood would remain a staple of DM&IR operations until the late 1980s. The Wales Branch north and east of Two Harbors was built specifically to tap the timber resources. Located at the end of the branch was Sawbill Landing, shown in 1961. (Basgen photo)

per plants in Northern Wisconsin including Wausau, Rhinelander, Tomahawk, Nekoosa and Wisconsin Rapids. Pulpwood was also loaded at almost every town across the iron ranges. Pulpwood was shipped to St. Regis Paper at Sartell, Minnesota, Blandin Paper at Grand Rapids and via interchange to D&NE at Saginaw, Minnesota to the large Northwest Paper mill at Cloquet, Minnesota.

Shipments off the Wales Branch started to decline in 1965 with the shut-down of the Tomahawk Timber sawmill at Forest Center. Congress sealed the fate of the Wales Branch in 1978 with the establishment of the Boundary Waters Canoe Area Wilderness which would control future logging in this area. The Wales Branch was pulled up to Norshor Junction, the connection with Reserve Mining, in June 1983 ending most of the pulpwood shipments off the branch. Pulpwood shipments began to decline on the rest of the system during the 1970s and by the late 1980s most shipments were being made locally by truck.

BENTONITE

Development of the taconite industry in Minnesota was spurred along by the continued reduction of available natural ore. As the natural ore reserves began to be depleted geologists and mining companies began to look for an alternative ore.

The first attempt at mining taconite was in 1870 by Peter Mitchell near Babbitt, Minnesota. Numerous attempts, some successful, were used to process taconite into a high iron content product, but because of the huge supply of natural ore, taconite was only a small player. This all changed in 1951.

Prior to 1951, many attempts at beneficiation of taconite were attempted. Most were failures because once the impurities in taconite were removed, only a high-iron content dust was left over. A binder was needed that could be added to the taconite to create a shippable product that blast furnaces could use. E.W. Davis, "The Father of Taconite," had experimented with the processing of taconite at the University of Minnesota Mines Experiment Station beginning in 1918. The continued breakage of the "pellets" was solved by Percy L. Steffensen of Bethlehem Steel. Mr. Steffensen found that by adding small amounts of bentonite clay to a balling drum feed succeeded in making the taconite balls tougher. It is unclear why adding 15 to 20 pounds, per ton of taconite concentrate, would make the balls able to withstand rougher treatment without breakage. Another benefit of the bentonite and taconite mixture was that damp balls could be heated to a high temperature quickly without exploding or breaking apart.

Bentonite clay is found in the following areas in the western United States: Bell Fourche, South Dakota, and Colony, Wyoming area, Upton and Osabe, Wyoming area about 50 miles directly south, Kacyee and Casper, Wymong area, Lovell, Greybull and Worland, Wyoming area and two different areas in Montana near Malta and Vananda.

Strip mine development of some of these areas began as early as 1950. Besides being used for taconite processing, bentonite is also used for: oil well drilling mud, cosmetics, kitty litter, and a binding agent in foundry sand molds and as a binder for animal feed pellets. Bentonite is named for the one-time Fort Benton, now Benton, geological stratum located in the eastern Wyoming Rock Creek area. Montana bentonite has higher silica content and this would play a significant part in later taconite pellet production.

The development of taconite reserves on the Iron Range went hand in hand with the need for a continuous supply of bentonite clay. Seizing this opportunity, Hallett Minerals was organized during the spring of 1964 to mine, market and process bentonite clay. At that time, Reserve Mining and Erie Mining were producing pellets and both Eveleth Taconite and Minntac taconite plants were under development. The need for processed bentonite was at hand.

From 1964 to 1965 Hallett Minerals acquired almost 6,000 acres—about 9 square miles—in an area two to five miles south of Vananda, Montana. Initial estimates showed enough bentonite was available for about 10 years. Besides developing the strip mine, Hallett installed a field dryer and railroad siding on the Milwaukee Road main line at Vananda. The bentonite was strip mined and then processed in the field dryer to reduce the moisture content from 28 percent to as low as 15 percent. Once this was done the bentonite was screened to about 1-inch size and then loaded into railroad hoppers cars.

During December 1964, R.N. McGiffert of Hallett Dock Co. inquired with the DM&IR about leasing property near Burnett, Minnesota, for use as a possible construction site for a bentonite processing plant. This inquiry was forwarded to the DM&IR Engineering Department so that lease costs could be developed for the land involved.

Construction of a processing plant at Burnett, with an annual capacity of 150,000 tons of finished bentonite, was begun in 1965. Burnett is located 25 miles north of Duluth on County Highway 7 alongside the DM&IR Proctor-to-Virginia, Minnesota main line. The site for the processing plant was chosen for its central location to proposed and under-construction taconite plants.

At the July 19, 1967, DM&IR operations meeting held at Gilbert, Minnesota., R.J. Hoch, DM&IR manager of sales, reported that: "An attempt is being made to establish a rate on semi-processed clay from Wyoming and Montana for the Hallett Minerals Co. proposed plant at Burnett, Minnesota. Complications have arisen due to the fact that origin carriers (MILW) refuse to establish a rate lower than the one already existing for a move to Chicago."

At the September 22, 1965, DM&IR op-

Hallett Minerals bentonite processing plant at Burnett, Minnesota late 1967. (MRHS collection)

erations meeting held at the Wolvin Building in Duluth, J.P. Keeney DM&IR manager of pricing, reported that: "Arrangements have been completed with the Milwaukee Road for establishment of a $6.50 per ton rate (predicated on the movement of 100,000 tons annually in multiples of 10 cars) on semi-processed clay from Montana to the Hallett Minerals Co. plant at Burnett. DM&IR division of this rate will be 10 percent and cars will be received at Duluth. Rates for movement of clay from Burnett to taconite companies have not been established."

At the December 29, 1965. meeting at the Wolvin Building, R.J. Hoch was reporting on clay and bentonite movements: "Tonnages continue to increase due to increased plant capacity at both Reserve and Erie Mining companies. The Eveleth Taconite Co. will receive 850 tons of bentonite in 1965; their annual consumption of bentonite will be 17,500 tons when the pellet production rate reaches the 1.6 million ton-per-year level. Estimated tonnage for the Hallett Minerals Co. bentonite processing plant at Burnett is 100,000 tons of crude clay inbound and 80,000 tons of processed bentonite outbound to taconite plants on the range."

At the August 1, 1966, operations meeting at Gilbert, Mr. J. P. Keeney manager of sales reported that: "Hallett Minerals Co. expects to be in production by September 1; outbound rates are being established. Another major construction firm is interested in locating a bentonite processing plant on the Great Northern line near Hibbing. Our Traffic and Law Departments protested in Minneapolis a motor carrier application by William Yonkovich for handling in motor service material that had a prior rail movement. GN has been working with the motor carrier. This movement, however, would be restricted to a 20-mile radius of Chisholm. The reason for GN support of the motor application appears to tie in with the establishment of this new bentonite plant." To this author's knowledge this proposed plant near Hibbing was never constructed.

The first shipment of unprocessed bentonite from Vananda, to Burnett was made in late February 1967. Bentonite was loaded into MILW 91000, 95000, 96000 series hopper cars at Vananda and interchanged to the DM&IR at Rice's Point for movement to Burnett. The first shipment of processed bentonite was made from Burnett on March 16, 1967, and delivered by DM&IR to Reserve Mining Co. railroad at Norshor Jct. for final movement to the Silver Bay, Minnesota taconite plant.

The processing of the bentonite at the Burnett plant consisted of the following: the unprocessed screened bentonite arrived in open hoppers and was bottom dumped out of the hoppers via an under-track unloader. Once dumped, a conveyor took the bentonite to a covered, 6,000-ton, storage area. The bentonite was then reclaimed with a Michigan 125 front-end loader and fed to a bucket elevator that lifted it to a 160-ton capacity surge bin. From here the bentonite was fed through a Raymond 6669 roller mill that ground and flash-dried the material at the rate of 25 tons per hour. The bentonite was now a very fine powder with the moisture content reduced from 15 to 5 percent. The finished product was then either bagged for truck shipment, or placed in one of two concrete silos having a capacity of 2,000 tons each. From the silos the bentonite powder was loaded directly into covered hoppers. To provide quality control an on-site lab was staffed by two men who tested the incoming crude bentonite. Much like taking samples of iron ore before shipping, the bentonite had eight samples per carload taken when loaded at Vananda to allow for consistent quality control of the finished product.

Taconite development on the iron range went hand in hand with the amount of bentonite handled by DM&IR. In 1963, bentonite movements on DM&IR were 176,000 tons to Erie, Reserve and Eveleth Taconite, who produced 15,896,835 tons of pellets. By 1974, this movement totaled 453,000 tons to Erie, Reserve, Minntac, and Eveleth, who produced 36,052,976 tons of pellets. Of the 453,000 tons of bentonite handled in 1974, 92,044 tons, a little more than 20 percent originated at Burnett.

A Milwaukee Road traffic study in December 1975 for the calendar year 1974, showed 2,137 open-hopper carloads of bentonite being shipped from Vananda, Montana to Duluth for interchange with the DM&IR. During the same period, the Burnett plant shipped a total of 1,104 covered hopper loads of processed bentonite.

According to Bill McGiffert, who was the general manager of the Burnett plant when it opened, it was cheaper to ship unprocessed bentonite in open hoppers to Burnett for final processing. Railroad rates were much more expensive when shipping finished products than unfinished. Unprocessed bentonite was considered unfinished product and thus was cheaper to ship from Montana to Burnett. Hallett Minerals felt that they would be able to compete with other shippers of bentonite from South Dakota and Wyoming because of the cheaper railroad rates, Hallett's higher degree of quality control, and a consistent and controlled covered hopper fleet supplied by the DM&IR. This was one of the main reasons for building the plant at Burnett.

Hallett found a good customer in Reserve Mining. Almost 80 percent of the output at Burnett in 1974 — 879 carloads — was shipped to Reserve. Bill said that Reserve found a good fit for the bentonite out of Burnett and it "worked well with their process." The remaining taconite plants received very small amounts of Burnett bentonite. "Cub" Kubis, who worked at the plant from its initial construction in 1964 until its demise in 1986, recalled that Burnett shipped bentonite by rail to the Minntac, National, Butler and Eveleth taconite plants. He also said bentonite was trucked to Inland (Minorca). Trucks were also used to ship bagged bentonite to various locations in Wisconsin.

The remaining 20 percent of the output for 1974 was 89 carloads for Erie Mining, eight carloads for Minntac, 70 carloads for Eveleth Taconite and 138 carloads for Bruce Lake/ Atikokan, Ontario plants. Bill said, "Even with the higher quality product, the other taconite plants chose to use Burnett bentonite to fill in only when they encountered a shortage and needed a quick delivery," Throughout the entire plant's history, Reserve Mining remained the largest customer.

R.C. Anderson, retired engineer in the taconite industry, recalled; "The quality of (Burnett) bentonite was on the low end of things, which resulted in higher usage. Price kept them going until the amount used became a factor. One of the first things I worked on (Eveleth Taconite at Fairlane) was to get the best bentonite quality even at a higher price because we were under pressure to keep the silica content down on finished pellets. Bentonite is 60 percent silica. The more bentonite you use, the worse the pellet metallurgical quality is."

Beside the rail shipments to Burnett, Hallett also was handling bulk shipment of bentonite over Hallett Dock 6. On September 2, 1968 the steamer *E.B.Barber* of the Algoma Central fleet took on a cargo of 9,959 net tons of crude bentonite shipped from Vananda destined to Wabush Mines at Pointe Noire, Quebec. Loading was complete on September 4, 1968.

During 1974 Hallett Minerals undertook a study to move the plant from Burnett to the Hallett Dock 6 location. Bill McGiffert said one of the reasons for considering the move was expansion of capacity and the ability to re-use the coke oven bins, at the adjacent Zenith Furnace property, as feed bins for the bentonite processing. However, the cleanup necessary at the old Zenith Furnace property negated the cost savings and expanded production opportunities, and did not take place.

Also during 1974, Hallett, with the need for additional bentonite, proceeded with further exploration in northern Montana. North of Glasgow, in Valley County, a drilling and test site was established and operated by Brazil Creek Bentonite. Bill said that "good quality bentonite was found at this location" but only a small amount was shipped to Burnett during 1975. These shipments were loaded in Burlington Northern open hoppers and shipped from Malta, ontana via Minot and Grand Forks to Duluth.

Hallett Minerals continued to operate the plant until 1981, at which time the plant was sold to Federal Bentonite. At that time, production at the plant was basically what it was during 1974, about 1,100 carloads per year. Hallett's decision to sell the plant proved to be a good one. In 1986, Reserve Mining declared bankruptcy and the plant was shuttered. The loss of this customer caused Federal Bentonite to close their plant in September 1986.

The tracks over County 7 at Burnett have been paved over and like many mines on "The Range," the processing plant has faded into history.

COAL

Coal was a major commodity transported by the DM&IR and predecessor roads. Prior to 1907 it was handled over various coal company docks at Duluth and Two Harbors. DM&N contracted in 1907 with Barnett & Record Co. to build an 1,800-foot long by 604-foot wide coal dock with three vessel-unloading towers at Duluth. After the dock was completed all "corporate" i.e. U.S. Steel, Oliver Iron Mining and DM&N coal was received over these docks. D&IR had a similar coal dock at Two Harbors, providing the same service for corporate coal.

Besides locomotive and company coal for stations, it had a number of major consumers including coal yards, lumber yards, power plants at Two Harbors, Proctor, Virginia and mining companies. It was also handled to a number of smaller communities' power plants and shipped via D&NE at Saginaw, Minnesota to Northwest Paper at Cloquet, Minnesota.

Once U.S. Steel built the large American Steel and Wire plant at Steelton, coal was hauled from Duluth docks to the plant for steel and coking purposes. After World War II the DM&IR and DW&P alternated hauling coal each year from the Duluth docks to Virginia Water and Light public utilities at Virginia, Minnesota. DM&IR hauled the coal on odd-numbered years and DW&P on even-numbered years, beginning each October.

DMIR announced on December 29, 1965, that the 50/50 sharing of coal movement to Virginia, Minnesota with DWP had been arranged to begin on April 1, 1966. This traffic had averaged about 65,000 tons annually in 1960. Coal was also hauled to Hibbing, Minnesota's public utilities power plant. During 1960 this averaged about 58,000 ton annually, divided equally between GN and DM&IR. Virginia Water and Light discontinued inbound rail shipments of coal on July 24, 1999, when the plant started using coal trucked from docks at Duluth.

Taconite plants used large amounts of electricity and the construction of Erie Mining's taconite plant in 1949 taxed the electrical power grid of the iron range. Additional taconite plants being considered in the early 1950s brought with them the need for a new power plant centrally located on the iron range. Beginning with the May 10, 1953, arrival of

DM&IR, former D&IR, coal dock at Two Harbors, in July 1957. Company coal and consumer coal both were received at this dock from lake boats. Larry cars on top moved from the dock to storage area located to the left of the photo. (D.P. Holbrook collection)

10,000 tons of eastern coal at the Pickands Mathers coal dock in Duluth, coal started to be transported between Duluth docks and Colby, Minnesota to the new Minnesota Power and Light Co. (MP&L) Laskin power plant. In 1953 they received 153,000 tons of coal.

On April 1, 1960, the DM&IR became party to a through rail-lake-rail rate on coal from Ohio and West Virginia, Minnesota to Colby, Minnesota. This coal was handled mostly in battleship Q2-, Q5- and Q7-type cars instead of ore cars. MP&L began discussions in 1968 to shift from purchasing high-sulfur Eastern coal to low-sulfur Western coal. During May 1969, coal from Peabody Coal Co.-owned Big Sky Mine in Montana was shipped in the first unit coal trains operated by Northern Pacific and Great Northern to the Clay Boswell steam-generating plant located on the GN at Cohassett, Minnesota.

Initially this was one unit train every six days. During 1969 coal started being trans-loaded at Cohassett and shipped in GN (later BN) and DM&IR hopper cars from that location to the Colby, Minnesota MP&L plant. By 1989 BN unit trains were being operated to Keenan yard for interchange to DMIR. These trains would then be split at Keenan and delivered to Colby as half a unit train. When unloaded the empties would be returned to Keenan and be reassembled into a unit train for BN crews to depart Keenan.

Both Eveleth Taconite and Minntac, for a period of time, used eastern coal at their taconite plants. Minntac in 1979 was receiving 1,165 loads of eastern coal per year. EVTAC received a 10-car test shipment on October 18, 1978, and continued receiving coal during the winter of 1978, however, freezing of the coal created difficulty in unloading and coal shipments were discontinued to EVTAC during the summer of 1979.

The DM&IR Duluth limestone dock was retired in 1963 and torn down in 1972. Coal and limestone business after this date was handled by Interlake Iron Co., Berwind Coal and other docks in Duluth. By 1972 these movements had been consolidated at Hallett Dock 5 at Duluth and Dock 6 at West Duluth.

REEFER OPERATIONS

What was an iron ore railroad doing with 75 refrigerator cars? Less than carload lot (LCL) perishable business contributed greatly to the development of the Vermillion and Mesabi iron ranges. Until improved highways began to be constructed during the 1940s, rail was the only way to get perishable traffic from Duluth to these locations. Some of the DM&N and D&IR reefers were used to load provisions for the various logging railroads operating in Northern Minnesota. These cars were interchanged off-line to Duluth & Northeastern, Duluth & Northern Minnesota and other logging carriers.

Besides shipping perishable traffic to on-line points, the DM&N had found a niche business in shipping perishables to Northern Minnesota, to points on the Canadian Northern Railway. Perishable and merchandise LCL traffic arrived by rail and highway and was loaded into reefers and box cars. DM&N shared the Northern Pacific freight house in downtown Duluth until the 1938 merger. The D&IR constructed their freight house just northeast of downtown Duluth at Endion, Minnesota. Ice was harvested or manufactured locally and loaded into the reefers via an elevated platform. As traffic declined during the 1950s an icing truck was used at Endion and stored when not in use in a vacant roundhouse stall.

Prior to completion of the Canadian Northern to Duluth on June 30, 1912, and beginning in 1908, the DM&N hauled perishable from Duluth to Virginia, Minnesota and delivered it to the Duluth, Rainy Lake & Winnipeg at Virginia, who forwarded the traffic on to the Canadian Northern at Ft. Frances, Ontario. The Duluth, Rainy Lake & Winnipeg name was changed to Duluth, Winnipeg & Pacific in 1910. After June 1912, this traffic was delivered to the Duluth, Winnipeg & Pacific at Duluth, thus continuing to earn revenue for the DM&IR

Service in 1911 was provided as follows: daily Monday-through-Friday service was provided to all points on the DM&N Alborn Branch including Coleraine, except Mt. Iron and Proctor on an overnight schedule. Cars loaded at Duluth (Endion Yard) on Monday night were spotted early the next morning. Proctor and Mt. Iron, were served on Monday, Wednesday and Friday. Mainline points between Proctor and Mt. Iron were spotted on Monday, Wednesday and Thursday.

Service to on-line points on the Canadian Northern (Duluth, Winnipeg & Pacific, DW&P) were provided on Tuesday and Thursday. Service to points in Rainy River Valley on Canadian Northern Railway (Beaudette, Pitt, Graceton, Williams, Roosevelt, Swift, Warroad and Longworth, Minnesota) was provided on Mondays.

By the early 1930s, abandonment of the logging railroads and the use of system ownership reefers by Canadian Northern, had decreased the demand for reefers supplied by the DM&N and D&IR. By 1940 only 37 reefers remained from the fleet of 75. The remaining cars were being used mostly for LCL shipments to on-line DM&IR points.

By 1958 the expense of operating the cars in LCL service was being felt. Total LCL tonnage had fallen below 12,000 tons. DM&IR management authorized a study to determine the cost savings of moving LCL operations to trucks. The study showed that there was insufficient traffic to make trucking LCL profitable. Losses continued to mount and effective Oc-

Far from home rails, express box car UP 9197 is spotted at the Virginia, Minnisota DM&IR freight house during the summer of 1960 along with DM&IR 5270 and one Class F1 5224-5260 box car. The majority of LCL was transloaded and consolidated at the main DM&IR freight house in Endion, but occasionally a foreign boxcar load of LCL had multiple stop-offs at DM&IR stations and it was not transloaded. (Jim Maki Sr.)

tober 1, 1962, instructions were issued to reduce LCL service. Perishable traffic was only to be accepted at Endion, Minnesota freight house three days a week. Non-perishable shipments would continue to be accepted five days per week, but LCL service would only be provided on Monday, Wednesday and Thursday. By October 31st, the reduced service had had the desired effect; LCL traffic was on a downward trend.

In an effort to stem the continuing losses on LCL the DM&IR began to slowly eliminate service. On December 3, 1962, LCL service to Chisholm, Minnesota was discontinued with this traffic being drayed (trucked) from Hibbing. February 1, 1963, the Biwabik, Minnesota, LCL was discontinued with this traffic being handled at Aurora. On March 1, 1963, Eveleth and Gilbert LCL service was moved to Virginia, and by the end of 1963, service to Tower, Minnesota had also been discontinued. LCL loadings in 1962 totaled 8,432 tons dropping to 6,216 tons in 1963. Even with the station closings and reduced service, the DM&IR was still losing money on LCL.

During 1964, 29 refrigerator cars remained in service. A study to use "dry ice" instead of ice to cool perishable traffic was commissioned. The cost savings were favorable and DM&IR AFE 9882 was approved on January 21, 1965, for converting all 29 cars. This was later changed to 23 refrigerator cars be converted from ice to "dry ice" by the addition of racks inside the car to hold the dry ice and insulated curtain walls. This project started on February 1, 1965, and was completed on June 18th. Seventeen reefers had racks installed on both ends and six had racks installed on one end only. The cars with racks on one end could be used for perishable on one end of the car and non-perishable traffic on the other end. The ice truck at Endion was retired during the summer of 1965.

May 1965 found the DM&IR still losing about $100,000 per year on LCL. Service again increased to five days per week for merchandise and three days per week for perishable traffic. Use of the six reefers with end racks in one end must have been successful and a study was commissioned in October 1965 to

A Interstate Branch transfer crew has cut off the empty GN ore cars sitting on the main and is going up into BN (former GN) Saunders, MN yard to the left of the photo to check condition of yard before shoving the empty cars into yard. Mixing of iron ore for specific boat cargos caused DM&IR to interchange iron ore with GN(BN) at Saunders, WI and NP/Soo at Pokegama during the ore shipping season. Empty cars would also return via these junctions. 1975 photo. (George LaPray)

study the use of "dry ice" racks in box cars. This would allow the cars to be freely used for either perishable or merchandise traffic. No known box cars were ever converted.

LCL service to Embarrass, Iron Jct. and Fairbanks, Minnesota, was discontinued during 1965. In November 1965 a meeting was held between DM&IR and the Minnesota Railroad and Warehouse Commision (MRRWC) notifying them of their intent to go to a 6,000-pound minimum for LCL shipments. MRRWC advised the DM&IR to submit their proposal.

By late 1966, the 6,000-pound minimum had all but eliminated LCL. Most LCL was shipments had been in small lots from 100 to 2,000 pounds and the 6,000-pound minimum did not allow for small shipments. The development of the highway system in Northern Minnesota and the 6,000-pound minimum put an end to LCL shipments on the DM&IR. LCL service was discontinued in 1967.

INTERCHANGE PARTNERS

Most of the iron ore was shipped through the docks at Two Harbors and Duluth, but starting in 1920s natural ore began to be interchanged to Northern Pacific and Soo Line at Pokegama, Wisconsin, and Great Northern at Saunders, Wisconsin to allow steel companies to mix ore for the correct chemical grade for their furnaces. These carriers also interchanged ore to DM&IR for mixing. All-rail ore movements to Granite City, Illinois, and eastern mills began in the early 1930s with a peak movement occurring during World War II to supplement shipping on lake boats. Lake ship-

The shortline Duluth & Northeastern connected with DM&IR at Saginaw, Minnesota. D&NE served the large Northwest Paper Co. and Wood Conversion Co. plants at Cloquet, where it connected with Great Northern and Northern Pacific. Soo and C&NW would use DM&IR to get traffic to and from D&NE at Saginaw. Much of the pulpwood used by Northwest Paper Co. originated on DM&IR. DM&IR also handled interchanged pulpwood from Duluth, Winnipeg & Pacific at Virginia, Minnesota, for NWP. D&NE 16, a 2-8-0, was purchased new by D&NE from Baldwin in 1913 and is on display in the city park in Cloquet. The train is heading north from Cloquet to Saginaw during September 1963 with a string of DM&IR rack flats on the head end. (L.A. Hastman collection)

ping came to a halt each winter and if needed steel mills would have ore shipped all-rail during the winter months.

Traffic was interchanged to:

Great Northern: Chisholm, Minnesota, Coleraine, Minnesota, Hibbing, Minnesota Virginia, Minnesota, Sauders, Wisconsin, and Duluth, Minnesota (Endion and Missabe Jct.)

Northern Pacific and CMStP&P: Pokegama, Wisconsin, Duluth, Minnesota(Endion and Missabe Jct.)

Soo Line: Ambridge, Wisconsin, Saunders, Wisconsin and Duluth, Minnesota(Endion and Missabe Jct.)

Chicago & Northwestern: South Itasca, Wisconsin and Duluth, Minnesota (Endion and Missabe Jct.)

Duluth & Northeastern: Saginaw, Minnesota

Duluth, Winnipeg & Pacific: Virginia, Minnesota, Duluth, Minnesota (Endion and Missabe Jct.)

Reserve Mining: Babbitt, Minnesota, Norshor Jct. Minnesota (not a common carrier)

Erie Mining: Emco, Minnesota (not a common carrier).

D&NE at Saginaw, Minnesota was an important interchange partner as Soo and Omaha used DM&IR to handle merchandise traffic to and from D&NE. This consisted of mostly empty box cars for paper loading and chemicals. Much of the pulpwood used by the large Northwest Paper Co. plant located on D&NE at Cloquet, Minnesota originated on DM&IR, much of it later coming from the branch to Forest Center, Minnesota. Pulpwood and wet pulp was also received from DW&P at Virginia, MN and handled by DM&IR to Saginaw for movement to Cloquet by D&NE. Coal for Northwest Paper originated at Duluth Docks served by DM&IR. ■

DM&IR 7128, former D&IR 8028, on display at Lake Superior Museum of Transportation at Duluth, on July 10, 1984. The car was restored by Proctor shop employees. Compare this photo with the in-service photo of 7128 and the built date now shows 1906. The painted door hardware was not used when car was in revenue service. (Bruce Meyer)

Various passenger cars were sold off to museum and tourist operations. Former D&IR 28 was donated to Lake Superior & Mississippi in mid-1980s and then went to Wisconsin Great Northern at Spooner, Wisconsin, and relettered for Duluth & Northern Minnesota Rwy. and named *Arrowhead.* The car is seen at Spooner; September 27, 2000. (D.P. Holbrook)

EPILOGUE

PRESERVED EQUIPMENT

Many pieces of DM&IR and predecessors D&IR and DM&N equipment have been preserved, some of it dating to the turn of the 20th century. Their preservation is a remainder of the constant winter shopping programs at Two Harbors and Proctor, extending their service lives and eventually allowing their preservation. Most of these cars are in excellent preserved condition; however, some have fallen on hard times.

Lake Superior Museum of Transportation, Duluth, Minnesota: DM&N 849 (ore car, former DM&N 4163), 68 (coach), 2124 (flat car), D&IR 6105 (gondola), 19 (coach), DM&IR C-9, C-12, C-205 (cabooses), 5124, 5132, 5380 (box cars), 7128 (refrigerator car), 33 (coach), *Minnesota II* (coach), W24 (combine), Missabe and Northland (business cars). One 3300-series box car was converted to passenger car at Iron World and then transferred to LSMT.

Lake County Historical Society, Two Harbors, Minnesota: D&IR 22 (caboose), 251(wood ore car).

Mid-Continent Railroad Museum, Baraboo, Wisconsin: D&IR 5537 (flat car), DM&IR C-74, C-183 (cabooses), 7122 (refrigerator car).

Iron World, Chisholm, Minnesota: DM&IR C-97 (caboose), 3325 (box car), 28034, 28036, 27651(steel ore cars). All three ore cars were donated on August 14, 1991, on AFE 12346. Two additional 3300-series box cars have been converted to passenger cars for the museum railroad.

Minnesota Museum of Mining, Chisholm, Minnesota: DM&IR 14536, 14613, 14954 (50-ton steel ore cars).

Cloquet Terminal Railroad, Cloquet, Minnesota: DM&IR 5200 (auto box car, on blocks), two 50-ton ore cars.

Soudan Underground Mine State Park, Soudan, Minnesota: D&IR 163 (50-ton steel ore car). Car is actually former DM&N 4727.

Tower Historical Society Railroad Museum, Tower, Minnesota: DM&IR C-73 (caboose), 81 (coach).

Wisconsin Great Northern Railroad, Spooner, Wisconsin: DM&IR 28, 32, 34 (coach), 112 (baggage/RPO). Some of these cars are not in their original as-built configurations.

Illinois Railroad Museum, Union, Illinois: DM&IR 84 (coach).

Lake Superior and Mississippi Railroad, Duluth, Minnesota: DM&IR 29, 85 (coach), W-184 and W-187 (caboose).

Lumberjack Steam Train, Laona, Wisconsin: DM&IR C-160 (caboose). ■

APPENDIX

A: GENERAL INFORMATION

In 1887 Congress passed the Interstate Commerce Act that created the Interstate Commerce Commission (ICC). Besides prohibiting railroads from unfair practices involving freight rates, it also established the ICC as a regulatory body. The reality was the ICC had very little power because of the vague language in the law. The passage of the act was the beginning of Federal regulation of the operating and business practices of the railroads.

The next regulation passed by Congress was the Safety Appliance Act in 1893. The act mandated automatic couplers on all railroad equipment used in interchange by January 1, 1898. It also mandated air brakes on a "sufficient" number of cars to allow the engineer to control the speed of the train without handbrakes being used to control speed. The act was effective January 1, 1898. The act was further amended in 1903, mandating 50-percent of cars in a train to have air brakes, and again in 1910, the ICC raised this percentage to 85 percent.

In 1910, Congress created the Safety Appliance Act, authorizing the ICC to establish railroad industry standards for sill steps, ladders, grab irons, running boards and hand brakes. The final standards were published in 1911, with full compliance mandated by 1916. The standards specified locations of safety appliances and critical dimensions. The Standing Committee on Safety Appliances of the ARA Section III Mechanical reported at their June 1919 meeting, that 100 percent of the D&IR roster of 7,151 cars had been equipped as specified by the act. The DM&N roster of 9,350 cars had 27 cars left to be equipped per the act. This amounted to 0.3 percent of the roster. By the time of the meeting only 3.8 percent of the nationwide fleet of freight cars remained to be placed into compliance with the Safety Appliance Act.

STENCIL GROUPS

Stencil group letters began to be applied to ore cars in 1952. Shown in an example of stencil group "B." The number 1 was added in the 1980s for unknown reasons. (S.D. Lorenz)

Each year, the DM&IR had a shopping process that allowed for one-quarter of the ore car fleet to be thoroughly shopped during the winter shopping season. The DM&IR had the advantage of shutting down most ore operations when the Great Lakes were frozen over. This allowed the shops at Proctor and Two Harbors to perform annual maintenance during the winter months. AAR requirements called for cleaning and testing air brakes systems every three years; this was changed to four years in 1958. Until 1952 this was tracked by individual car numbers and tracking ore car air brake testing was a problem. Short of having to look at each car closely it was difficult, at a glance to determine when a car had received its last yearly maintenance. Making matters even more difficult was that ore dust covering the cars all but obscured the lettering. Starting in June 1952 it was decided to add an identifying code called a "stencil group" each year to make it easier to determine what year a car was due for periodic air brake testing. Stencil group letters were established by dividing the ore car roster into thirds. DM&IR had applied for and received an exemption in 1958 for cleaning and testing the air brakes on the brand-new cars with AC brakes every five years. This allowed switch crews to identify easily each fall when switching out empty ore cars, which cars needed to be held at Proctor for movement to the car shop during the winter shopping season. Usually, most cars were completed during the November to April/May shopping season.

When "stencil group" letters began to be applied in late 1952 the DM&IR had 5,257 50-ton cars and 8822 70-ton cars on the roster. Most of the remaining 50-ton cars were scheduled for replacement and were only used during traffic surges. This caused these cars to have, for the most part, no specific stencil group letter assigned to a group of cars and cars within one block may have any of the four A, B, C, D, stencil group letters applied. Below left is a listing of total remaining 50-ton cars and early 70-ton cars with letters known to have been applied to various cars.

The stencil groups were assigned as follows to 70-ton cars in 1970:

umbers	Class	Total Cars	A	B	C	D
0-849	U3	89		Yes	Yes	
05-10554	U3	159	Yes			
555-11706	U4	118	Yes			Yes
000-14149	U6	130				
500-15499	U9	971				Yes
000-19999	U10	943				Yes
000-21000	U11	987	Yes	Yes	Yes	Yes
001-21026	U12	26				Yes
101-21126	U13	24				Yes
001-22125	U14	125				
001-23125	U15	125	Yes			
171-23177	U16	7			Yes	
900-34399	D	1				
400-34899	D3	18	Yes		Yes	
000-39699	D4	52			Yes	
250-41399	E6	135				
500-42299	E8	739				
300-42749	E9/E10	432	Yes	Yes		
750-43249	E11	483		Yes	Yes	Yes

Note D stencil groups did not start to appears until 1958-59 shopping season.

Car Numbers	Class	Stencil Groups - Total Cars in Groups							
		A	B	C	D	X	Y	Z	Z-1
23975-24163	U17				189				
24164-24999	U17	836							
25000-25149	U18				150				
25150-25499	U18			350					
25500-25999	U19		500						
26000-26499	U20		500						
26500-26999	U21		500						
27000-27499	U22			500					
27500-27999	U23		500						
28000-28332	U24		333						
28333-28499	U24			167					
28500-28999	U25			500					
29000-29012	U26			13					
29013-29499	U26				487				
29500-30499	U27				1000				
30500-30879	U28				380				
30880-30999	U28			120					
31000-32499	U29	1500							
32500-32999	U30			500					
33000-33166	U31					167			
33167-33333	U31						167		
33334-33499	U31							166	
33334-33499	U31								167
Totals		2336	2333	2150	2206	167	167	166	167

Stencil groups did vary a little over the years. When the Class U30, 32500–32999 were delivered they were placed in stencil Group B before being placed in service and later shifted to Group C as 50-ton cars were retired and testing requirements changed. Another example was 190 Class U18 cars that had been in storage in 1969 were shifted from stencil Group C to Group A. Internal correspondence on June 30, 1958 indicates that 148 cars in Class U18 were in stencil group A; these would be shifted to stencil group C by 1970. Prior to 1970 there were other instances of shifting of stencil groups because of retirement of 50-ton cars and many 70-ton cars being in deep storage during the business downturn during the mid to late 1960s. After 1970, the cars remained in the stencil groups shown above until stencil groups were discontinued in 1982. Note that cars with accident damage or extensive repairs could have the stencil group changed, but this was an exception.

For AB equipped cars the Stencil Group testing years were:

A: 1952, 1955, 1958, 1962, 1966, 1970, 1974, 1978

B: 1953, 1956, 1959, 1963, 1967, 1971, 1975, 1979

C: 1954, 1957, 1960, 1964, 1968, 1972, 1976, 1980

D: 1961, 1965, 1969, 1973, 1977, 1981

AC brake-equipped cars were delivered in 1957. Five-year testing of AC brake-equipped cars 33000–33499 were:

X: 1961, 1966, 1971 Y: 1962, 1967, 1972

Z: 1958, 1963, 1968, 1973

Z1: 1959, 1964, 1969, 1974

Years 1960, 1965, 1970 were non-inspection years for AC-equipped cars. Each testing year shown started in May of the year shown and ended on April 30 of the following year.

Intra-company correspondence on August 25, 1958 indicated that DM&IR had been notified that AB brake testing was being changed from every three years to every four years, adding stencil Group D to cover the fourth year. At the time DM&IR owned 9,821 70-ton ore cars and 3,689 50-ton ore cars. The four-year plan called for 3,377 cars to be tested during the winter 1958-59 shop season, 3,377 during 1959-60, 3,377 during 1960-61 and 3,379 during 1961-62. The first "D" stencil group cars began to appear during the 1958-59 shopping season. For unknown reasons, a stencil group B with a small number 1 was introduced in the 1980s.

A new stencil group "E" began to appear in April 1981 for Class U23 through Class U29 cars. By Dec 31, 1981, 21 Class U24, 308 Class U25, 30 Class U26, 76 Class U27 and 8 Class U28 cars had been placed in stencil group "E". The need for the new "E" stencil group was created when FRA extended the periodic air brake testing period from four years to six years on Jan. 1, 1981 on AB, ABC-1 and ABD-1 valves. DM&IR added the "E" and "F" stencil groups effective with this change. By Oct. 1982 42 Class U24, 610 Class U25, 60 Class U26, 150 Class U27 and 16 Class U28 regular ore cars had been placed in stencil group E.

Notation on November 30, 1982 DM&IR Equipment Lists states that Stencil Group tracking was being discontinued. It does not appear any cars ever received the stencil group F.

B: PAINTING AND LETTERING

Standards for information and stenciling were established in 1901 by The Master Car Builders Association (MCBA). This included the size and location of lettering but these were often ignored. Both D&IR and DM&N deviated from the 1909 MCB standard MCB-26A which required the reporting marks on the left end of car. Both carriers placed reporting marks to the left, but the number to the right on house cars prior to 1901. This changed over time and both carriers began to paint cars with reporting marks and number to the left when facing the car. Ironically, DM&N Class Q1 4500–4699 twin open hoppers delivered in

1916 from Pullman had the reporting marks and number stenciled at the very right hand while facing the car. Starting in 1921 these began to be repainted with the reporting marks and number stenciled on the left-hand side while facing the car. Minor revisions were made over time, but these standards remained in effect until the October 20, 1920, when complete revision was made by the American Railway Association (ARA), requiring reporting marks and numbers to be displayed to the left when facing the car. DM&N again deviated from recommended practice when Class P2 3300–3399 single-sheathed box cars were delivered in 1923. The DM&N monogram and car number were placed in the left panel with reporting marks placed on the right-hand portion of the car. In 1934, the ARA became the Association of American Railroads (AAR), and with minor revisions those standards remained in effect until today. Further revisions in the 1970s changed the Clean, Oil, Test, and Stenciled (COTS) so they had to be shown in a black and white outlined box, originally with a single box, then two boxes and later three boxes to the right-hand end of every car.

Recommended ARA—later AAR—practice starting in 1918 was for horizontal 1-inch lines above the reporting marks and below the car number. D&IR and DM&N did not follow this practice. After the January 1, 1957, Mechanical Division meeting of AAR, the standards calling for 1-inch horizontal lines above the reporting marks and below the road numbers on the side of the car were eliminated. After the merger, DM&IR applied these lines until 1967 on box cars, refrigerator cars, gondolas, open hoppers and covered hoppers.

D&IR AND DM&N PAINT

Not until the advent of color photography are we able to determine paint colors, except when they were specified on builder's orders. Freight cars, other than refrigerator cars, for both D&IR and DM&N appear to have been painted either red oxide or black. Photos of D&IR equipment by 1917 show equipment being painted black and builders' orders from this time frame specified red oxide as a base color and black for the final exterior coat of paint. Refrigerator cars appear to have been painted either yellow or white. D&IR specified red oxide for the Class E10 and E11 ore cars built in 1916.

D&IR specifications for a 35-ton box car in 1917 shows cars to be painted with a base coat of Illinois Steel Co. #550 brown followed by a second coat of paint consisting of Illinois Steel Co. #290 black. A letter to ACF specified that Class E9 ore cars be painted the same as the 1917 boxcar specifications. D&IR specification for a proposed order of refrigerator cars in 1917 specified Illinois Steel Co. #950 yellow for side walls, fascia and doors. Illinois Steel Co. #550 brown for end walls, fascia and buffer blocks. Illinois Steel Co. #370 graphite black was to be used for all safety appliances, underframe, and grab irons, corner bands and other metals parts. Specification for a 50-ton gondola and a 30-ton flat car in 1917 showed it being painted with first coat of Illinois Steel Co. #370 graphite black followed by a second coat of Illinois Steel Co. #290 black. DM&N Class U4 ore cars show specifications to be painted Illinois Steel Co. #845 black. DM&N Class U10 ore cars 19000–19999 show specifications to builder for cars to be painted Illinois Steel Co. #370 graphite black.

No additional DM&N car order specification have been found, but it appears DM&N followed D&IR painting specifications once U.S. Steel took over both roads in 1901.

DULUTH & IRON RANGE LETTERING

The first Duluth & Iron Range monogram was designed in 1915. The octagonal monogram was designed by Algol Johnson who was the mechanical engineer for the D&IR at Two Harbors. The indian chief head was selected as having tourist appeal and could assist in increasing passenger traffic in the Lake Vermilion and Ely districts. This monogram, in various color schemes, was used until 1926 when the indian head within the circle was replaced with the words Duluth & Iron Range Railroad. At that time the "Get the Safety Spirit" was changed on the outside circle to "The Safety Spirit."

The first box cars were lettered with just D&IR reporting marks with RR below them to the left of the door and road numbers on the right hand side of the car. Around 1896 the "RR" below the reporting marks was dropped. The author has been unable to locate any freight cars or cabooses with the D&IR indian head monogram. About 1910, the road number was placed directly below the reporting marks on the left side.

Wooden ore cars had simple D&IR lettering on the bottom side sill with the number applied to the center of the top chord. Early on it was noted that the ore dust was making numbers almost impossible to read. Two Harbors shop began to cut out raised metal numbers that were applied at the same location. Even when weathered, the standout metal lettering was still readable.

Starting in 1905, steel ore cars from 3900 up to 12499 were delivered with "Duluth & Iron Range" spelled out on the side of the car with road number below and no reporting marks. Starting in 1916, with 12500-series steel ore cars were delivered with just D&IR reporting marks followed by periods with the road numbers below. Added in 1916 was a very small stacked "Duluth and Iron Range" on the side of the cars. Some of the steel ore cars received the octagonal D&IR logo after 1915.

DULUTH, MISSABE & NORTHERN LETTERING

The Class L and N box cars, 3000–3062, purchased between 1892 and 1899, were lettered with a small arched Duluth, Missabe and Northern to the right of the door. Starting with Class N1 boxcars 3063–3072, delivered in 1903, lettering was a large stacked Duluth, Missabe and Northern to the left of the door with the road number to the right of the door on cars with left-opening doors. Cars with right-opening doors had the lettering swapped.

The stacked lettering applied to vehicle cars 3073 and 3074 had 10-inch high DULUTH, MISSABE, NORTHERN lettering. The "AND" was 6-inch high. Separation between the lettering was 8-inch. Numbers on 3073 were 10-inch high. DM&N stencils and numbers on the door were 4-inch high with 5-inchseparation.

Class K refrigerator cars had a simple road number to the left of the door with "Summer and Winter Car" lettering on the doors. The Class J cars starting with 5003 received the arched DM&N lettering to the right of the door. Starting with Class J1 cars, they received the same lettering as the Class N box cars.

Flat cars prior to 1906 were delivered with DM&N Extended Gothic centered on the car with the number placed between the far right pair of stake pockets. About 1906 the numbers were shifted to the left and were placed to the right of the seventh stake pocket counting from the right. The characters changed from Extended Gothic to Railroad Roman about 1906. Reporting marks remained in same location. On August 18, 1937 the reporting marks were moved to right of the third stake pocket and the number to the right of the fourth stake pocket.

Wooden ore cars had simple DM&N initials on the bottom side sill with the number applied to the center of the top chord.

The first steel ore cars 8000–8004 were delivered with DM&N reporting marks centered on the carbody with the number to the right of the reporting marks. Beginning with Class U1 cars starting at 8005 cars were delivered with Duluth, Missabe, (on the top line) and (on the second line) and Northern (on the third line) stacked lettering on the side of the cars with road numbers below.

The first Duluth, Missabe and Northern round monogram was designed in 1910.

Starting in 1910 a large DM&N ball monogram was introduced and this replaced the stacked lettering. This continued on all box cars until the Class P2 3300–3399 series were delivered with a small ball monogram on the left hand of the car in 1923 because of the single sheathing on this group. The large ball monogram was used on double-sheathed box cars until the late 1920s. A number of box cars were repainted with the large DM&N round logo, replacing the stacked DM&N to the left of the door. Class U12 ore cars delivered in 1925 were the first series of ore cars painted with the 20-inch DM&N round stencil monogram, 7-inch numbers. Three-inch stenciling was used for class and capacity.

The Duluth, Missabe and Northern monogram was redesigned in 1917. A.V. Rohweder was the supervisor of safety and he suggested to F.E. House, the federal manager at the time, that the current monogram be modified using the slogan "Safety First." A new round monogram, with Safety on the top and First on the bottom of the circle encompassing the Duluth, Missabe and Northern name, was designed by Steve Raetz, mechanical engineer at Proctor. It appears that freight cars only received the DM&N ball monogram without the Safety First surrounding the ball. The color scheme was generally red and green, but sometimes black-white. It was in general use until 1937 when the D&IR and DM&N were merged into the Duluth, Missabe & Iron Range.

DULUTH, MISSABE & IRON RANGE PAINT & LETTERING

After the consolidation, George Bohannon, mechanical engineer, redesigned the monogram basically following the DM&N monogram and placing the new name in the center of the monogram. Between 1937 and early 1942 a Duluth, Missabe & Iron Range monogram was used without the "Safety First" slogan. The Safety First slogan was added in the outside circle in 1942 on new Class U19 ore cars. The monogram had various color schemes over the years, mostly green and red, but sometimes black and white and, starting in 1959, maroon and gold. Class Q5 hopper cars, 2700–2949 delivered in 1952 received a modified Safety First DM&IR monogram with much smaller Safety First, Duluth, Missabe and Iron Range lettering with closer spaced outside circles. It is unclear why the monogram was modified, but the experiment was not repeated. Initials on freight equipment began to be changed from D&IR and DM&N to DMIR on August 21, 1937. The ampersand on reporting marks was also dropped on this date.

Ore cars were painted iron oxide brown from 1925 until the Class U31 cars were delivered in 1957. Specifications on Class U23 through U30 ore car orders specified cars and trucks to be painted with "iron oxide lamp black" paint with white paint for lettering. Class U31 cars were painted black with white lettering to identify these cars as being non-interchange-approved account being equipped with experimental AC-type brakes. This was supposed to assist in identifying the cars to keep them on line, but iron ore dust and weathering quickly turned the cars brown. Ore cars repainted at Proctor from 1955 to 1962 received the iron oxide lamp black paint. During 1959 a painting program of 70-ton ore cars was accelerated with 1,536 cars completed by the end of September 1959. Management correspondence indicated the paint color as "ore car brown." This ore car repainting program

A	24"	30"	36"
B	8 3/8"	10 3/8"	12 5/8"
C	5"	6 3/4"	8 1/2"
D	1/2"	7/8"	1"
E	1/2"	11/16"	5/8"
F	2 1/8"	2 5/8"	2 3/4"
G	2 1/4"	2 1/2"	3"
H	1 1/2"	2 1/8"	2 1/4"
J	5 3/4"	7 1/2"	8 7/8"
K	3 1/4"	3 3/4"	4 1/8"
L	1/2"	11/16"	13/16"
M	7/16"	9/16"	9/16"

was terminated in 1962 account cost-cutting measures caused by the economic downturn on the Iron Range. Starting in the spring of 1966 the only ore cars fully repainted were rebuilt taconite and crude taconite cars, which were painted maroon with yellow lettering. Occasionally a returning leased car was painted by an outside shop and re-stencilled with non-standard stenciling but this was the exception.

DM&IR monogram Tower, Minnesota, August 12, 2014. (D.P. Holbrook)

The DM&IR monograms on ore cars were 24-inch diameter, 36-inch on box cars and gondolas, 30-inch on covered hoppers and cabooses.

During 1952, painting of cabooses was done every 3 to 5 years depending on type of service car had been subjected to.

Initially, DM&IR cabooses were painted yellow with black lettering and roofs. Starting in 1949, perhaps earlier, Missabe Division assigned cabooses had roofs and lettering painted maroon, and Iron Range Division assigned cabooses retained black roofs and lettering. Safety First stencil monograms were introduced with the delivery of the first extended vision cabooses in 1952. This monogram was initially located on the extended vision cupola below the windows. It was moved to below the cupola starting on Februry 24, 1960.

Increased car loadings of coke from AS&W at Steelton caused DM&IR to purchase 500 rebuilt 77-ton open hoppers, DM&IR 12000–12149 and EJ&E 74000–74349 series, from Penn Central. Cars were rebuilt at PC Samuel Rea shops with Bethlehem Steel-supplied kits from offset side hoppers originally owned by NYC and B&A. DM&IR 12105 former NYC 905298 rebuilt August 6, 1975, in Duluth. September 20, 1975. (Doug Buell)

Fifty specially designed quad hopper cars Class Q2 4700–4749 were acquired by DM&N in 1937. 250 additional cars Class Q5 2700–2949 built in 1952 by Pressed Steel Car and 100 Class Q7 2950–3049 built in 1957 by Pullman Standard were acquired to protect coal movements. DM&IR 2795 is at Virginia, Spring 1962. The Class Q5 hoppers would be the only cars acquired with this distinctive monogram. (Jim Maki Sr.)

DM&IR 12030, former B&A 910089 rebuilt July 2, 1975, one of 500 triple hoppers, DM&IR 12000–12149 and EJ&E 74000–74349 acquired in 1975. All were former B&A/NYC triple hoppers rebuilt with Bethlehem Steel supplied kits at PC Samuel Rea Shops. Butler, WI August 7, 1976. (D.P. Holbrook collection)

At least two wood cabooses, C190 and C191 received the stencil monogram when rebuilt for T-Bird service in 1965. During the winter shopping season of 1959-1960 four cabooses, C-38, C-93, C-136, C-161 were painted in a lighter shade of yellow with maroon roofs and lettering but no monograms. A limited number of cabooses assigned to Missabe Division received this paint scheme with Iron Range Division cabooses retaining the black roof and lettering. The first extended vision cabooses, C-200 through C-219 were delivered

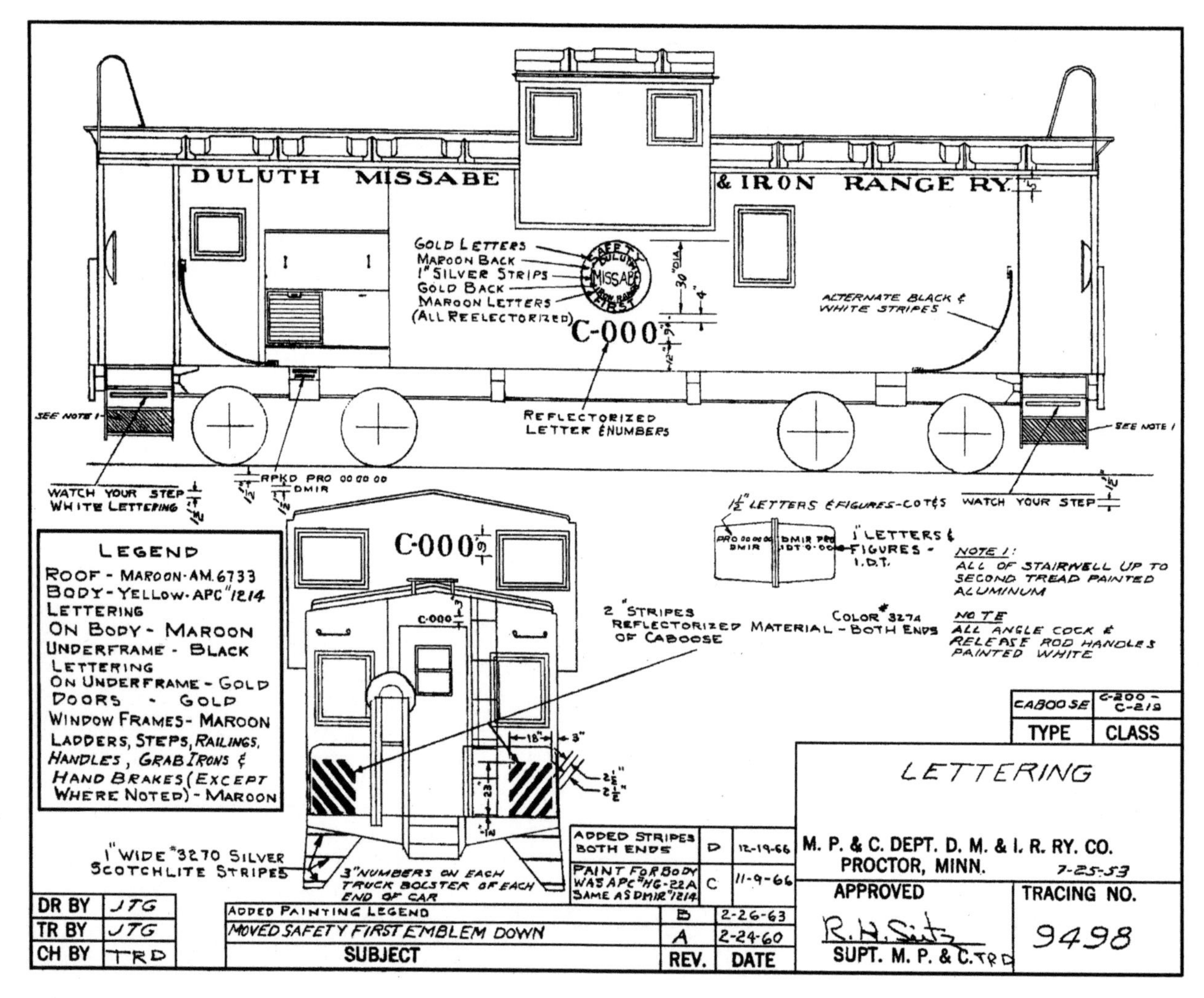

with black roofs and lettering. Starting in early 1960, all cabooses began to be repainted into a standard yellow carbody, with maroon roofs, step platforms, ladders and lettering. This paint scheme was further modified for steel cabooses only on December 19, 1966, when safety stripes for higher visibility were added on either side of the doors on both ends. Beginning in 1954, bottom step and riser of end stairways were painted silver allowing them to be easier identified at night. In late 1960, the carbody color was AP 1214 yellow with roof and lettering being AM 6733 maroon.

The car body color was changed to APC HG-22A yellow on November 9, 1966. This yellow was supposed to be the same as AP 1214 but photos show this to be a slightly more gold shade of yellow. Version 2, maroon, gold and white Scotchlite monograms were introduced in 1959 with the first application on cabooses in late 1960. Cabooses prior to this had the car number centered on the carbody. Effective with the application of the Version 2 monograms it was moved to be centered below the cupola. Initially in spring of 1960 all cabooses had window frames painted the carbody color. Two cabooses, C-161 and one other, were changed to maroon window frames in October 1960 for inspection during a business car trip. Not until late 1960s would this become a standard on newly painted cabooses.

On wood cabooses the Scotchlite monogram and road numbers were applied to a metal placard that was installed below the cupola on the side of caboose. On steel cabooses this was applied directly to the carbody below the cupola replacing the cupola stencil monogram. At least one caboose, C-209 when rebuilt from

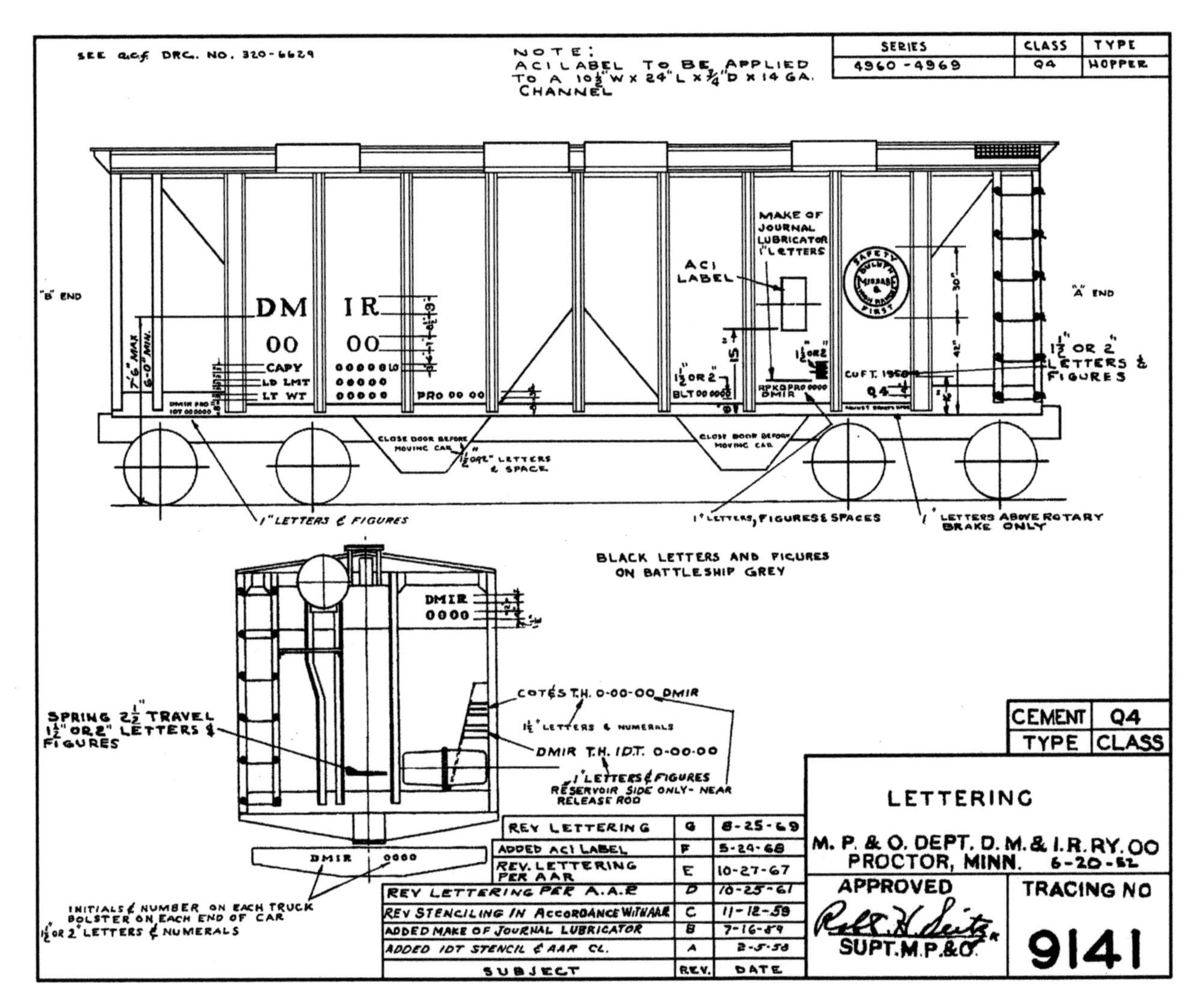

SUBJECT	REV.	DATE
REV LETTERING	G	8-25-69
ADDED ACI LABEL	F	5-24-68
REV. LETTERING PER AAR	E	10-27-67
REV LETTERING PER A.A.R	D	10-25-61
REV STENCILING IN ACCORDANCE WITH AAR	C	11-12-59
ADDED MAKE OF JOURNAL LUBRICATOR	B	7-16-59
ADDED IDT STENCIL & AAR CL.	A	2-5-58

extended vision to side-window caboose, had an original Version 1 variation of Scotchlite monogram. Version 2 with a yellow instead of gold center was introduced during 1969 and used on a few cabooses. Version 3 green and white Scotchlite monograms started to be applied to cabooses during the winter 1973-1974 shopping season. At the same time Scotchlite diagonal stripes were applied along the bottom edge of carbody. A large red Scotchlite placard was added to the ends of the cupola with the road number applied.

The "Arrowhead" logo was developed in 1968. A small version of this was applied on a few wood cabooses C-43, C-44, C-52, C-56, C-58, C-136, C-138, C-143, C-144, C-180, C-192, C-193 located where the car number was displayed and the number superimposed on the logo. Cabooses C-44 and C-180 when repainted had the car number stenciled on the sides and cupola in roman lettering and it appears to be the only cabooses so lettered. Cabooses C-43, C-55, C-58 and C-138 received gothic numbers on the Arrowhead logo but retained roman lettering on both ends of the cupola. Most, if not all, other Arrowhead logo cabooses had gothic numbers on both the side and cupola. On bay windows C-204, C-205, C-218, C-219 a small version was applied without the car number. It should be noted that some cabooses retained the black roofs and lettering until being retired.

Refrigerator cars were painted yellow with black underframes, roof, ends, trucks, grab irons and appliances. The only known exception was 7063 being painted in maroon with gold Scotchlite reporting marks, numbers and a maroon, gold and white monogram.

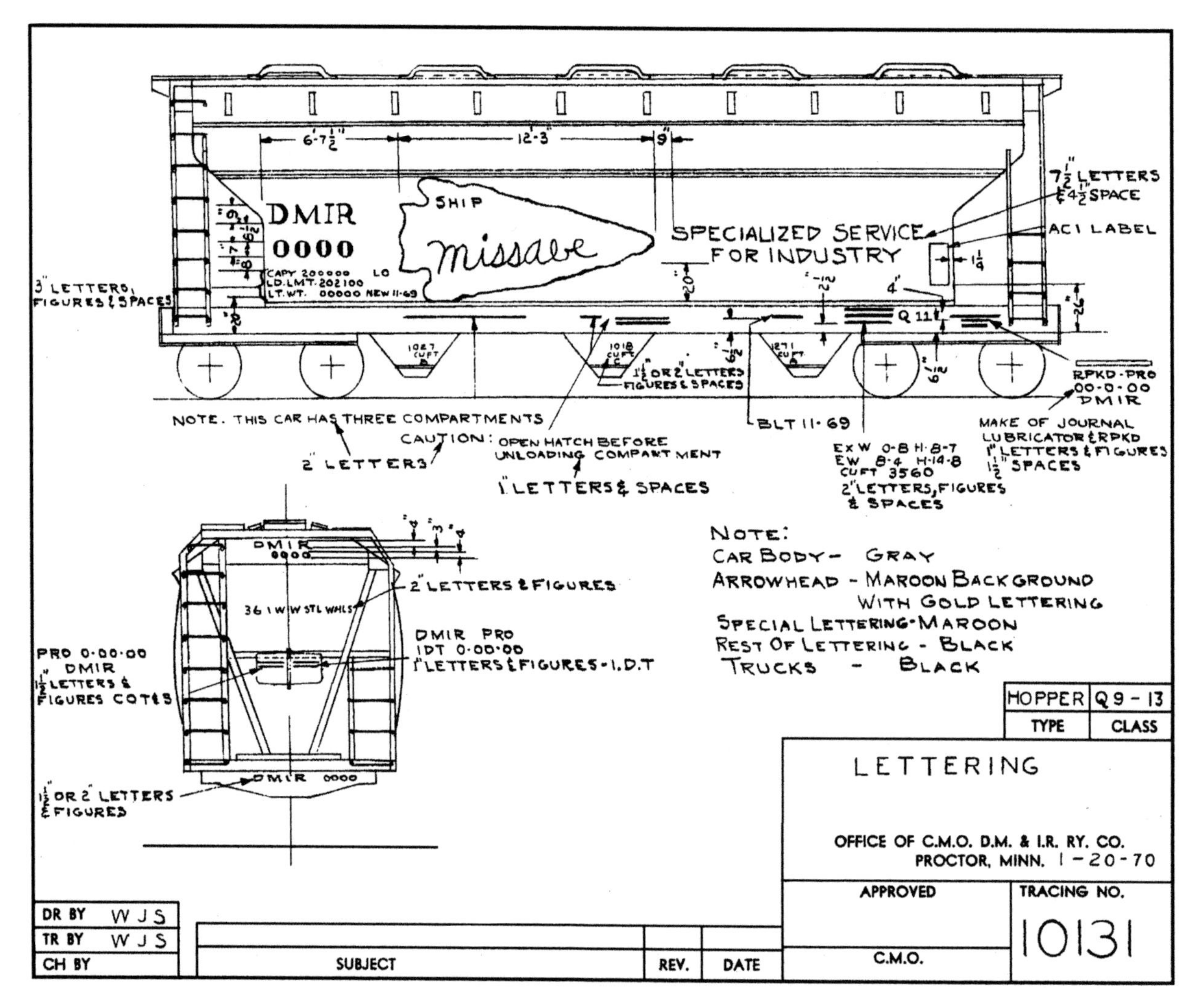

In 1971, class Q5 hopper cars were to be painted ore car brown with white lettering except the first car, 2808, was painted maroon and gold for publicity purposes. A letter was issued on April 4, 1971, instructing shop forces to repaint the first car but all remaining cars were to be left in primer aluminum and placed in service with reporting marks and number account of a severe hopper shortages. These cars would then be shopped during the summer of 1971 as time was available. For a short period of time from April 1971 until October 1971 the Q5 hoppers rebuilt were in aluminum colors. All remaining rebuilt hoppers between 1971 and 1973 would be painted maroon with gold lettering.

Taconite cars were not initially painted when side extensions were added. An August 19, 1965, letter to general car foreman at Proctor specified that rebuilt taconite cars start to be painted with USSSS No. AM-6733 Maroon paint and that No. 95-014 Dulux gold paint be used for all stenciling.

The final DM&IR monogram was the "Arrowhead" logo, which was developed in 1968 and applied to Class Q9 through Q13 covered hoppers and a handful of wood cabooses and steel bay-window cabooses. Covered hoppers were delivered from ACF without the Arrowhead logo which Proctor shops applied. Logo was Dupont Maroon #88-2336 with Gold #95-014 used for the lettering on the logo.

Class V2 gondolas were delivered painted black with white lettering. Beginning in 1958, gondolas and flat cars would be painted maroon with gold lettering.

A special T-Bird logo (see photos) was applied on crude taconite service ore cars 40000–40279 instead of a DM&IR monogram.

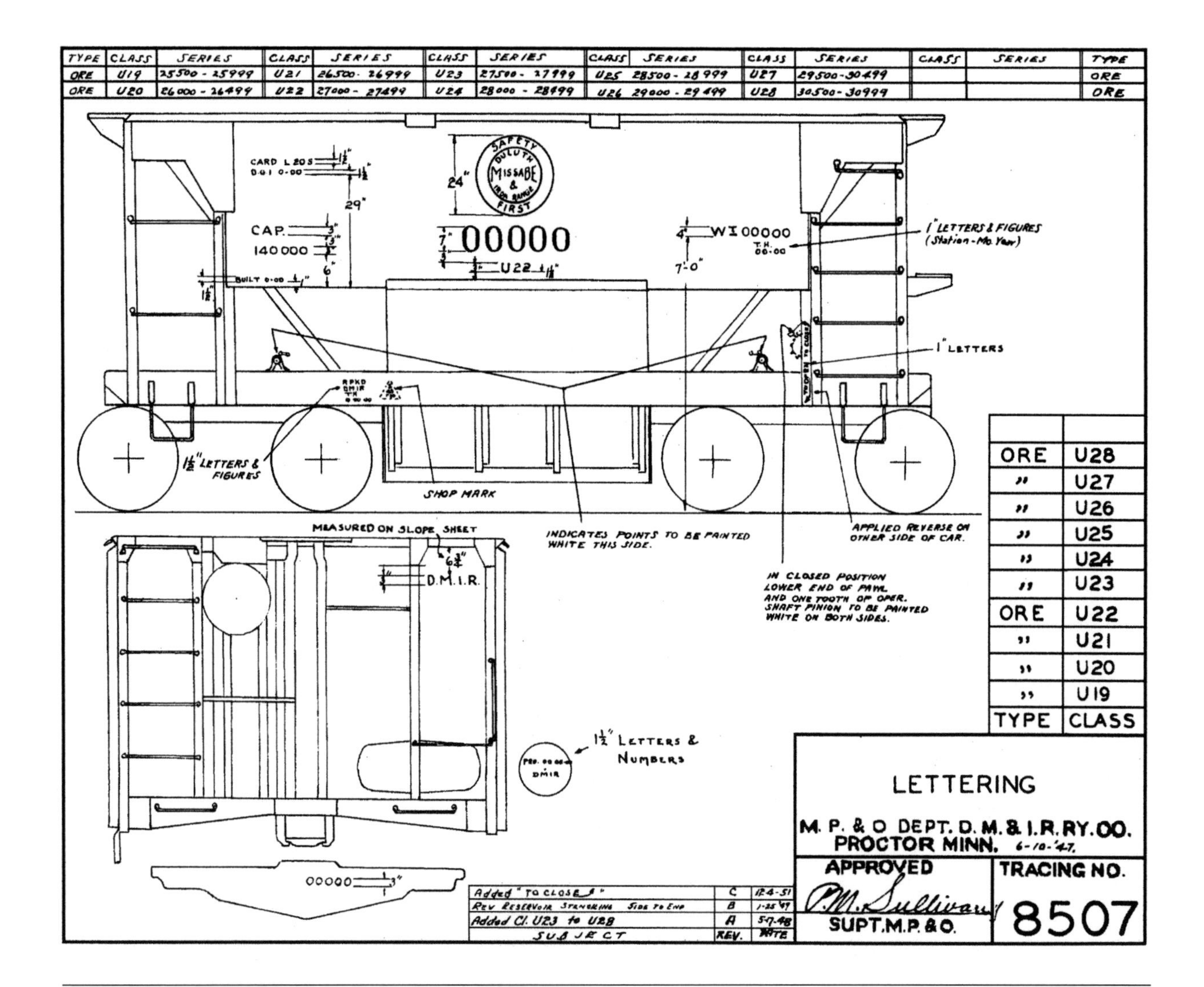

SCOTCHLITE LETTERING AND MONOGRAMS

Scotchlite monograms arrived in 1955 with the application of a green and red monogram with white lettering and circles separating the green and red colors. It was applied to SW9s, SD9s, RSD-15' and the RDC. Unique to these applications the monogram wording was: "Duluth, Missabe, and Iron Range." This was modified in 1959 with an "&" replacing the word "and" in the monogram. All future use of the monogram on freight cars, cabooses, locomotives and MofW equipment would be this version of the monogram.

Starting in 1959 a program was set up to use a new maroon and gold color scheme for painting all work equipment and merchandise cars as painting became necessary. Scotchlite reflective monograms, numbers and reporting marks began be used during 1959 on freight cars and cabooses although some cars did received non-reflective stencil monograms. Through October 31, 1960, 137 box, gondola and boarding cars were painted, using the new maroon and gold color scheme. White lines above and below reporting marks and numbers were no longer recommended practice after 1951, however DM&IR did not discontinue this practice until 1967. Modern portable airless spray painting equipment had been placed in service at Proctor and Two Harbors car shops for use in the car painting program.

By September 30, 1962, an additional 132 merchandise and work cars had been painted in the new maroon with Scotchlite monograms, reporting marks and numbers. Cabooses received yellow carbodies and maroon trim. A year later found an additional 57 merchandise and 27 cabooses painted in the "new" scheme.

The Scotchlite monograms can be broken down to five variations:

Version 1 – Monograms developed in 1955 had two white outlined circles around the monogram with the center of the monogram being red and the outer band being green with all lettering in white. Company name on the monogram was: "Duluth, Missabe, and Iron Range." Used on SW9s, SD9s, RSD-7s, RDC and former B&LE F7A/Bs. Unique on these monograms were a white "dot" on the green outer band on either side of the word "Missabe." These Version 1 monograms had a wider white outline circles. No photos have been found showing its use on equipment other than locomotives.

Version 1a – Company name was changed by changing "and" to "&" and dropping the white dots on the monogram in 1959 with the first use on EMD SD18 locomotives and later SD38s, SD-Ms and SD40-3s. Used on freight equipment from winter 1973-74 until CN merger. This monogram was used mostly on cabooses, but was also used as a replacement monogram on 5900-series box cars, two gold-painted taconite ore cars, 52931, 52454 and MofW equipment.

Version 2 - Monograms had two white outlined circles around the monogram with the center of the monogram being maroon and the outer two bands being gold with all lettering in white. This type of monogram was used on box cars, gondolas and MofW equipment from 1959 until mid-1960s.

Version 2. Scotchlite monogram with gold border, maroon center and white circles and lettering. DM&IR 3106 Virginia, 1960. (Jim Maki Sr.)

Version 1a. Scotchlite monogram with green border with red center and white circles and lettering. September 27, 2000. (D.P. Holbrook)

Version 3 - Monograms had two white outlined circles around the monogram with the center of the monogram being orange and the outer band being maroon. Lettering in center of monogram was in maroon and the outer circle lettering was in gold. This monogram was used mostly on cabooses from 1960 until the winter of 1972-1973.

Version 3. DM&IR Scotchlite monogram with orange background and lettering on caboose C-91 Virginia, July 27, 1969. (Owen Leander)

Version 4 – Monogram matched Version 2 except the center of the monogram was changed from orange to yellow. This began to be used during late 1966 and intermixed with Version 2 monograms until the supply was exhausted.

Version 4. Scotchlite monogram with maroon border, yellow center, white circles and yellow and maroon lettering. (D.P. Holbrook collection)

Cost-cutting measures during the economic downturn on the iron range during the 1960s saw Scotchlite monograms and lettering dropped from freight cars in 1968 with the above exceptions.

DM&IR REPORTING MARKS AND NUMBERS

Reporting marks were normally 9-inch high lettering and road numbers 7-inch high. Reporting marks and end numbers were normally 4-inch on all cars except ore cars which had 3-inch high DMIR and 4-inch high end numbers. These were normally stenciled in the upper right corner of each end including cars equipped with pulpwood racks. Photos indicate most DM&IR ore cars only received reporting marks and no end numbers.

Surprisingly, DM&N starting in 1925, and later DM&IR ore cars, had a monogram with road number below, but no reporting marks. FRA caught up with this fact by 1975 on regular ore cars and 1995 on mini-quads and required DM&IR to apply reporting marks. Main objection was that AAR Interchange Rules required cars to carry reporting marks when used in interchange service. The delay in stenciling the mini-quads was because they were used for on-line service only until the start of the all-rail ore movement using these cars in 1995.

Solving this problem, the DM&IR stenciled DM to the left of the monogram and IR to the right of the monogram on mini-quads and most regulars had the monogram painted out and DMIR stenciled over the old monogram location. A few cars without the reporting marks are still in service in 2015. T-Bird cars rebuilt in 1976 did not have the DM&IR ball monogram but did have DMIR reporting marks above the number, complying with FRA car stenciling requirements for cars used in interchange. Ironically, none of the T-Bird cars was ever used in interchange service.

Required by AAR and later FRA rules, DM&N and DM&IR did not apply reporting marks on ore cars. FRA began to enforce the rules starting in 1975 on regular ore cars and 1995 on mini-quad sets, "DMIR" reporting marks were stenciled above the number and in some cases the monogram was painted out when the reporting marks were applied. DM&IR 53324 Proctor, October 4, 1998. (D.P. Holbrook)

DM&IR WEIGHT, DIMENSIONAL AND MECHANICAL DATA

Weight data stenciled on the left side of the car consisted of: nominal capacity, load limit, light weight and "NEW or reweigh location and date (month and year). The MCBA in 1911 agreed that the date weighed should also include the location. This was stenciled on DMIR cars in 3-inch high lettering. Reweigh locations consisted of: STN for Steelton, PRO for Proctor, D for Endion Yard at Duluth, and T.H. for Two Harbors. Dimensional data was stenciled in 1½- to 2- inch high letters.

Class lettering was normally 4-inch high letters which were placed near the lower right corner of the car. On ore cars it was placed directly below the road number.

Iron ore was shipped as "long tons" (2,240 pounds was a long ton). Normal stenciling on ore cars showed the light (tare) weight of the car to the right-hand side of carbody. Beginning in winter 1952 the long ton decimal equivalent number began to be stenciled below the light weight shown. An example would be 15.04 long tons for a car with a lighr weight of 33,700. This stenciling was precipitated by the development of computers being used for figuring weights on loaded cars. DM&IR, GN and NP all adopted the decimal light weight stenciling on ore cars during the 1950s.

DM&IR ADDITIONAL LETTERING

During the mid-1950s a "CT" stencil on side of car was added to "regular" ore cars to indicate cars were "clean and tight" for shipments of special grade of magnetite concentrate that was produced for the cyclones at Pilotac. The stencil was a reminder for cars that were suitable for this loading. Made from premium grade taconite it was called "coarse grind." It was shipped to Arcturus and other mines for use as heavy media with the heavy media cyclone processing of natural ore.

Various other markings appeared over the years. "P" was located to left of monogram on ore cars for cars equipped with plywood slope sheets. A number of cars had the letter "X" placed on both sides of the monogram during the late 1950s to indicate they were being used to test how paint and lettering reacted in the infra-red thawing sheds. Initially appearing on the side sill and later to the right of the car number on the sides was a circle with a "D" inside the circle. This indicated car had tight doors. An angled slash through this logo indicated that the door seals had been removed. A white dot or oval above the right sill step indicated the sills had been patched. A "spider" logo on the side sill indicated "supports for rear lug casting" had been repaired. Starting in 1995 white stars began to appear on the side sheet of the car. This indicated car had passed FRA inspection and was okay for service, progressively more and more stars were added each year the car was inspected.

Stencil group letters began to be applied in 1953 at the center of the side sill and were discontinued in November 1982.

A letter "X" indicating overage trucks and wheels, and later an "R" to indicate car was overage, were stenciled on the side of cabooses, adjacent to the car number on the side of the car starting about 1985.

FRA made changes to "Clean, Oil, Test and Stencil" requirements during 1971. New stenciling required the information to be standardized in black painted boxes with white outlines in a standard location on all cars. DM&IR began to apply the COTS painted boxes on 3,100 ore cars on December 2, 1974 and completed the project on February, 23, 1977.

"WHITE BALL" AND "WELD" LETTERING

Identification of the remaining K-brake equipped 50-ton ore cars became a significant problem once train makeup compliance was a concern. Beginning in 1955, DM&IR began to stencil two white circles (balls) on either side of the car number on each side of cars still equipped with K brakes. These circles earned the moniker "White Ball" on the K-brake equipped ore cars. On July 28, 1955, 1691 50-ton ore cars were still equipped with K-2 brakes and considered "White Ball" cars, broken down as follows: 557 Class U9 cars, 540 Class U10 cars, 144 Class E9 cars, 195 Class E10 cars and 255 Class E11 cars. These stencils were applied only to 50-ton ore cars (see Chapter 2, page 74, DMIR 42421).

During 1956, it was noted that there was a problem with ore leakage (loss) at the junction of the ends of the doors with the end slope sheets. A welded extension was developed, at a cost of $11.00 per car, which would reduce the loss of ore to a minimum. Cars rebuilt with the "welded" extension to the slope sheets had a white circle painted on the far left and right side panels with the script word "Weld" located in the middle of the circle to indicate that the rebuilding had been accomplished. This was later modified to a stenciled "welded sheets."

The white ball with two lines and the word weld was placed on 50-ton ore cars beginning in 1956 to indicate that interior side and slope sheets had been welded and slope sheet extensions added to prevent leakage of ore and taconite pellets. Cloquet, Minnesota September 26, 2002. (D.P. Holbrook)

ORE CAR SLOPE SHEET TESTS

The desire to reduce freezing ore in ore cars caused the DM&IR to try various solutions over the years. One of these was the application of plywood linings on slope sheets. Two cars were converted during September 1965. AFE 10033 for $1035 was approved on October 12, 1965, for the application and testing of plywood on end sheets of ten additional ore cars. Cars 33087, 33123, 33129, 33155, 33161, 33352 had 3/8-inch thick plywood installed and cars 33017, 33056, 33107, 3319, 33147, 33371 had ¾-inch thick plywood installed. The plywood was tested with various coatings. Car 33155 with fuel oil, 33129 with graphite lubrication, 33087 with brown ore car paint, 33123 with epoxy paint, 33161 with antifreeze, 33056 with 30 percent calcium chloride solution, 33139 with diesel lubricating oil, 33017 with Chemcote, 33147 with coal spray oil and 33107, 33352, 33371 with no coating. To indicate their test status a letter "P" was stenciled on both sides of car to the left of the DM&IR monogram.

On November 3, 1965, Mr. Halley (chief industrial engineer) reported that plywood end sheets installed on 70-ton ore cars showed no appreciable wear. The cars dump easily but this is partly due to favorable weather conditions.

Application of rubber belting to one standard ore car 33154 and one high-side taconite car 533309 to improve dumping was completed on Nov 5, 1965 under shop order 3354. Rubber conveyor belting was applied to ease the unloading of these cars during freezing temperatures. Rubber belting was also applied, in later years, to many of the T-Bird crude taconite service cars, as described in more detail in Chapter 4.

STEAMING AND INFRARED THAWING OF ORE

Discharging iron ore from ore cars was forever a problem. Coarse ore emptied through the bottom doors was not a problem; however, fine ore was subject to "bridging," causing the load to be stuck in the car. Car punchers were employed on the ore docks with long metal bars that were used to poke at the loads from the end platforms in the hopes of encouraging the load to discharge into the dock pockets. This was even more of a problem as washing plants developed on the iron range. These plants would wash the impurities out of the ore, but the ore still had a watery mix when loaded. Crews called these "mud, jelly or flat top" cars indicating the smooth almost level appearance of the loads. These were even more difficult to poke at and unload.

All of these cars were subject to freezing temperatures during the winter months causing the loads to become frozen in the cars. Steam lines were in place at both Proctor and Two Harbors that allowed steam engines to be hooked up to these lines to provide steam and lances were hooked up to the steam lines and placed through the side lance openings on the ore cars to unthaw the ore. The first ore steaming facilities at Two Harbors, with 75 car capacity, were constructed in 1917.

A later development was a three-car infrared radiation thawing building built at Two Harbors in 1957. This building was replaced in 1960 with a two-track building with a capacity of 36 cars. A similar, smaller shed, was built on Dock 5 in Duluth during 1964. Once thawed the cars would be moved direct to the docks for unloading. Close clearances within the shed caused Class U14 and U15 50-ton ore cars to be removed from ore service because they did not fit inside the building. During 1961 requests were made to place these cars in coal service to eliminate the problem. The Duluth shed was removed in the early 1970s and the Two Harbors shed removed in 1979. By this time shipping of natural ore had declined and the need had passed for thawing.

C: MECHANICAL AND AIR BRAKES, TRUCKS

The majority of freight cars in the United States built after 1910 were equipped with 5-foot 6-inch wheelbase trucks. The first wood ore cars, flat cars and box cars built for D&IR were delivered with 5-foot 4-inch wheelbase arch-bar trucks. D&IR would develop their own arch-bar truck and call it "D&IR Standard" in diagram books. DM&N Class P 200–379, Class Q 380–399 and Class O 400–1200 wood ore cars were delivered with 5-foot 6-inch wheelbase arch-bar trucks. Class R 1201–1400, Class S 1401–2000 and Class T 6000–7699 were all built with 5-foot 4-inch wheelbase arch-bar trucks. D&IR and DM&N ore cars trucks would remain 5-foot 4-inch wheelbase until DM&N Class U5 12000–12099 and D&IR 11400–11499 were delivered in 1910. All future ore cars would ride on 5-foot wheelbase trucks. A few ore cars would be delivered with unique 4-foot 2-inch wheelbase Pflager and Verona trucks. DM&N Class U13 21101–21126 and Class U16 23171–23175, former STDX 171–175, were built with Pflager 4-foot 2-inch wheelbase trucks. DM&N diagrams indicate two of these cars, 21102 and 21108, had "Special Pflager" trucks but author has found no further details. Two DM&N cars, Class U16 23176, 23177, former STDX 176, 177, were built with Verona 4-foot 2-inch wheelbase trucks. Class U14 and U15 were built with Dahlmen 2-level trucks, Class U17 with National B1, Class U19 with Buckeye and almost all other 70-ton ore cars would ride ASF Spring Plankless or ASF A3 Ride Control trucks. Many of these would later be modified with roller-bearing inserts.

ARCH-BAR TRUCK REPLACEMENT

On December 31, 1939, arch-bar trucks were prohibited except for six months on returning empties. Total prohibition on arch-bar trucks in interchange service was July 1, 1940. At that time many DM&IR cars still had arch-bar trucks. It should be noted that AAR allowed arch-bar truck-equipped cars to be acceptable in interchange from owner for loading or unloading within the same terminal switching district in which the interchange occurred. This allowed DM&IR to continue to interchange arch-bar truck-equipped ore cars to GN, NP and SOO within the confines of the Duluth and Superior Terminals.

Verona truck on STDX 23177. (Standard Steel Car Buildes photo, D.K. Retterer collection)

Pflager truck on DM&N 21126. (Bob's Photo)

Arch-bar truck under former DM&N Class U5 ore car. Cloquet, Minnesota, September 26, 2002. (D.P. Holbrook)

AAR began the process to outlaw arch -ar trucks in interchange service during 1931. DM&IR was concerned about making this change and a letter was issued on April 3, 1936, requesting information on car requirements for interchange to determine how many cars needed to be converted. On that date DM&IR had 1,281 interchange cars that were affected. Prior to this letter, on April 15, 1935, a decision had been made to not convert arch-bar trucks on flat cars or refrigerator cars for interchange service.

On May 27, 1937, a total of 10,917 ore cars were on the roster with 6,015 equipped with arch-bar trucks and 4,902 with cast steel side frames. Replacement Andrews U-section side frames were purchased and conversion from arch-bar to Andrews U-section side frames started July 1, 1937, at the rate of seven cars per day. This initially included DM&N: 47 Class P box cars, 97 Class P1 box cars, 55 Class Q gondolas, 26 Class V1 gondolas; D&IR: 25 Class H gondolas and 100 Class G gondolas. It was determined that 454 cars would remain with arch-bar trucks and be restricted to on-line service only. A letter from Supt. Motive Power and Cars to C.E. Carlson president on February 28, 1938, specified that these were D&IR: 78 Class F box cars, four Class M stock cars, 113 Class I flat cars and eight Class R refrigerator cars; DM&IR: 44 Class V flat cars, 158 Class V1 flat cars, 11 Class Y flat cars and 38 Class J1, J2, J3 refrigerator cars. Ore cars and cabooses were not included in this list; they were not interchanged.

Many of these cars would remain in on-line service into the 1970s with arch-bar trucks. Cars that retained arch bars began to be stenciled "For On-Line Service Only." Duluth & Northeastern, since the late 1930s, waived the overage requirement for cars interchanged to their railroad and DM&IR continued to use arch-bar equipped cars for pulpwood shipments to D&NE into the early 1960s.

By January 1, 1935, cars with wooden underframes were outlawed in interchange service. Any remaining D&IR and DM&N cars of this type were either retired before this date or remained on the roster, restricted to on-line service only.

ROLLER BEARINGS

In 1955, Superintendent of Motive Power and Cars Robert H. Seitz recommended the installation of roller bearings on 20 ore cars to test reliability and the reduction in hot boxes associated with solid-bearing trucks. Twenty sets of Timken Roller bearings were ordered on September 11, 1955, for application to 20 ore cars on AFE 7923 at a cost of $18,150. Car numbers were: 32531, 32547, 32552, 32591, 32611, 32699, 32740, 32771, 32778, 32782, 32791, 32811, 32814, 32824, 32833, 32840, 32841, 32874, 32896, and 32951.

The installation of roller bearings was completed on March 26, 1956. At the time of installation all cars had white stripes painted on either side of the DM&IR monogram and the left stripe had the words "Roller Bearing." The cars when rebuilt retained journal box covers and the stripes and lettering were applied to assist carmen when oiling solid-bearing cars so they would not oil the roller-bearing cars.

Timken roller bearings were placed on 20 of the 1953-built Class U30 70-ton ore cars by March 1956. The roller bearings were inserts into standard sideframes that retained the journal boxes but lacked the lids. Worried about carman oiling these cars, distinctive white stripes were applied on either side of the monogram with the words ROLLER BEARINGS in the left stripe to make these highly visible to mechanical and operating department employees. This paint scheme would be retained when all were rebuilt with sideboards for taconite service and repainted in maroon and yellow. (Basgen photo, D.P. Holbrook collection)

Twenty ore cars were equipped with Timken roller-bearing wheelsets in late 1955. They received distinctive white stripes on either side of the monogram with the left stripe lettered ROLLER BEARINGS to assist operating department in identifying the cars. All were rebuilt for taconite service in 1964 and 1965. Repainted maroon with gold lettering, the stripes were retained as shown on DM&IR 52531 at Rainy Jct. Yard, Virginia, July 27, 1969. Stripes were removed when cars were repainted with T-Bird logos or rebuilt as part of mini-quad sets. (Owen Leander)

Beginning in September 1964, these cars began to be rebuilt to high-side taconite cars and during the rebuilding they were painted maroon with yellow lettering and retained the distinctive stripes and ROLLER BEARINGS lettering with the stripe color changed from white to yellow. Initially, the rebuilt taconite cars had the number 5 added to the original numbers, creating a six-digit reporting number. This process was discontinued in late 1966 when the first number 5 was eliminated and the number 3 was changed to 5.

K BRAKES TO AB BRAKES CONVERSIONS

Air brakes were essential for operation of the heavy tonnage ore trains on DM&N and D&IR. ICC reported in 1887 that the D&IR was the first railroad to completely equip all the locomotives and cars with automatic air brakes. By 1890 a common ore train was 50 35-ton capacity cars with about 2,500 trailing tons. K brakes were recommended practice by Master Car Builders in 1913 and made standard practice in 1917. Wood cars of the period were averaging a fully loaded to empty weight ratio of 2:1. Steel cars were 3:1 to 3.5:1 gross to tare loaded to empty weight ratios. These ratios led to the development of empty/load braking equipment during the 1919-1921 period.

D&IR and DM&N both adopted the empty/load braking systems. Longer train operation of 100 or more cars by 1920 was possible, but the K brake system did not provide a satisfactory system on these longer trains. The ICC proposed specifications for a new braking system in 1924 with development and road testing on SP in 1929-31 and PRR during 1932-33. The new AB brake system eliminated the violent slack action associated with the operation of 100-car trains. AB brakes were required on all newly built cars effective September 1, 1933, and on newly rebuilt cars effective January 1, 1937. The elimination of K brakes became a focus in the early 1950s.

This was an inopportune time for DM&IR because iron ore tonnage was increasing and a large part of the equipment roster still had K brakes. K brakes would be banned in interchange service January 1, 1954, but DM&IR would file multiple applications for an extension with ICC over the 1953–1956 period.

In 1953, the DM&IR had 2,890 50-ton ore cars equipped with K brakes. An ICC order of August 3, 1953, in docket 13528, Investigation of Power Brakes, extended the time limit for application of AB brakes to DM&IR 50-ton ore cars to December 31, 1954. During 1954, 69 of the 50-ton ore cars were rebuilt with AB brakes.

On November 3, 1954, application was made to the ICC requesting permission to use 1,500 of the K-brake equipped 50-ton ore cars in ore service until December 15, 1956. Five hundred were to be used in short transfer movements in trains of 60 cars or less at a maximum speed of 25 mph and 1,000 cars to be used in regular service. It was planned to retire the remaining 1,321 cars not equipped with AB brakes by December 15, 1956. As of July 28, 1955 there were 557 U9, 540 U10, 144 E9, 195 E10 and 255 E11-class 50-ton ore cars still equipped with K brakes.

The ICC issued an order to supplement, docket 13528, dated April 11, 1955, granting the DM&IR permission to use these cars in ore service *only*, and then in trains not exceeding 100 cars in length, until December 31, 1955. Faced with a growing shortage of serviceable equipment, the DM&IR ordered 500 sets of second-hand AB brake in May 1955 (on AFE 7822 at a cost of $143,000) and another 500 second-hand sets in August 1955 (on AFE 7861 at a cost of $136,500).

The anticipation that ore tonnages for the 1956 ore-shipping season would be as least as great as 1955, and because new ore cars could not be built and delivered before the fall of 1956 or the spring 1957 (due to a steel shortage) caused the DM&IR to look at the 50-ton ore car situation. It was decided to install AB brakes on 800 of the "strongest" of the remaining 50-ton K-brake equipped ore cars (on AFE 7925 at a cost of $426,400). By the 1956 ore shipping season approximately 3,300 50-ton ore cars were equipped with AB brakes. This included 424 old-style cars: Class U3 8405–10554 Class D 4400–4899, Class U4 10555–11704, and 86 Class U4 100–849, 4-door cars for use in other than ore service.

It was thought, however, that this installation would not provide an adequate ore car supply to mining companies during the 1956 season, nor would it result in the desired efficiency in ore operations. An additional application to the ICC was made in late 1955 for permission to use the remaining 706 50-ton K-brake equipped ore cars, from April 1 until November 1, 1956, on an emergency basis in trains of less than 100 cars in non-interchange service, until additional AB brake application could be made. The 680 remaining K-brake equipped cars were outlawed on November 1, 1956, and further relief from the ICC order was not forthcoming. K-brake equipped cars continued to be used MofW service after that. During the 1956-57 winter shopping season 375 of the remaining 680 K-brake equipped cars were converted to AB brakes.

These efforts were anticipated to relieve some of the tight car supply during the 1957 season, due to the fact that the 500 new Class U31 ore cars, 33000–33499, would not be delivered until the fourth quarter of 1957 and not available for the 1957 ore shipping season.

CAR SHOPPING

Normally ore was shipped starting on April 1 and continuing through November until freezing temperatures caused shipping to be discontinued. During 1952, this left an average of 83 working days on which to complete necessary repairs on 5,000 50-ton and 9,000 70-ton ore cars. Repairs were broken down into the following three categories:

Light repairs, including the repairs to air mechanism to all the 50-ton cars and to one-third the ownership of 70-ton cars.

Light repairs, but not including repairs to air mechanism to two-thirds of the 70-ton cars.

Repairs to either 70-ton or 50-ton cars which have been wrecked or badly damaged.

Item No. 1 was mandatory and had to be completed. Repairs listed in Item 2 which had been completed every year since the acquisition of the 70-ton cars, could be dispensed with during the winter months if necessary and provisions made to complete mandatory item during the normal ore shipping season. Item 3, which comprised the smallest number of cars, was completed to the extent possible.

Repairs to ore cars were apportioned to cars in categories 1 and 2 for repair at Proctor to one car to be repaired at Two Harbors. All category 3 repairs were always done at Proctor. To compensate for this, all major repairs to commercial, work equipment and passenger equipment was done at Two Harbors. Cabooses in service on the respective divisions during the summer months were repaired during the repair season at the shop of the division on which they were in service.

During 1952, winter ore car repairs comprised: clean and test air brakes, repack journal boxes on 50 percent of cars, necessary truck and body work. Seasonal ore car repairs comprised: repacking journal boxes on 50 percent of cars, greasing dumping mechanisms and hand brakes, necessary truck and body work, adjust piston travel.

During the winter shopping season of 1958-59, ore cars shopped at Proctor or Two Harbors were being stenciled on side sills with a "star" insignia with the figure "S" "59" and either "T.H." or "Pro" to indicate the point from which shopped. Cars shopped for seasonal repairs were stenciled on side sills "59 P" to designate cars shopped at Proctor and "59 T.H." for cars shopped at Two Harbors.

As of 1952, merchandise cars were shopped for heavy repairs as follows:

Class V, V1, K, K1, K2, K3, V and Y every 8 to 12 years; Class Q, G1, G2 and V1 every 12 to 15 years; Class P, P1, F, F1, M, P, J1, J2,

J3 and R every 15 to 20 years and Class H, H1, Q1, Q2, Q3, Q4, P2, F2, V2 every 20 to 25 years.

Cabooses were shopped for extensive repair (not rebuilding) at 15- to 20-year intervals.

During 1963 lading strap anchors were applied to 37 interchange flat cars in order to comply with modified AAR requirements. In 1965 safety door hangers were applied to 100 DM&IR interchange box cars to comply with requirements of the AAR

Early in 1975, the Code of Federal Regulations, Part 215, Title 49 had the following changes made affecting the DM&IR:

Clean, Oil, Test and Stencil (COTS) information, called Consolidated Stencils, were to be applied to 3,100 ore cars during 1975 with the remaining ore cars, commercial cars and cabooses required to be completed by December 31, 1976, required by paragraph 215.11(c). These stencils consisted of a black-painted box with a white outline that contained all the COTS information.

Ore car repacking would go from a 48-month to 30-month cycle. The 30-month cycle was contingent on journal boxes being stabilized. Required completion date was December 31, 1978, per paragraph 215.97.

Waiver Stencils. During the changeover it was necessary to stencil every ore car with a waiver explaining lack of repack date compliance. Required completion date was December 31, 1975

Caboose wheel change-out. Wooden cabooses with wheels dated prior to 1927 had to have the wheels replaced.

These new requirements caused many problems for the mechanical department. The 48-month repack period had been standard since lubricator pads were adopted nationally in the late 1950s. The four-year interval had been selected by DM&IR to coincide with the air cleaning date and incorporated into the winter ore car shopping program for many years. By following that procedure, it had only been necessary to repack one quarter of the ore car fleet each winter. The captive DM&IR ore car fleet had made it possible to use the 48-month repack, although special agreements had been necessary, when cars moved off line. These new FRA standards caused the BN to refuse ore cars in interchange with repack dates over 24 months old. DM&IR was granted a limited waiver on January 24, 1975, that was valid until December 31, 1978.

AC AIR BRAKES

At the time the order was placed for Class U31 ore cars, further development in the AB brake systems was under way. The updated AB control valve was introduced in 1955, and classified as an "AC" brake system. The AC system incorporated diaphragm-operated piston, rubber "O" rings, and spool valves that replaced slide valves wherever possible. The AC system performed the same function as the AB valve. The weight of the removable portions was reduced by 40 percent. Except at two locations, spool valves with rubber seal rings replaced the AB slide valves. The speed of a release on the air brakes on 150 cars was reduced from 50 to nine seconds.

This eliminated ring leakage and friction variables and were used first in conversion kits for the AB portions and finally in the ABD-1 service portion. When the new parts were used in the emergency portion, the entire valve was designated the ABD control valve (which was interchangeable with the AB valve). This made it possible to accelerate air brake releases, more positive response to pressure changes and more consistent operation.

The AAR began looking in 1955 for a controlled environment to test the newer AC brake system. At the same time as the AC brake system was developed, high-friction COBRA composition brake shoes were introduced in North America.

What better a controlled environment than the iron ore railroads of Northern Minnesota? When Erie Mining began purchasing their freight car fleet they also committed to become the "test bed" for the Westinghouse AC brake system. All equipment delivered to Erie Mining was equipped with AC brakes, including cabooses. At the same time, DM&IR elected to receive the 500 U31 class ore cars with AC brakes. Additional railroads participating in the test include the Great Northern, with 50 cars equipped with the AC valve, and the Rock Island, with 50 "Covert-A-Frate" flat cars.

During July of 1957 evaluation tests began on Erie Mining. The combination of the AC valve and the pressure-maintaining feature on the Erie Mining locomotives did away with the use of retainer valves and permitted a given brake pipe reduction to be held constant for a long period of time. The test train consisted of five engines, 92 loaded taconite cars and three cabooses. The cabooses were all equipped with a speedometer and an air gauge that indicated brake pipe and brake cylinder pressures. Recorders were distributed throughout the train to indicate draw-bar action. At the time of Erie Mining's startup the roster of AC-equipped equipment was 389 pellet cars, 5 cabooses and 4 diesel fuel tank cars of 22,000-gallon capacity. At the time, these ACF tank cars were the largest welded tank cars in the world.

ASF A-3 Ride Control trucks on Class U26 ore car. Note the L-shaped empty-load sensor attached to the end of the bolster, allowing the automatic adjustment of the braking capacity based on the empty or load status of the car. (GATC photo, D.P. Holbrook collection)

What made the AC brake unique at the time was that it was *not* permitted in interchange service. Air brake cleaning and inspection at the time of delivery was on a four-year basis. ICC waived these requirements and allowed DM&IR to do the cleaning and inspection on AC cars on an every five-year basis. The remaining AC-equipped cars were converted to standard AB brake assemblies between April 1972 and May 1973.

EMPTY-LOAD BRAKES

DM&IR was an early proponent of empty-load brakes with the DM&N Class U12 95-ton ore cars delivered in 1925 being the first cars equipped. Empty-Load brakes were made necessary when the loaded to empty weight ratio exceeded 4.2 to 1. It provided for a braking force that manually, and later automatically, changed depending on if the car was loaded or empty. Without the system cars would have sufficient braking power when empty, but insufficient when loaded. Later versions incorporated a device which measured spring deflection on the trucks to determine if car was loaded or empty.

By February 1935, 309 cars were equipped with empty load brakes. All ore cars constructed after 1935 were delivered with empty-load brake systems.

Development and testing of composition brake shoes showed that the empty-load brake systems were no longer needed. Cost Authority 6001 was approved October 20, 1966, to equip 2,399 70-ton ore cars with single-capacity brakes and composition brake shoes. During the conversion these cars were changed from empty-load brakes to single-capacity brakes. Once the conversion was completed a stencil, reading, "Single Capacity Brake, E1-friction Shoes" was applied on the side sill directly below the load cylinder on both sides. This was accomplished during the normal winter shopping program.

Cost Authority 6003 was approved on August 4, 1967, for the same modification to an additional 1,786 70-ton ore cars.

Cost Authority 6007 was authorized on July 15, 1968, for replacement of cast-iron brake shoes with Cobra heavy-duty composition brake shoes. Cobra shoes were introduced in 1956 and it eliminated the need for 12-inch cylinders on 90- to 100-ton cars, clasp brakes on freight cars and empty-load braking systems; it also extended wheel life with fewer thermal problems. Cars involved were 473 Class U26, 954 Class U27, 363 Class U28 and one Class U31 car, a total of 1,791 cars. By the early 1970s, DM&IR had eliminated empty-load brake systems.

PLAIN BEARING LUBRICATORS

DMIR was an early tester of lubricator pads to replace waste packing in plain bearings. By late 1955, 12 types of journal lubricator pads were being tested on 1,517 ore cars and 120 hopper cars. These included pads from Absco, Acme, Centra-Feed, Crown, Journa-Pak, Hennessy Miller, Premier, Rollins, Southland, Unipak and Wikit. Hennessy type pads were unsuccessful and discontinued in 1956. Lubricator pads were installed on Class Q7 2950–3049 open hoppers and Class Q8 4970–4989 covered hoppers, and when delivered from builder, AAR had mandated that journal box lids on these cars be painted yellow to allow easier identification by carmen, rip tracks and mechanical department of cars equipped with lubricator pads.

By 1960 DM&IR had equipped 2,400 cars with lubricator pads and all cars were completed by November 1961. AAR issued a directive mandating that all waste-packed bearings had to be changed to lubricator pads no later than January 1, 1962. On September 22, 1965, a program was instituted for replacement of Acme lubricator pads which were no longer approved by the AAR. An accelerated winter repair program was begun on December 9th to replace the Acme pads with 1,320 sets installed through December 28, completing the changeover.

PLAIN BEARING TO ROLLER-BEARING CONVERSIONS

The first conversion from plain or solid bearings to roller bearings took place with the 30 70-ton steel ore cars converted in 1956. By 1986 the maintenance of solid-bearing cars was becoming an expense that could be avoided. Cost Authority CA-6147 was approved on September 24, 1986, for the modifications of 260 ore cars from solid bearings to roller bearings. Timken provided gauges, jigs and technical assistance at no charge to the DM&IR to assist in the conversion.

As of June 1986 there were a total of 1,716 mini-quad cars in pellet service with 6 x 11-inch solid-bearing wheelsets. Journal boxes had to be oiled every 15 days and lubricator pads changed every four years. The conversion was accomplished by modifying the existing side frames to accept roller-bearing inserts. The project was started in 1987, and by January 1989 720 cars had been converted (280 T-Bird, 429 Mini-quads and 11 side-dump cars). National Transportation Safety Board (NTSB) recommended in late 1988 that modified plain-bearing boxes that had been converted to roller bearing be prohibited in interchange service. The conversion program was terminated in 1989.

LOW TO HIGH AIR HOSES

All D&IR ore cars were delivered with high-mounted air hoses. DM&N initially received ore cars with low-mounted air hoses, but during 1903 DM&N moved low-mounted air hoses to high-mounted locations to make it easier for crew members to couple them together. Trucks on ore cars stuck out past the ends of cars, making the space between cars very narrow with insufficient space for crew members to connect air hoses. After 1903 the high-mounted air hoses on ore cars became standard. ■

In the days of steam thawing of frozen ore loads, workmen had to insert a steam lance into a hole in the side of the ore car. It was a labor-intensive and demanding job. This photo was taken at Proctor, before rushing the thawed loads to the dock. (DMIR photo, Richard Laurens collection)

BIBLIOGRAPHY

PERIODICALS

The Commercial West , Saturday March 29, 1902
Railroad Gazette, Various issues 1880–1908
Railroad Age Gazette, Various issues 1908–1910
Railway Age Gazette, Various issues 1910–1918
Railway Age, Various issues 1918–1995
Missabe Railroad Historical Society publication *Ore Extra,* Vol. 1.1-29.1
Official Railway Equipment Register, 1895–2002
Railway Age, "Using Bulk Cement on Railway Construction Work," June 18, 1915
Skillings Mining Review, 1930–2000
Railway & Engineering Journal Vol. 63 Oct. 1889
Directory of Iron & Steel Works, U.S. and Canada, Tenth Edition,1890
The American Engineer: An Illustrated weekly Journal, Vol 13-14 (May 18, 1887, pg. 178) "Movement afoot to re-organize and reestablish Northwestern Car & Manufacturing"
Railway World Vol 31 Sept 24, 1887 page 920
Supreme Court records and briefs, 1884-1887
Minnesota History Spring 2015, "The Mail is Coming!" by David A. Thompson

DOCUMENTS & REPORTS

Duluth & Iron Range Annual Reports, 1884–1914
Duluth, Missabe & Northern Annual Reports, 1893–1918
Duluth, Missabe & Iron Range Annual Reports, 1940–1976
D&IR, DM&N, DM&IR AFE Files for years 1919-2001 at the MRHS archives.
D&IR Diagram books: 1893, 1910, 1921, 1929, 1931
DM&N Diagram books: 1895, 1900, 1905, 1905, 1906, 1908, 1917, 1931
DM&IR Diagram books: 1939, 1943, 1953, 1959, 1976, 1978, 1989, 1996, 1998
Proceedings Railway Systems and Procedures Association 1953–1960
Ship Masters Association Directory 1959, International Ship Masters' Association of the Great Lakes
Proceedings of the Air Brake Assoc. 1950–1985.
Minnesota Warehouse Commission annual reports, 1884–1940
ICC Valuation Docket No. 368 for Duluth and Iron Range Railroad Co. dated March 28, 1930
ICC Valuation Docket No. 369 for Duluth, Missabe and Northern Railway Co. dated July 5, 1929
Obsolete American Securities Vol 1, 1904

BOOKS

Duluth and St. Louis County, Minnesota; Their Story and People, Vol. 1
King, Frank A. , *The Missabe Road,* Golden West Books, 1972
King, Frank A. , *Locomotives of the Duluth, Missabe & Iron Range,* Pacific Fast Mail, 1984
Schauer, Dave C.,*Duluth, Missabe and Iron Range Railway,* Morning Sun Books, 2002
Leopard, John, *Duluth, Missabe & Iron Range Railway,* MBI Publishing Co., 2005
Dorin, Patrick C., *The Elgin, Joliet and Eastern Railway,* Signature Press, 2009
The Forest for the Trees. How Humans Shaped the North Woods, Minnesota Historical Society Press, 2004
History of the Saint Croix Valley, Vol 1.

Here is one more view of the No. 2 ore dock, with a classic Lakes ore boat, the *Benjamin F. Fairless,* alongside, Loading chutes are down as cars on the dock discharge ore. As the large lettering on the ship's side proclaims, the *Fairless* was part of U.S. Steel's Pittsburgh Steamship Company fleet. (Wayne C. Olsen collection, courtesy Richard Laurens)

The text of this book is set in a modern digital version of Caslon, the widely used face designed by William Caslon in London about 1720, and one of the great typefaces of the Western world. The version used here is Adobe Caslon, created for Adobe Systems by Carol Twombly in 1989, faithfully following the original Caslon specimens at the St. Bride Printing Library in London. This is widely regarded as the finest digital Caslon.

The titling face is Evogria, a strong, assertive and highly visible font designed by Situjuh Nazara specifically for headlines, posters and other display applications. It first appeared about 2004. He has designed dozens of free and commercial typefaces in the last decade, under the name 7NTypes.